Date Due

Due 12-30-87		
MAY 1 9 1988 JUN 1 4 1988		JUL 1 4 1995
JUN 29 1988		MAY 2 0 1996
FEB 2 1989		
DEC 2 0 1989 MAR 22 1990		
MAY 2 1990 DEC 1 1 1990		
MAY 2 0 1992		
OCT 1 7 1994		
NOV 7 1994		

Earthquakes and the Urban Environment

Volume III

Author

G. Lennis Berlin
Associate Professor
Department of Geography
Northern Arizona University
Flagstaff, Arizona

CRC Press, Inc.
Boca Raton, Florida

Library of Congress Cataloging in Publication Data

Berlin, Graydon Lennis, 1943-
  Earthquakes and the urban environment.

  Bibliography: p.
  Includes index.
  1. Earthquakes. 2. Earthquakes and building.
I. Title.
QE539.B48   551.2'2   77-16131
ISBN 0-8493-5175-1

Direct all inquiries to CRC Press, Inc., 2000 N.W. 24th Street, Boca Raton, Florida, 33431.

© 1980 by CRC Press, Inc.

International Standard Book Number 0-8493-5173 (Volume I)
International Standard Book Number 0-8493-5174 (Volume II)
International Standard Book Number 0-8493-5175 (Volume III)

Library of Congress Card Number 77-16131
Printed in the United States

ACKNOWLEDGMENTS

I would like to express my deepest gratitude to the many people who contributed substantially to the creation of this book. The entire manuscript was reviewed by James R. Underwood, Jr., Chairman-Department of Geology, Kansas State University and John Kelleher, Seismologist, U.S. Nuclear Regulatory Commission. Their authoritative criticisms and suggestions are largely responsible for any success of this book. Because it was not possible to make all the changes suggested, the author is solely at fault for any errors or omissions that may remain. The following individuals reviewed portions of the manuscript related to their research specialities: Jack Barrish (Jack Barrish Consulting Engineers), James H. Dieterich (U.S. Geological Survey), Ajit S. Virdee (California State University at Sacramento), James H. Whitcomb (California Institute of Technology), and Peter I. Yanev (URS/John A. Blume & Associates, Engineers). The many individuals who provided photographs vital to the completion of this book are acknowledged in the figure captions.

Two colleagues at Northern Arizona University deserve special thanks. Dominic J. Pitrone provided countless constructive suggestions and invaluable data collection assistance, and Howard G. Salisbury, III, Chairman-Department of Geography and Planning, was untiring in his efforts to accommodate my many special requests. Cartographic assistance was provided by Michael Schramm and Nat Garcia. Carolyn Waller and Virginia Hall typed the entire manuscript.

The talented editors at CRC Press, namely Sandy Pearlman, Jeffrey Eldridge, Terri Weintraub, Beth Frailey, Barbara Perris, and Gayle Tavens, contributed a great deal to the refining of much of the text. These individuals also were most understanding concerning the many delays caused by the author.

My grateful thanks go to my wife Judy for her encouragement and forbearance over the several years that it took to complete this project.

Graydon Lennis Berlin
Flagstaff, Arizona
November 1978

THE AUTHOR

Graydon Lennis Berlin was born in St. Petersburg, Pennsylvania, on May 21, 1943, and educated in the public schools there. He received a B.S. degree in 1965 from Clarion State College (Earth-Space Science), an M.A. degree in 1967 from Arizona State University (Geography), and a Ph.D. degree in 1970 from the University of Tennessee (Geography). He began his educational career in 1968 as an Assistant Professor of Geography and Research Associate at Florida Atlantic University. He joined the faculty at Northern Arizona University in 1969, attaining the rank of Associate Professor of Geography in 1975. Between 1969 and 1978, Dr. Berlin was also a Research Geographer on a part time basis with the U.S. Geological Survey. At the present time he is the Director of the Advanced Training of Foreign Participants in Remote Sensing Program, a joint venture of Northern Arizona University and the U.S. Geological Survey.

Dr. Berlin is a member of several national professional organizations and was elected Chairman of the Geography Division of the Arizona Academy of Science in 1974. He is also a member of Gamma Theta Upsilon, the National Professional Geographic Fraternity and a Full Member of Sigma Xi, the Scientific Research Society of North America. Dr. Berlin is a biographee in American Men and Women of Science, Who's Who in the West, and the Dictionary of International Biography. He is the author of more than 30 journal articles and government reports.

DEDICATION

To Judy, Jodi, Mom, and Dad

TABLE OF CONTENTS

EARTHQUAKES AND THE URBAN ENVIRONMENT
G. Lennis Berlin

TABLE OF CONTENTS

Chapter 1

PLANNING FOR SEISMIC HAZARDS

I. LAND USE PLANNING

Land use planning has been used only recently as a method to mitigate losses due to earthquake hazards. In California and Alaska, for example, urban growth in recent decades has moved ahead with little regard to seismic hazards (Figure 3 in Volume I, Chapter 1). According to Mader,[1117] urban planners in the past have "generally not been sufficiently aware of seismic problems and therefore have too often simply ignored them." In addition, elected government officials "are often unaware of seismic problems or, if aware, find it difficult to deal with vague events such as earthquakes in the day-to-day realism of political decisions."[1117] Unlike the situation in existing urban areas where development patterns are already established, proper land use plans and decisions can reduce risks from seismic hazards in regions currently experiencing the urbanization process. Mader[1118] notes that although "experience is lacking, there are a number of promising approaches to dealing with seismic problems in land use planning."

A. Land Use Plans: A Sampling from California

In California, especially as a result of the February 9, 1971 San Fernando earthquake, several key pieces of legislation have been enacted to achieve seismic hazard reduction via land use planning. The California Legislature enacted *Senate Bill (SB) 351* in 1971 requiring all cities and counties to prepare and adopt a *seismic safety element* in their long-range *general plans* (Government Code Section 65302).* The element must consist of an identification and appraisal of seismic hazards including surface faulting, ground vibrations, ground failures, tsunamis, and seiches. In addition, an appraisal of mudslides, landslides, and other processes related to slope stability must be considered simultaneously with seismic hazards. The Governor's Earthquake Council has stated that the intent of the law is to require cities and counties to consider *all* seismic hazards in planning programs to reduce deaths, injuries, property damage, and economic and social dislocations resulting from future seismic events.[1119] With the passage of this legislation, seismic safety became a state concern with the burden placed on local and county governments to deal with seismic problems when making planning decisions.[1117]

The first seismic safety element study was prepared for the tri-cities of El Cerrito, Richmond, and San Pablo in western Contra Costa County.[1120] It was designated a model study by the Governor's Office of Intergovernmental Management and the California Council on Intergovernmental Regulations; the latter organization distributed the study to all cities in California. The essential parts of the tri-city study are (1) detailed findings of potential earthquake hazards, existing land uses, and disaster implications, (2) policies to regulate existing development and guide future development, and (3) specific recommendations for action by the three cities to reduce the impact of minor and major earthquakes.[1120]

In 1973, the California Division of Mines and Geology (CDMG) published and dis-

---

* California law also requires a general plan to include a land use element, a circulation element, a housing element, a conservation element, an open-space element, a noise element, a scenic highways element, and a safety element. The law further requires that zoning and subdivision of the land be consistent with the general plan.[1118]

tributed a detailed set of guidelines to serve as an additional aid for the preparation of seismic safety elements. The publication, entitled *Guidelines to Geologic/Seismic Reports,* was prepared by the Southern California Section of the Association of Engineering Geologists.[1121]

In early 1977, a committee of the California Safety Commission conducted a statewide survey and found that approximately 80% of all cities and counties had adopted seismic safety elements.[1118] Most of the elements contain background material prepared by geologists and planners and a policy statement to be included as a part of the general plan.[1118] According to Mader,[1117,1118] (1) the elements are of varying quality, ranging from those that brush the topic to those that treat the subject in great depth, (2) a variety of approaches have been used to deal with seismic safety,* and (3) the modifications to land use plans in response to the seismic data have varied. Mader[1118] summarizes the impact of Senate Bill 351 to date.

. . . It is clear, however, that effects of the legislation have been felt state-wide and have led to local identification of seismic problems and formulation of policy, and are leading toward significant impacts on land use decisions. The newly adopted elements have not, however, been in effect for sufficient time to judge their real impact. The State has by means of this legislation told local government to take seismic safety into consideration in general plans. The State has not yet said it will judge the adequacy of the local response.

California has taken a far-reaching approach in directing local government to cope with one particular seismic hazard — fault rupturing.[1117] The *Alquist-Priolo Special Studies Zones Act* of 1972** (amended in 1974 and 1975) requires the State Geologist (Chief, California Division of Mines and Geology) and the State Mining and Geology Board to assist "cities, counties, and state agencies in the exercise of their responsibility to provide for the public safety in hazardous fault zones."[1122] The act is designed to provide a means for reducing personal and property damage resulting from movement along an active fault. The legislation applies to new real estate developments and structures designed for human occupancy, with the exception of single-family wood-frame dwellings, in designated hazardous zones.[1123] Basically, a habitable structure must be sited so as to avoid "undue hazards" that could be created either by surface faulting or by fault creep, and geologic studies are required along specified fault traces as a prerequisite to construction projects.[1117,1123] The following is a summary of the official responsibilities and functions required by the act.

The State Geologist has the continuing responsibility to delineate *Special Studies Zones* that encompass limited areas centered on potentially hazardous faults. A zone boundary defines an area that the State Geologist believes warrants detailed geologic investigations to determine the presence or absence of hazardous faults.[1124] The State Geologist must revise existing zones and delineate new zones as additional geologic and seismic data become available; the zones are delineated on 7.5 minute (1:24,000) topographic maps. *Preliminary Review Maps* of new and revised zones are issued 1 July each year, and *Official Maps* are released 1 January of the following year. Once Preliminary Review Maps are released, cities, counties, and state agencies affected by Special Studies Zones have 90 days in which to review the maps and to submit comments to the State Mining and Geology Board.[1123]

Under Phase I of the CDMG program, Special Studies Zones were compiled for the San Andreas, Calaveras, Hayward, and San Jacinto faults. Phase I zoning was completed in 1974. Phase II of the program has been extended to the following faults: Antioch, Buena Vista, Elsinore-Chino, Fort Sage, Garlock, Kern Front, Manix, Mal-

---

*   Mader[1118] discusses the different approaches used in the seismic safety elements of Santa Barabara County, Santa Clara County, San Jose, and San Francisco.
** Formerly known as the *Alquist-Priolo Geologic Hazards Zones Act.*

ibu Coast-Raymond, Newport-Inglewood, Owens Valley, Rogers Creek-Healdsburg, San Gregorio, Sierra Madre-Santa Susana-Cucamonga (includes San Fernando), Whittier, and White Wolf.[1123] As of January 1, 1978, a total of 261 Official Maps had been issued by the State Geologist.[1125]

The State Mining and Geology Board has the responsibility of formulating policies and criteria to guide affected cities and counties, serving as an Appeal Board for disputes that cannot be handled at the local level, and advising the State Geologist. The following is a summary of policies and specific criteria adopted by the board.[1123]

Policies

1. Specifies that the Act is not retroactive.
2. Suggests methods relating to review of Preliminary Maps prior to issuance of Official Maps.
3. Policies and criteria apply only to areas within Special Studies Zones.
4. Defines *active fault* (equals potential hazard) as a fault that has had surface displacement during Holocene time (last 11,000 years).

Specific Criteria

1. No structures for human occupancy are permitted on the trace of an active fault. (Unless proven otherwise, the area within 50 feet [15.2 m] of an active fault is presumed to be underlain by an active fault.)
2. Requires geologic reports directed at the problem of potential surface faulting for all real estate developments and structures for human occupancy.
3. Requires that geologic reports be placed on open file by the State Geologist.
4. Requires cities and counties to review adequacy of geologic reports submitted with requests for development permits.
5. Permits cities and counties to establish standards more restrictive than the policies and criteria.
6. Sets fees for building permits at 0.1 percent of estimated assessed valuation of proposed structure.
7. Defines a) structure for human occupancy, b) technically qualified geologist, and c) new real estate development.

Affected cities and counties (1) are responsible for the local implementation of the act, (2) approve or disapprove sites for every new real estate development or structure designed for human occupancy within Special Studies Zones, and (3) collect fees for building and development permits to cover administrative costs. State agencies have the responsibility for siting state structures safely within Special Studies Zones.[1123] As noted by Hart,[1123] in many cases where the existence of a fault hazard is unclear, a local jurisdiction must decide whether or not a proposed development or habitable structure is an acceptable risk based upon the site investigation made by a geologist licensed in California.

A special agency, the *San Francisco Bay Conservation and Development Commission (BCDC),* was created by the California Legislature with authorization to prepare and enforce a comprehensive plan for controlling development along the shoreline of San Francisco Bay. BCDC shares jurisdiction with cities and counties over land use decisions and, with minor exceptions, a permit from BCDC is required for all projects within its area of jurisdiction.[1118] As noted by Mader,[1118] the commission "in effect, holds veto power over any project proposed in conflict with the San Francisco Bay Plan."

California also has adopted legislation for the siting of new public school and hospital buildings to insure that the proper foundation and geologic conditions are assessed for seismic safety (also discussed in Volume 2, Chapter 3). For example, under the *Hospital Safety Act* (Senate Bill 519), the California Division of Mines and Geology is assigned the responsibility of determining the adequacy of geologic/seismic reports for new hospital sites prepared by certified engineering geologists and submitted to the Office of Architecture and Construction[952] (Figure 1). CDMG uses *Guidelines to Geologic/Sesmic Reports,*[1121] *Recommended Guidelines for Determining the Maximum Credible and the Maximum Probable Earthquakes,*[1126] and *Checklists for the*

FIGURE 1.   The Fairmont Hospital in San Leandro, California is one of many hospitals located within the active Hayward fault zone. A major earthquake along this fault would very likely put all of these hospitals out of commission. In 1973, the California Legislature enacted the *Hospital Safety Act* to prevent the construction of new hospitals within fault zones. (Courtesy of James L. Ruhle and Associates, Fullerton, Calif.)

Review of Geologic/Seismic Reports[1127] in the review process for determining if the appropriate foundation and earthquake conditions were considered and properly evaluated in the analysis of seismic safety.[952]

As noted by Mader,[1117] *environmental impact reports (EIR)* can be effective in defining potential seismic hazards, and because they are required for a wide variety of projects, they can cover developments "not otherwise subject to adequate governmental review."[1117] However, Mader[1117] further observes that because environmental impact reports "are for information purposes, they do not insure that problems which are uncovered will be treated properly in development plans." In California, the *Environmental Quality Act* of 1970, with amendments, requires all state and local agencies to prepare, or to have prepared by contract, an environmental impact report on any project that could have a significant effect on the environment.[1128,1129] The geologic problems that must be considered in the preparation and review of an EIR are (1) earthquake hazards, (2) loss of mineral resources, (3) waste disposal, (4) slope and/or foundation instability, (5) erosion, sedimentation, flooding, (6) land subsidence, and (7) volcanic hazards. Earthquake hazards include (1) fault movement, (2) liquefaction, (3) landslides, (4) differential compaction/seismic settlement, (5) ground rupture, (6) ground shaking, (7) tsunamis, (8) seiches, and (9) flooding from the failure of dams and levees.[1130] The California Division of Mines and Geology is responsible for the review of geologic analyses in the reports to determine if the appropriate geologic factors have been adequately considered.[1129,1131] To assist in the review process and to provide guidance to public agencies, EIRs are evaluated on the basis of checklists and guidelines contained in the CDMG publication *Guidelines for Geologic/Seismic Considerations in Environmental Impact Reports,*[1130] which identifies the geologic conditions present in the report area and a list of elements that should be included in the report.[1129]

Zoning ordinaces can be used effectively for dealing with seismic hazards at the local level. The Town of Portola Valley* has adopted several innovative procedures for reducing the impact of seismic hazards. For example, the town's *subdivision ordinance* requires a geologic report to be submitted with the application for a subdivision to

---

*   Portola Valley occupies a site within the San Andreas fault zone southeast of San Francisco.

insure that the proposed site is physically suitable for the residential development. The report must be approved by the Town Geologist prior to development.[1117,1118] In addition, a *fault setback ordinance* was enacted because the San Andreas and other active faults pass through the community. One requirement of the ordinance is that no structure designed for human occupancy can be located closer than 15.2 or 30.4 m to a fault trace with a known or inferred location, respectively.[971] Additional requirements of this ordinance are described in Volume II, Chapter 3.

B. Possible Land Use Plans and Building Policies for Mitigating Earthquake Hazards

Nichols and Buchannan-Banks[1132] describe various types of land use plans and building policies that can be implemented for developing and developed areas, respectively, to mitigate losses arising from surface faulting, ground shaking, ground failure, and tsunami and seiche effects.

1. Surface Faulting

1. Land uses that would be compatible with ground displacement should be recommended for future development. Compatible uses include undeveloped open-space, recreational areas such as golf courses, parking lots, and drive-in theaters. Only those uses essential to public welfare, such as utility and transportation facilities, should be considered in areas of extremely high risk.

2. Establishment of a fault hazard easement, requiring varying setbacks from active faults, can reduce the risk. The more critical the structure, the greater the setback. This type of approach is being used in Portola Valley, California.

3. For any development to be within or immediately adjacent to an active fault, detailed geologic studies should be required to demonstrate "that the construction would conform to standards of community safety and that an undue hazard to life and property would not ensue." This procedure is addressed by the Alquist-Priolo Special Studies Zones Act in California.

4. Where urban development already exists within active fault zones, jurisdictions can adopt policies leading to the removal of critical-engineered structures on the most accurately located active fault traces. Nonconforming building ordinances could be considered that would require "the eventual removal of structures in the greatest danger, starting with those that endanger the greatest number of lives — schools, hospitals, auditoriums, office buildings, and apartment houses, followed by commercial buildings, and perhaps eventually by single family residences."

2. Ground Shaking

1. In areas expected to be severely shaken in future earthquakes, low-density land uses might be appropriate for future development.

2. Building code criteria appropriate to ground vibrations can be adopted. For example, the most stringent regulations might be applied to thick, water-saturated areas having long fundamental periods of vibration that would closely match the natural periods of high-rise buildings.

3. Ordinances could require detailed geologic, soil, and engineering analyses for structures having high occupancies in areas suspected to have the greatest motion. A procedure similar to this is now required for public school and hospital construction in California.

4. Hazardous building and parapet abatement programs can be initiated. The former program was first implemented in Long Beach, California, and the latter in Los Angeles.

3. Ground Failure

1. In areas where instability can be expected, open-space or other nonoccupancy uses could be implemented.
2. Analyses could be required to demonstrate that a hazardous condition has been eliminated by site preparation work or special engineering design before occupancy unit plans are approved.
3. In developed areas where a severe instability problem exists, consideration should be given to the implementation of a hazardous building ordinance or to the initiation of nonconforming use procedures.

4. Tsunami and Seiche Effects

1. Stringent controls should be applied to all land uses within areas that could be subject to tsunami and seiche runup or potential areas of inundation downstream from water-retaining structures that are sited within active fault zones and landslide-prone areas.
2. Controls could include the following: (1) allow only those land uses that are essential (docks, warehouses) in possible inundation areas, and warn owners and occupants of the potential hazards, (2) prohibit new construction in undeveloped inundation zones, (3) initiate warning systems and evacuation plans, and (4) eliminate potentially dangerous dams.

Because neither the location nor the magnitude of *tectonic land changes* can be predicted, very little can be done to minimize the effects before the deformation occurs.[1132]

C. Earth Science Data for Identifying and Assessing Seismic Hazards

It is essential to have a broad-based earth science information inventory if land use policies and regulations are to be used effectively to minimize earthquake hazards. However, Linville[1133] observes that in most seismic areas "hazards have not been identified on a scale that can be significant for planning." In other instances, earth science data may be available, but they "must be translated from scientific and technical language into a form that can be used effectively in the decision making process."[1134] Nichols and Buchannan-Banks[1132] have compiled a list of earth science data types and sources that are likely to yield the most useful data for identifying and assessing the potential seismic hazards of a given area.

1. Bibliographic research of geological and geophysical data, seismicity, historic earthquake records, including accounts of damage from shaking, faulting, and tsunamis.
2. Interpretation of remote sensing data, including conventional aerial photographs. Both the earliest and most recent photography at different scales should be examined.
3. Regional geologic maps, generally at a scale of 1:62,500 or larger. These commonly will have been prepared by the U.S. Geological Survey or state surveys.
4. Special-purpose detailed geologic maps of fault traces and zones, landslides, unconsolidated deposits subject to liquefaction, settlement, and subsidence . . . These may have been prepared only by private consultants for individual sites although the federal and state surveys have been preparing such maps for large areas in recent years.
5. Repeated geodetic measurements over long periods (decades) to detail possible horizontal and vertical land or sea changes. These are normally conducted by public agencies to establish mapping control and design of facilities.
6. Geophysical surveying to determine such things as depth to bedrock, seismic velocities, earth structures, magnetic properties of rocks, or shear wave properties. Surveying may have been conducted for research or design of specific facilities.
7. Measurements of fault creep and earth strain. Normally undertaken for research.
8. Seismometer arrays to determine seismic activity, fault location, type and attitude, and likely hypocenters; conducted for research.
9. Strong-motion instrumentation data from different geologic environs and representative buildings; collected for design and research.

10. Trenches across critical faults to determine location and type of displacement and to secure samples that permit age dating of past movement to determine frequency of fault displacement. Trenches dug and examined both for research and site exploration.
11. Subsurface exploration to locate water levels and barriers and to obtain samples for determination of soil properties needed in computing ground response characteristics. Such data are largely collected by consultants as part of site exploration studies.
12. Detailed topographic maps and submarine profiles are needed to estimate slope stability, prepare possible inundation maps, and evaluate tsunami runup potential. Where available, these normally have been prepared by federal agencies but are being made increasingly for design of large coastal installations.
13. Empirical or theoretical modeling of ground response in typical geologic/soil environments. Such work has been done in universities, by consultants, and by government researchers in the United States and some other countries.
14. Preparation of relative risk maps. These have largely been prepared for specific, large land development projects.

The types and amount of data will vary with the area in question, depending upon the geologic complexity, earthquake history, the type and distribution of land use (existing, planned), the level of planning development, and the amount of funding available.[1132]

1. The San Francisco Bay Region Environment and Resources Planning Study

Because of the recognized need for incorporating earth science data in planning efforts, the U.S. Geological Survey and the Office of Policy Development and Research of the Department of Housing and Urban Development initiated a pilot program in 1970 to develop an essential earth science information base that could be used to relate geologic hazards with land use planning and decision-making efforts. The experimental program centered on the nine-county San Francisco Bay region.[1135,1136] The philosophy and intent of the *San Francisco Bay Region Environment and Resources Planning Study* has been summarized by Borcherdt.[1134]

Although the study focuses on the nine-county . . . San Francisco Bay region, it bears on a different issue that is of national concern — how best to accommodate orderly development and growth while conserving our national resource base, insuring public health and safety and minimizing degradation of our natural and manmade environment. The complexity, however, can be greatly reduced if we understand the natural characteristics of the land, the processes that shape it, its resource potential, and its natural hazards. These subjects are chiefly within the domain of the earth sciences: geology, geophysics, hydrology, and the soil sciences. Appropriate earth science information, if available, can be rationally applied in guiding growth and development, but the existence of the information does not insure its effective use in the day-to-day decisions that shape development. Planners, elected officials, and the public rarely have the training or experience needed to recognize the significance of basic earth science information, and many of the conventional methods of communicating earth science information are ill suited to their needs.

The study is intended to aid the planning and decision-making community by (1) identifying important problems that are rooted in the earth sciences and related to growth and development in the bay region, (2) providing the earth science information that is needed to solve the problems, (3) interpreting and publishing findings in forms understandable to and usable by nonscientists, (4) establishing new avenues of communication between scientists and users, and (5) exploring alternate ways of applying earth science information in planning and decisionmaking.

The study encompasses four program elements: (1) topographic, (2) geologic and geophysical, (3) hydrologic, and (4) planning. The basic products of the study are maps and three types of reports: (1) basic data contributions, (2) technical reports derived from the basic data for a technical audience, and (3) interpretive reports which represent final derivations for a nontechnical audience, such as elected officials.[1137] The investigations and products for the geologic and geophysical elements (active faults, slope stability and engineering behavior of bedrock areas, physical properties of unconsolidated deposits, seismicity and ground motions) are presented in Appendix G.

Cooperative funding for the San Francisco Bay study was continued through 1975 by the two federal agencies. The U.S. Geological Survey ended the study as a separate, formal project on June 30, 1976. By early 1977, more than 100 maps and reports covering a wide range of topics had been published, and many of the methods devised and tested in this study have been adapted and used elsewhere.[1136]

2. Seismic Zonation — San Francisco Bay Region

The geologic environment influences the severity of a seismic event by determining "(1) the potential location and size of damaging earthquakes, (2) the potential for rupture of the ground surface by faulting, both slow creep and sudden movement, (3) the potential for damaging levels of ground shaking on different geologic units at various distances from the source of the earthquake, (4) the potential for flooding from dam failures, tsunamis, seiches, and tectonic changes of land level, and (5) the potential for shaking-induced ground failures such as landslides and those related to liquefaction."[1138] A recent U.S. Geological Survey study demonstrates that it is feasible to assess the above potential earthquake effects for the purposes of *seismic zonation** for a portion of the San Francisco Bay region using *existing* geologic and geophysical knowledge. Seismic zonation represents the necessary base from which land use plans can be developed to minimize future earthquake losses.[1138]

In the USGS study,[1138] methodologies are described for constructing the basic tools needed for seismic zonation in the San Francisco region: (1) an active fault map,[45] (2) data for estimating bedrock motion at the surface,[513] (3) differentiation of sedimentary deposits and a map showing qualitative estimates of ground motion,[535,1139] (4) a map showing areas of potential inundation by tsunamis,[1140] (5) a map showing liquefaction potential,[1141] and (6) a map showing landslide susceptibility.[552]

Borcherdt et al.[630] then developed a methodology for the "composite application" of the basic tools described in the above paragraph to predict the geologic effects of potential earthquakes. To illustrate the strategy, a demonstration profile was selected perpendicular to the San Andreas fault from Sky Londa to the southern tip of the Coyote Hills, along which a $M_L = 6.5$ earthquake was postulated (Figure 2).

The demonstration profile includes five geologic units on the basis of physical properties.[630]

1. Bay mud; most recently deposited soft clay, silt, and minor sand; contains more than 50 weight percent water;
2. Holocene alluvium; poorly consolidated clayey silt, sand, and gravel; contains less than 40 weight percent water;
3. Late Pleistocene alluvium; primarily same material composition as Holocene alluvium, but contains less water and is more consolidated; in some places overconsolidated (soil-engineering sense);
4. Pliocene and early Pleistocene deposits; primarily continental Santa Clara and marine Merced Formations consisting of semiconsolidated and consolidated sandstone, siltstone, and mudstone; and
5. Pre-Tertiary and Tertiary bedrock; includes Franciscan Formation, consisting mostly of sandstone and shales with lesser amounts of radiolarian chert, greenstone, limestone, and serpentine; marine sandstone and shale of Eocene, Miocene, and Pliocene age; and Page Mill Basalt, consisting of lava flows and pyroclastic rocks of Miocene age.

The stratigraphic relations of the five units are illustrated in Figure 3.

The effects of the postulated earthquake (surface faulting, ground shaking, flooding, liquefaction, and landsliding) are dependent upon the distribution of the geologic units with respect to the San Andreas fault. Generalized predictions for each geologic effect along the profile are described below and shown in Figure 3.[630]

Surface Faulting

On the basis of past observation, the postulated magnitude 6.5 earthquake probably would be associated

---

*   Delineation of geographic areas with varied potentials "for surface faulting, ground shaking, liquefaction, and landsliding during future earthquakes of specific size and location."[1138]

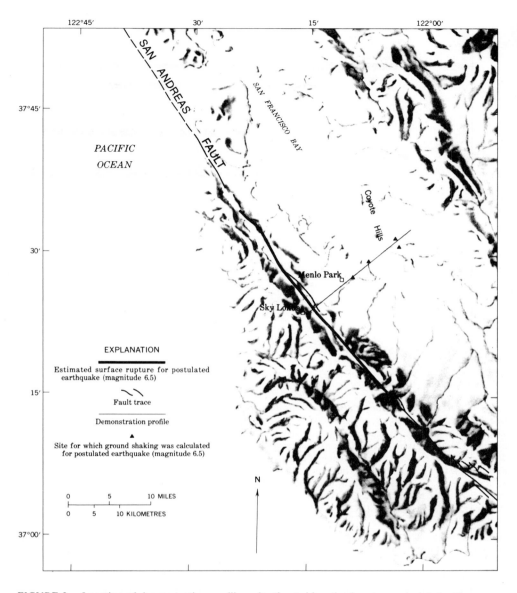

FIGURE 2. Location of demonstration profile and estimated length of rupture associated with a postulated earthquake of local magnitude (M_L) 6.5 on the San Andreas fault, southwestern San Francisco Bay region. (From Borcherdt, R. D., Brabb, E. E., Joyner, W. B., Helley, E. J., Lajoie, K. R., Page, R. A., Wesson, R. L., and Youd, T. L., Studies for Seismic Zonation of the San Francisco Bay Region, Borcherdt, R. D., Ed., U.S. Geological Survey Professional Paper 941-A, U.S. Government Printing Office, Washington, D.C., 1975, 89.)

with right-lateral displacement along the San Andreas fault that may be as great as 1 m . . . The length of estimated surface rupture is 40 km . . . plus or minus about 10 km . . . The main zone of surface rupture will range in width from a few metres . . . to several tens of metres . . ., but small fractures and permanent ground distortion may extend to much greater distances. Locally, branch and subsidiary faults, such as the Black Mountain fault, the Cupertino fault, and the Cañada fault, may also move, but movements on such lesser faults are much more difficult to predict. If sympathetic surface movements do occur along these lesser faults, they are expected to be less than on the main fault rupture.

Ground Shaking

In general, strong shaking (50-125 cm/s) . . . could be expected from the postulated earthquake for all

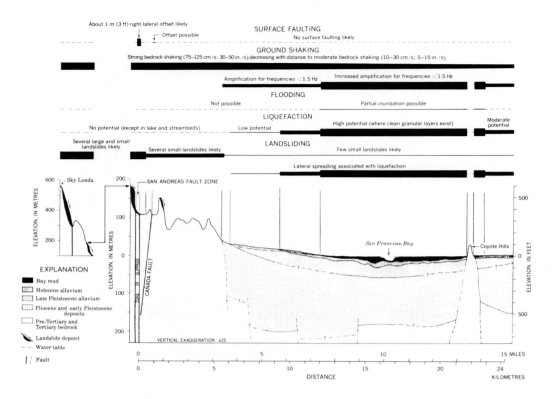

FIGURE 3.   Predicted geologic effects of a postulated earthquake ($M_L = 6.5$) on the San Andreas fault (see Figure 2 for location of demonstration profile and estimated length of surface rupture). The severity of each earthquake effect is indicated qualitatively by thickness of underlining and quantified to the extent permitted by the current state- of-the-art for seismic zonation on a regional scale. The severity of the predicted earthquake effects generally depend on the type of underlying geologic material. Geologic cross section compiled by K. R. Lajoie. (From Borcherdt, R. D., Brabb, E. E., Joyner, W. B., Helley, E. J., Lajoie, K. R., Page, R. A., Wesson, R. L., and Youd, T. L., Studies for Seismic Zonation of the San Francisco Bay Region, Borcherdt, R. D., Ed., U.S. Geological Survey Professional Paper 941-A, U.S. Government Printing Office, Washington, D.C., 1975, 91.)

surface bedrock sites along the profile west of the bay plain . . . The model calculations suggest that a substantial amplification of bedrock shaking in the frequency range below 1.5 hertz could be expected for all parts of the demonstration profile underlain by alluvial deposits, with increased amplifications for the parts underlain by bay mud. The predicted amplifications are large enough to suggest that ground shaking for frequencies below 1.5 hertz may be stronger at the sites underlain by bay mud and alluvium than at sites underlain by bedrock much closer to the fault. Manmade structures with natural periods coinciding with those of the underlying unconsolidated geologic deposits are particularly susceptible to damage.

Flooding

. . . For the postulated earthquake, the most probable cause of inundation by water is the failure of dams or dikes. Flood water from such failures could originate from either San Francisco Bay or upland reservoirs.

For the postulated earthquake, the likelihood of a large vertical offset of the sea floor or large submarine landslide necessary to generate a tsunami seems remote . . .

The postulated earthquake . . . probably is not large enough to generate a seiche in San Francisco Bay. However, seiches could be generated in the Upper and Lower Crystal Springs Reservoirs.

The tectonic setting of the San Francisco Bay region suggests that large tectonic changes of land level (such as occurred during the 1964 Alaska earthquake) are very unlikely to accompany the postulated earthquake. Only minor local changes in vertical elevation of about 0.3 m . . . were produced by the 1906 (San Francisco) earthquake.

Liquefaction

. . . Sediments with the greatest potential for liquefaction are the clay-free granular layers within the bay mud unit. Holocene alluvium has a generally moderate potential with locally high potential in some recent channel and overbank deposits. Late Pleistocene alluvium is generally dense and has a low potential for liquefaction. In addition, much of the Holocene and the late Pleistocene alluvium is normally above the water table and thus has, at most, a seasonal potential for liquefaction.

The most common type of ground failure expected to result from liquefaction along the profile is that of lateral-spreading landslides . . . Areas of the profile with the highest potential for this type of ground failure from the postulated earthquake are underlain by bay mud along the western margins of San Francisco Bay.

Landsliding

The postulated . . . earthquake could be expected to generate several landslides along the demonstration profile . . .

If the postulated earthquake were to take place during a wet season and high ground-water levels, many landslides could be expected along the profile. A few large (more than 150 m . . . in maximum dimension) landslides are likely on the steep slopes between Sky Londa and the San Andreas fault. In this area, existing landslide deposits could be reactivated. Several small landslides could be generated in the area between the San Andreas fault and the western margin of the bay plain. Lateral-spreading landslides associated with liquefaction could be expected near the margins of San Francisco Bay.

If the postulated earthquake were to occur during a dry season and low ground-water levels, the amount of landsliding is expected to be much less. Some landsliding still could be expected between Sky Londa and the San Andreas fault. A few small landslides probably would occur in the other hilly areas along the profile, and lateral-spreading landslides associated with liquefaction still could be expected near the margins of San Francisco Bay.

3. Fault Detection

Identifying active faults (e.g., movement occurring in Holocene time or the past 11,000 years) is one of the most important tasks for assessing seismic risk, especially in areas undergoing development for critical-engineered facilities such as nuclear power plants, dams, refineries, and high-occupancy structures. One of the most promising methods for showing previously unrecognized *lineaments,* a portion of which may represent potential faults, is *remote sensing,* which is defined as the collection of information about a target without physical contact. Target data are measured by methods using various bands of the *electromagnetic spectrum.* Bands of electromagnetic energy used in remote sensing are *ultraviolet (UV), visible light, near infrared (IR), thermal infrared,* and *microwave* (Figure 4). If the data are acquired by a photographic method, the scene representation is called a photograph, if by a nonphotographic method, the representation is called an image. Remote sensing observations are commonly made from aircraft and satellite platforms.[1142] It must be stressed that remote sensing does not replace field investigations. Rather, in its role as a reconnaissance tool, remote sensing may define suspect targets for concentrated field investigations and reveal geologic relationships that might otherwise have been obscure or unknown, even in thoroughly mapped areas. Various remote sensing techniques and applications are described in a number of recently published books.[1142-1149]

A lineament is defined as "a mappable, simple or composite linear feature of a surface whose parts are aligned in a rectilinear or slightly curvilinear relationship and which differs distinctly from the patterns of adjacent features and presumably reflects a subsurface phenomenon."[1150] Lineaments that represent lines or zones of structural displacement are faults or fault zones. In other cases, lineaments may represent zones of crustal weakness where displacement has not occurred (e.g., joints, bedding planes).[1142] Detailed field investigations are required to establish the validity of lineaments mapped as inferred or suspect faults and to determine which of the identified faults might be active. Usually, only a small percentage of lineaments turn out to be faults.

Standard black and white photographs, obtained from suborbital altitudes, have

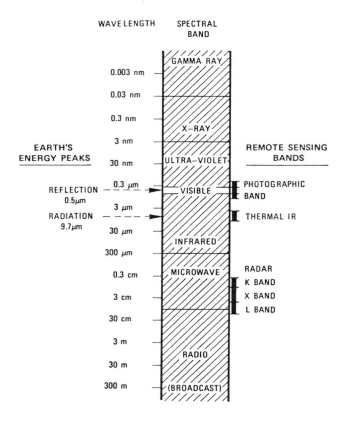

FIGURE 4.   *Electromagnetic spectrum* showing bands employed in *remote sensing*. Electromagnetic energy in space has a constant velocity of 3×10^8 m/sec (i.e., the speed of light). *Wavelength* is the distance between successive wave crests or other corresponding points in a wave. Metric units used for wavelength measurements include nanometers (nm), 10^{-9} m; micrometers (μm), 10^{-6} m; centimeters (cm), 10^{-2} m; and meters (m). (From *Remote Sensing: Principles and Interpretation* by Floyd F. Sabins, Jr., W. H. Freeman and Co. Copyright © 1978.)

long been used for geologic mapping, and color, color IR, and black and white IR photographs are being used increasingly. Photointerpretation techniques and fault identification examples are well documented in a number of books[1142,1145,1149,1151-1160] and professional journals such as *Photogrammetric Engineering and Remote Sensing.* Surrogates or identification clues used for locating suspect faults on photographs and images include straight stream channel segments, straight contacts between erosional and depositional features, straight valleys in hard rock areas, linear anomalies of natural vegetation and lakes, topographic scarps, sag ponds, distinct hue or tonal changes on opposite sides of a lineament, lineaments detectable in both rock and adjoining surficial materials, and offsets in drainage channels, topographic features, and lithologies. Studies using suborbital photographs for identifying and mapping faults in California, Utah, and Texas are included for discussion.

Clark[1161] reports on the utility of small scale black and white aerial photographs* for several fault investigations in California. For example, in a Garlock fault (Figure 5 in Volume I, Chapter 2) study, 1:120,000 scale vertical photographs were used to

---

*   A *small scale aerial photograph* (e.g., 1:120,000) covers a larger ground area in more limited detail than a *large scale aerial photograph* (e.g., 1:20,000), which depicts a small ground area in considerable detail.

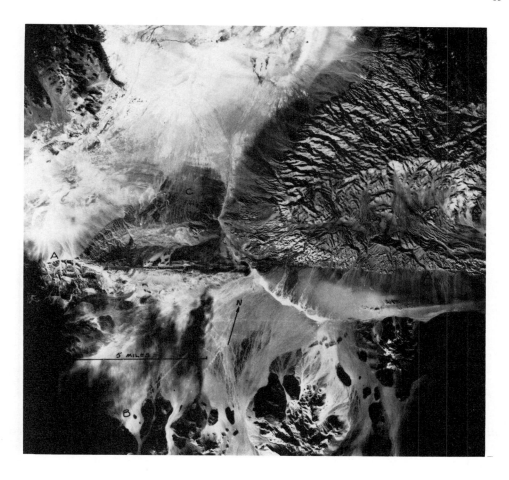

FIGURE 5. Small scale photograph of the left-lateral Garlock fault (A-A′) near Searles Valley, California Cloud shadows obscure detail of volcanic rocks in the southwest corner at B. Shorelines of Pleistocene Lake Searles are evident at C. (From Clark, M. M., Geologic Utility of Small-Scale Airphotos, U.S. Geological Survey Open File Report 1319, 1969, Plate 3.)

help locate the most recent breaks along its 242-km length. The photographs also where extremely useful for revealing its probable continuity in places where surface evidence of the fault was missing for several kilometers (Figure 5). As noted by Clark, had the Garlock fault and the surrounding geology been unknown, ''a geologist would have been able to easily discover it and determine most of its extent from a cursory inspection of a few 1:120,000 scale photos, if not from orbital photos at 1:600,000 scale or so.'' The Coyote Creek fault, located in northwestern Imperial Valley between the Santa Rosa and Vallecito mountains, was the site of the Borrego Mountain earthquake of April 9, 1968 (M_L = 6.4). The 1968 surface rupture extended the known position of the fault approximately 19 km to the southeast. However, an inspection of several 1:145,000 scale photographs received in December 1967 revealed two lineament segments that were identified as a probable extension of the fault. The earthquake occurred along this extension some four months later.

Cluff and Slemmons[1162] used *low-sun angle aerial photographs* to evaluate the active Wasatch fault zone (dip-slip) in Utah. The low solar illumination angle technique captures shadows during their maximum development and enhances subtle geomorphic features, in this case fault scarps, that are usually obscure on *high-sun angle aerial*

FIGURE 6.   High-sun angle photograph acquired about midday along the Wasatch fault just south of Bells Canyon, Utah. Arrows point to fault scarps. Note that the shadows indicating the location of the fault are at their minimum development. Compare with Figure 7. (From Cluff, L. S. and Slemmons, D. B., in *Environmental Geology of the Wasatch Front, 1971,* Utah Geological Association, Salt Lake City, 1972, G4. With permission.)

*photographs** (Figures 6 and 7). The optimum enhancement of scarps and other geomorphic features occurs "when the sun is on the upthrown side of the scarp, is shining nearly perpendicular to the trend of the scarp, and is at an altitude that is below the projected slope of the scarp surface."[1162] The special black and white photographs were interpreted stereographically to locate probable fault traces and fault-related surface features such as tilted blocks and landslides. The lineaments were then divided into three classes to demonstrate a hazard heirarchy appropriate for evaluating future site developments along the fault zone.[1162]

Probably all the Class I lineaments are well-defined topographic features that mark the most recent surface fault ruptures, produced by rapid fault displacement associated with strong earthquakes. Most Class I ruptures are undoubtedly the result of repeated fault displacements that are concentrated along previously established planes of weakness. Therefore, the Class I faults are the most likely candidates for significant future movements.

The Class II features are probable surface faults. They have little vertical relief and may be secondary fault-related features associated with ground failure or graben development.

The Class III features are possible surface faults. They have little or no vertical relief. Most of them appear to be related to the Class I and II fault features; however, some Class III features may represent erosional faultline features or shoreline features and this should be taken into consideration in more detailed investigations. The Class III features are shown because we feel they are possibly fault related and are important enough to be considered for further investigation and evaluation. It is apparent that our confidence level decreases from Class I to III.

In the Houston-Galveston area of Texas, there are a number of active faults associ-

---

*   Most aerial photographs are acquired between 10:00 a.m. and 3:00 p.m. to minimize shadow development. This type of illumination is desirable for topographic mapping to insure an unobstructed terrain.[1142]

FIGURE 7. Low-sun angle photograph acquired at 9:00 a.m. along the Wasatch fault just south of Bells Canyon, Utah. Arrows point to fault scarps. Note that the shadows indicating the location of the fault are at their maximum development. Compare with Figure 6. (From Cluff, L. S. and Slemmons, D. B., in *Environmental Geology of the Wasatch Front, 1971*, Utah Geological Association, Salt Lake City, 1972, G5. With permission.)

ated with tectonic movements, the production of petroleum and natural gas, and subsidence caused by the widespread withdrawal of ground water. Clanton and Amsbury[1163] found that the most useful tool for mapping active faults in the affected region was small scale color IR photographs. Black and white aerial photographs acquired prior to urban development were also useful. The following surrogates were used by Clanton and Amsbury to map potentially active faults on the color IR photographs: fault scarps, sag ponds along fault traces, more luxuriant vegetation and wet ground on the downthrown side of faults, and modifications to drainage patterns on opposite sides of a fault. Two classes of faults were mapped and field checked. Where the photolinears coincided with failures in roads, runways, and buildings, the faults were categorized as active because of the existing structural damage. Straight-line scarps were used to identify inactive faults or faults that are active but cause no structural damage because of their rural setting.

Remote sensing from satellite platforms[1142-1148,1164-1175] offers a synoptic perspective for investigating regional geologic phenomena and is proving to be exceptionally useful for identifying new lineaments in many different geologic environments.[1142,1176-1221] A number of these features have been found to represent important faults. In other cases, observable lineaments have been used to extend the ground-determined extent of previously known faults. Several studies have been included for discussion.

Bechtold et al.[1182] successfully used *Landsat multispectral scanner (MSS) images** for the regional inventory of relatively small faults which cut alluvium in a belt along the Colorado River in the Basin and Range Province of southern Nevada, southeastern California, and northwestern Arizona. Faults previously reported by Longwell[1222] were identifiable on the images, and interpreted faults were confirmed during subsequent field reconnaissance. Many of the newly recognized faults occur along range fronts in intermontane alluvium, a setting making them difficult to recognize in ground-based mapping. The faults generally trend northward and probably represent late Tertiary to Recent east-west extension. Bechtold et al. concluded the study by noting that (1) the Landsat images have a sufficient resolution for conducting surveys of Recent faulting in semi-arid areas and (2) use of the images "can have important applications in the planning and design of engineering projects such as dams, highways, and nuclear power generating stations."[1182]

Abdel-Gawad and Silverstein[1183] demonstrated the effectiveness of Landsat images for identifying faults showing evidence of geologically recent breakage in several areas of the southwestern U.S. and Sonora, Mexico. Several faults not previously known were identified. A California example is shown in Figures 8 and 9. One approach used by Abdel-Gawad and Silverstein to assess the application of Landsat images to the problem of identifying potentially active faults was to plot the epicenters of historic earthquakes on overlays corresponding to the geometry of Landsat images and to correlate the seismicity pattern with "observations on fault continuities, intersections, and inferred evidence of recent breakage." The following quotation summarizes their findings.

Considering the relation of seismicity to the fault pattern, study of . . . [the] imagery shows that in most areas where earthquake clusters occur, there is usually ample evidence of recent faulting. The opposite, however, is not necessarily true: there are many areas and even major fault segments characterized by geomorphic features indicating recent faulting, such as stream offsets and sharp lineaments in the alluvium, which are seismically quiet. This indicates . . . that characterization of earthquake hazardous areas largely on the basis of the present distribution of earthquake occurrences can be quite misleading and may result in unpleasant surprises. Based on our analysis of ERTS [now Landsat] data over large areas in California, Nevada, Utah, and Sonora, it appears feasible to utilize ERTS imagery to identify many more potentially active faults than are presently known. Maps showing the location of potentially active faults could be prepared and in our opinion would be far superior to available maps and certainly would contribute to a more reliable earthquake safety planning program.

Gedney and Van Wormer[1184,1197] have demonstrated the usefulness of Landsat images for delineating structural features in Alaska that had never been recognized on the ground. In one study,[1197] the authors used MSS image mosaics of south-central and central-interior Alaska to map features that had the appearance of being faults for comparison with epicenter maps of recent earthquake activity to determine which of the features might be seismically active. In both regions, most of the epicenters fell on or near lineaments that had not been mapped previously as faults. Two of the apparent faults pass close to Anchorage. The Trans-Alaskan pipeline crosses several of the major lineaments mapped in the central-interior region. Gedney and Van Wormer[1197] point out that because "so little is known of the seismicity of these areas

---

*   The Landsat series of remote sensing satellites[1142-1148,1160,1168-1171] were inaugurated on July 23, 1972 with the launch of Landsat-1 (formerly known as the Earth Resources Technology Satellite-ERTS). The satellites are in a near-polar, sun-synchronous orbit at an average altitude of 900 km. The multispectral scanner (MSS) sensitivities on Landsat-1 and -2 are as follows: band 4, 0.5 to 0.6 μm (green); band 5, 0.6 to 0.7 μm (red); band 6, 0.7 to 0.8 μm (near IR); and band 7, 0.8 to 1.1 μm (near IR). The MSS on Landsat-3 also has a thermal IR channel with a 10.4 to 12.6 μm sensitivity. Each MSS image measures 185 km on a side.

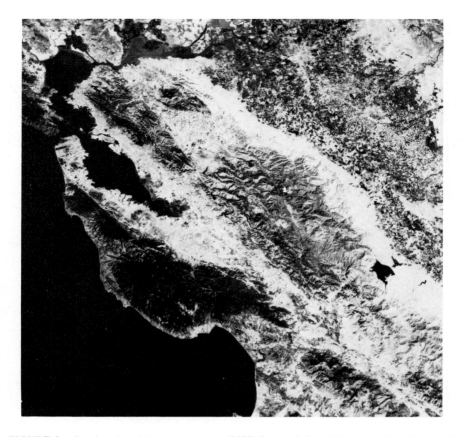

FIGURE 8.   Landsat-1 multispectral scanner (MSS) image (0.5 to 0.6 μm sensitivity) acquired over the Coast Ranges-Central Valley area of California on October 26, 1972.

over long periods, we must regard each of these lineaments . . . as potential sites for future earthquakes.''

Landsat images are being used to evaluate proposed nuclear power plant sites in the U.S. and several foreign countries.[1203,1204,1212] Eggenberger et al.[1204] describe their use of Landsat images in this endeavor.

> The imagery has been used by the authors on both reconnaissance and detailed levels. On the reconnaissance level the LANDSAT imagery enables the investigator to independently compile regional geologic and tectonic maps. In areas of detailed geologic mapping, the LANDSAT derived maps are used to evaluate the quality of mapping and to pinpoint areas where more detailed studies may be required. In areas where sufficiently detailed geologic maps are not available or where geologic mapping has not been conducted, the LANDSAT imagery provides an excellent base for the initiation of regional geologic studies and provides clear evidence for large scale structures. The large scale structures are further studied by conventional aerial photography and by standard geologic field methods.

In Europe, known ground truth data in the technical literature and data derived from the interpretation of Landsat images were used to identify a possible extension of an existing seismically active fault. It has been recommended that a proposed plant site which lies on the extension be withdrawn from consideration.[1204]

Side looking airborne radar (SLAR) systems,[1142-1147,1223-1226] which utilize *active** mi-

---

* SLAR is an active system because it supplies its own source of electromagnetic energy to ''illuminate'' the earth's terrain. For this reason, SLAR is entirely independent of solar lighting, and missions can be conducted day or night.

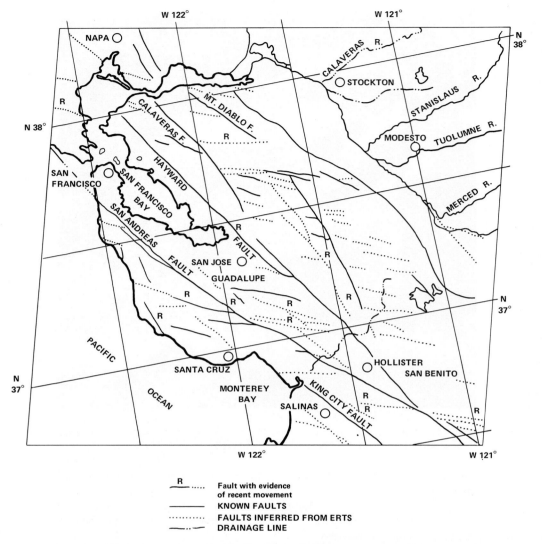

R ..... Fault with evidence
of recent movement
——— KNOWN FAULTS
·········· FAULTS INFERRED FROM ERTS
—··— DRAINAGE LINE

FIGURE 9.   Fault map of the area depicted in Figure 8. (Adapted from Abdel-Gawad, M. and Silverstein, J., in Symposium, Significant Results Obtained from the Earth Resources Technology Satellite-1, Vol, 1, Freden, S. C., Mercanti, E. P., and Becker, M. A., Eds., NASA SP-327, U.S. Government Printing Office, Washington, D.C., 1973, 444.)

crowave energy, are well suited to lineament detection because SLAR's comparatively low angle of "illumination" produces a shadowing effect (*radar shadowing*) that often accentuates subtle fault and fracture systems aligned obliquely or perpendicular to the antenna (Figure 10). In addition, the ability of SLAR to display large areas in a small image format and the relatively low resolution of radar images tend to suppress vegetation signatures that can hamper the recognition of geologic structure (Figure 10). Because of the long wavelengths associated with microwave energy (Figure 4), SLAR essentially has an all-weather capability. This parameter is especially important for wet tropical areas that never have been satisfactorily photographed from the air because of persistent cloud cover.[1227]

A number of studies have demonstrated the usefulness of radar images for structural mapping.[1227-1251] Snavely and Wagner[1233] were among the first to demonstrate the use-

A

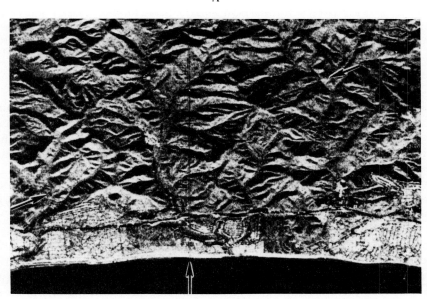

B

FIGURE 10.   Comparison of radar image and photomosaic of fault near San Clemente, California (A) Aerial photomosaic acquired on December 1953. Fault is indicated by arrows. (B) K-band (0.86 cm wavelength) radar image acquired on November 1965. Oblique radar "illumination" causes highlights and shadows that enhance the fault trace. Antenna look direction is indicated by the arrow at the bottom of the image. The strength of radar reflections returned from ground targets and received at the antenna determine the brightness on the image. Light tones represent strong returns, and dark tones represent weak or no returns. (From Sabins, F. F., Jr., in *Geology, Seismicity, and Environmental Impact*, Moran, D. E., Ed., Association of Engineering Geologists, Los Angeles, California, 1973, 151. With permission. Courtesy of F. F., Sabins, Jr., Chevron Oil Field Research Co., La Habra, Calif.)

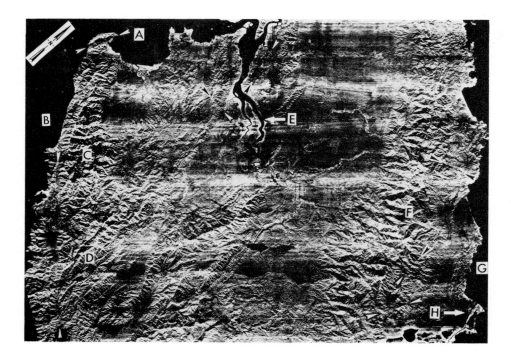

FIGURE 11.   K-band (0.86 cm wavelength) radar mosaic of Darien Province, Panama, acquired by the Westinghouse Corporation in 1967. The mosaic was compiled by the Raytheon Company. A near total cloud cover was experienced during the overlights. Annotated features are: (A) Gulf of San Miguel, (B) Pacific Ocean, (C) Serrania del Sapo, (D) Cordillera de Jurado, (E) Rio Tuira, (F) Serrania del Darien, (G) Caribbean Sea, and (H) Atrato Delta. Small arrows locate several of the faults identified by Mac-Donald.[1242]

fulness of radar images for geologic mapping in regions where bedrock is obscured by dense vegetation. Image analysis and field evaluations for a 16-km-wide strip of the Oregon coast from the California border to the Columbia River indicated that the radar imagery "provides a very clear rendition of the topography and in essence 'defoliates' the thick cover of vegetation," and "fault traces can be discerned on the imagery by juxtaposition of areas with different tonal rendition and by associated linear topographic features."[1239] Field mapping has shown that several of the interpreted lineaments represent previously unrecognized faults.[1233,1239]

The U.S. Army Engineer Topographic Laboratory initiated a radar acquisition and mosaicing project in the late 1960s to improve the geologic maps for the persistently cloud-covered and rugged Darien Province of Panama (Figure 11). MacDonald[1242] extrapolated from the mosaic a wealth of terrain information that was previously unknown. For example, MacDonald was able to develop the following interpretive fault categories: (1) normal fault inferred from radar images, (2) high-angle reverse fault, fault trace inferred from radar images, movement from corroborative data, (3) fault trace inferred from radar images, movement cannot be determined, and (4) strike-slip fault, movement inferred from radar images. The most conspicuous fault on the radar mosaic, and not noted on published maps, is a large strike-slip fault paralleling the Pacific coast (Figure 11). This fault is thought to be a continuation of a fault present in the Intendencia de Choco of Colombia. The total length of the fault in Panama and Colombia is approximately 300 km. Several of the faults identified by Mac-Donald[1242] are annotated in Figure 11.

Gelnett[1250] conducted a study for the National Aeronautics and Space Administration and the U.S. Army Corps of Engineers to determine the usefulness of various remote sensing systems for the identification of regional and specific structural features that could affect the design and location of major engineered structures. The Butler Valley Dam Project in northern California* was selected as a demonstration site because (1) it is located in a mountainous region that is covered by dense evergreen forests, (2) the geology is extremely complex and largely unknown, (3) there is difficulty in deciphering geologic features due to the lack of exposures, and (4) it is a seismically active region. Remote sensing products used in the study included Landsat images, X-band radar images, and black and white (bands of the visible and near IR spectrum), color, and color IR photographs at several different acquisition scales. The primary objectives of the study were

(1) To refine the location of known regional faults and possibly detect previously unknown faults;
(2) To determine the presence or absence of a regional fault and fracture pattern that may be related to that found locally in the vicinity of the project;
(3) To assist in defining landslides existing along the proposed relocated road; and
(4) To evaluate the applicability of individual sensors for future use by other engineers and geologists.

The objectives were satisfactorily achieved, and the data "obtained from the remote sensor imagery added materially to and improved significantly upon existing knowledge of the geology in the project area as well as the entire study area." However, the radar images revealed substantially more structural detail at both the local and regional levels than any of the other systems. This was attributed to the ability of SLAR to record topographic and geologic detail while at the same time suppressing vegetation detail. In addition, the synoptic view enabled Gelnett to recognize discontinuous or intermittent fault traces.[1250]

An X-band radar image of the Butler Valley Dam Project area is presented in Figure 12. Lineaments verified in the field as actual faults or interpreted from the radar image as probable faults are shown in Figure 13. Most of the faults in the immediate vicinity of the dam site were identified by prior field reconnaissance, but their extent and trends were poorly defined. However, one prominent fault, approximately 10 km in length with a N10°E trend, was overlooked repeatedly in the field. This fault is marked by arrows on the right side of Figure 13.[1250]

Because of its ability to highlight lineaments that might represent active geologic faults, radar images are playing an important role in analyzing sites proposed for nuclear power plants.[1226] The Dames and Moore Company, for example, has used SLAR images for a number of siting projects in the U.S. The images are used as a complement to satellite data for lineament detection purposes.[1212]

A few studies have evaluated the usefulness of *thermal IR images***[1142-1147] for mapping faults.[1252-1262] The best results have been obtained with predawn images over unvegetated terrain. When detected, faults are most often expressed as offsets of rock units that have different radiant temperatures or temperature anomalies caused by the blockage of ground water. In the latter case, fault zones act as underground barriers to ground water movement, thereby trapping and concentrating water in surficial materials overlying the faults. Sabins[1252] points out that this could be a potentially useful technique for locating faults that are covered by alluvium without surface expression.

One of the most dramatic examples of applying thermal IR images to structural mapping occurred in southern California. Sabins,[1142,1154,1262] upon examining black and

* The study area encompasses 15,280 km² and is located approximately 360 km north-northwest of San Francisco in the Coast Ranges and Klamath Mountains.

** Thermal IR imaging systems detect and display radiant temperatures of targets. The sensors normally operate in the 8 to 14 μm band of the electromagnetic spectrum (Figure 4). Because all objects continually emit thermal IR energy, collection missions can be conducted day or night.

FIGURE 12.   Motorola X-band (3-cm wavelength) radar image of the Butler Valley Dam Project area in northern California.   The image was acquired through a 50% cloud cover.   Arrow marks the proposed dam site.   (From Gelnett, R. H., Airborne Remote Sensors Applied to Engineering Geology and Civil Works Design Investigations, Technical Report No. TR-17621, Motorola Aerial Remote Sensing, Inc., Phoenix, Ariz., 1975, 16.   Courtesy of R. H. Gelnett, Motorola Aerial Remote Sensing, Inc., Phoenix, Ariz.)

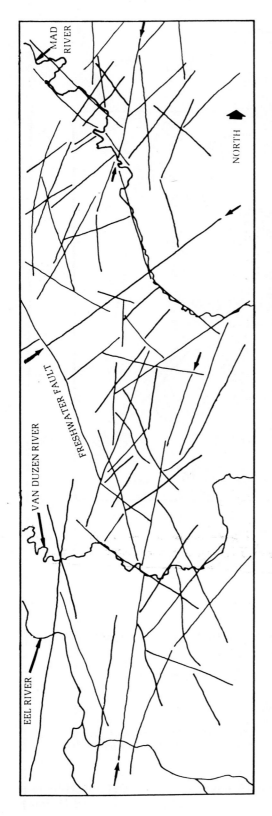

FIGURE 13. Lineaments mapped or interpreted as probable faults by Gelnett[1250] for the Butler Valley Dam Project area in northern California. Arrows mark the most prominent features. Compare with Figure 12. (From Gelnett, R. H., Airborne Remote Sensors Applied to Engineering Geology and Civil Works Design Investigations, Technical Report No. TR-17621, Motorola Aerial Remote Sensing, Inc., Phoenix, Ariz., 1975, 18. Courtesy of R. H. Gelnett, Motorola Aerial Remote Sensing, Inc., Phoenix, Ariz.)

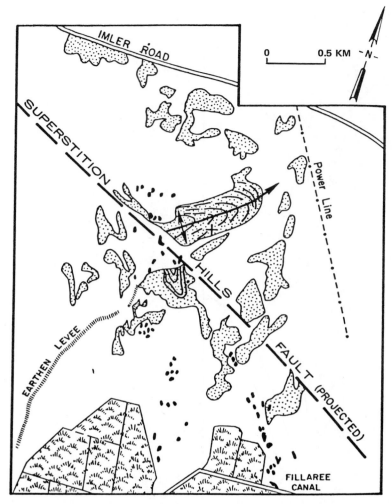

OUTCROPS OF DEFORMED TERTIARY LAKE DEPOSITS,
SHOWING TRENDS OF BEDDING.

RECENT WIND-BLOWN SAND COVER.

CULTIVATED AREAS.

STABILIZED SAND DUNES AND TUFA-COATED BOULDERS.

BRASS CAP MARKING SE CORNER SEC. 25, T.14S, R12E.

FIGURE 14.   Geologic interpretation map of predawn thermal IR image of the Super-
stition Hills fault area (southeastward extension) western Imperial Valley, California.
The thermal IR image is shown in Figure 15. (From Sabins, F. F., Jr., *Geol. Soc. Am.
Bull.*, 80, 402, 1969. With permission. Courtesy of F. F. Sabins, Jr., Chevron Oil Field
Research Co., La Habra, Calif.)

white photographs and a predawn thermal IR image of the Superstition Hills fault
area in the western Imperial Valley (Figure 14), concluded that the thermal IR image

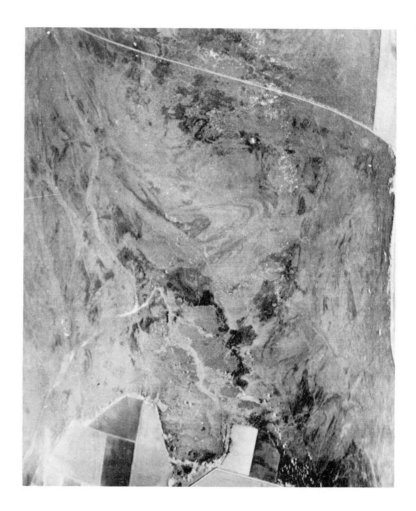

FIGURE 15. Predawn thermal IR image of the Superstition Hills fault area (southeastward extension) in western Imperial Valley, California. The image data were acquired in October 1963 and have a wavelength sensitivity of 8 to 14 μm. Light image tones represent relatively warm radiant temperatures, and dark tones relatively cool radiant temperatures. (From Sabins, F. F., Jr., *Geol. Soc. Am. Bull.*, 80, Plate 1-B, 1969. With permission. Courtesy of F. F. Sabins, Jr., Chevron Oil Field Research Co., La Habra, Calif.)

exhibited a far greater amount of geologic detail than the photographs (Figures 15 and 16). The major geologic features in the study area are the east-plunging Imler Road anticline (previously unknown) and the southeastward extension of the Superstition Hills fault (Figures 14 and 15). Besides truncating the anticline, the fault is delineated in the lower right portion of the thermal IR image by a lineament that is cooler on the east and warmer on the west (Figure 15). The trend of the lineament is parallel with and approximately 160 m east of a row of warm sand dunes. On April 9, 1968, the Borrego Mountain earthquake (M_L = 6.4) caused surface breaks (less than 3 cm of right-lateral displacement) along the Superstition Hills fault. The breaks closely correspond to the lineament visible on the image (Figure 15). Sabins[1142] states that the thermal IR image, "which was acquired four years prior to the earthquake, located an important fault that is obscure both on aerial photographs and in the field."

In addition to remote sensing techniques, improved seismographs are making it pos-

FIGURE 16.   Vertical aerial photograph acquired in May 1953 of the Superstition Hills fault area (southeastward extension) in western Imperial Valley, California. Compare with Figure 15. (From Sabins, F. F., Jr., *Geol. Soc. Am. Bull.*, 80, Plate 1-A, 1969. With permission. Courtesy of F. F. Sabins, Jr., Chevron Oil Field Research Co., La Habra, Calif.)

sible to record microearthquakes that may delineate a previously unknown active fault or fault segment. A recent example was associated with the May to June 1970 Danville, California earthquake swarm. A dense concentration of permanent and temporary seismographs revealed that the swarm sequence did not occur along the mapped fault trace of the Calaveras fault, but along a probable northwest extension of the Pleasanton fault that had not been previously mapped in the epicentral region.[1263] According to Lee et al.,[1264] this implies that not all future earthquake sources in the San Francisco Bay region can be identified by surface mapping techniques alone. Rather, delineating active faulting at depth by recording small earthquakes could help to identify areas of potential earthquake risk.

Once a fault has been identified in a small scale context, say for example from remote sensing interpretation and surface mapping, several subsurface investigation methods are available to more accurately determine a fault's location in a local environment. The more common methods are *trenches* and *boreholes* which represent *geological surveys*, and *geophysical surveys* which include *seismic refraction, electrical resistivity,* and *measurements of the earth's magnetic and gravity fields.*[1102,1265-1272] It is important to emphasize that geophysical methods alone are not sufficient to prove the existence or location of a fault, but rather, geophysical data should supplement geological information obtained from surface mapping, trenches, boreholes, and other sources.[1270,1271] It is also important to emphasize that if carbonaceous material is present along trench exposures, radiocarbon dating can be used to estimate the recency of movement along a fault. In addition, recent fault movement can be inferred from trench exposures if soil units or young alluvial deposits have been displaced.[1272]

Brabb[1273] notes that the number of active faults known in a region is related to the amount of geologic reconnaissance that has taken place. This was dramatically illustrated in the nine-county San Francisco Bay region where the U.S. Geological Survey and the U.S. Department of Housing and Urban Development cooperated in a pilot program (San Francisco Bay Region Environment and Resources Planning Study) to relate geologic hazards to regional and urban planning.[1134,1273] As a result of the field work completed for this project, the number of recognized active faults in the nine-county area more than doubled. Several of the faults were discovered as a result of investigations not designed for the explicit purpose of finding active faults (e.g., landslide, stream deposit, and coastal erosion studies). The following describes how one active fault was discovered in the course of a landslide investigation.[1273]

An active fault northeast of Alexander Valley, Napa County, was discovered by Carl Wentworth and Virgil Frizzell as they were looking for landslides on small scale aerial photographs. Their curiosity was piqued by a series of linear features that suggested fault movement. More detailed photographs showed places where streams had been damaged by fault movement and then filled with stream sediment. Dramatic proof for the activity of the fault came from Dorothy Radbruch-Hall who was independently mapping landslides on the ground. She observed two old redwood fences that had been offset in a right lateral sense about 30 centimeters by the fault. Subsequent work has shown that the fault is at least 35 kilometers long.

D. Reservoir-Induced Earthquakes

Although it is imperative that proposed sites for dams be searched for faults that might be seismically active, there is an additional problem of potential concern for planners, engineers, geophysicists, and geologists. There is a growing body of evidence indicating that the impounding of water in several large reservoirs has stimulated the release of stored tectonic energy by induced or enhanced local earthquake activity (*reservoir-induced earthquakes*).* To date, evidence ranging from suggestive to compelling has been presented for about 30 *seismic reservoir sites* in the U.S.,[1274-1284] Canada,[1285-1286] Japan,[1287,1288] Australia,[1288] New Zealand,[1289] People's Republic of China,[1290-1292] India,[1293-1300] Pakistan,[1301] Soviet Union,[1302-1306] Yugoslavia,[1307,1308] Greece,[1309-1311] Switzerland,[1312] Italy,[1313,1314] France,[1315,1316] Spain,[1288,1295,1316] Algeria,[1317] Zambia-Rhodesia,[1318-1321] and South Africa.[1288] In the U.S., reservoir-induced earthquakes have been reported at Lake Mead, Arizona-Nevada; Palisades Reservoir, Idaho; and Clark Hill Reservoir and Lake Jocassee in South Carolina. Several reviews describe the geology, reservoir parameters, and seismicity patterns for a number of the above worldwide sites.[1288,1289,1308,1316,1322-1328]

The following listing describes several characteristics of reservoir-induced earthquakes.

1.   Most of the earthquakes occur beneath a reservoir or within its immediate vicinity (within 25 km) at very shallow focal depths (less than 10 km).

2.   The stress system determined from fault plane solutions is consistent with the

---

*   Induced seismicity also can be caused by underground nuclear explosions, high-pressure fluid injection (discussed in Volume II, Chapter 2), and rock failures in mines.

pre-existing tectonic stress field in the region of a reservoir. The induced earthquakes are of the strike-slip or normal type; there are no reported cases of reservoir-induced earthquakes associated with thrusting mechanisms.[1329,1330]

3.   The seismic activity is usually associated with high water levels and rapid increases in water levels. Most often, the number of events increases as a reservoir level rises, and the largest earthquakes occur after the reservoir has reached its highest level.[1329] Reservoir surface area appears to be much less important than water depth.[1326]

4.   Foundation geology for affected reservoirs varies from granitic and metamorphic rocks to shale and sandstone.[57]

5.   Earthquakes have occurred in known seismic areas and in regions that, prior to impoundment, were apparently aseismic for some time. However in both cases, the rocks within which the earthquakes occurred probably were highly stressed before impoundment.[1329]

6.   Most earthquakes are of small magnitude, but four events have reached magnitudes greater than 6:
      Hsinfengkiang Reservoir, People's Republic of China, March 19, 1962, $M_L =$ 6.1
      Lake Kariba, Zambia-Rhodesia, September 23, 1963, $M_L = 6.1$
      Kremasta Lake, Greece, February 5, 1966, $M_L = 6.2$
      Shivaji Sagar Reservoir, India, December 10, 1967, $M_L = 6.4$

The largest local magnitude for a reservoir-associated earthquake in the U.S. is 5. This earthquake occurred at Lake Mead on May 4, 1939, approximately 10 months after the reservoir was filled to its operating level.

Two mechanisms have been proposed most often to explain the triggering of earthquakes at seismic reservoir sites. First, the extra weight of the water (i.e., *the load effect*)[1330] could increase the local vertical strain in reservoir rock that might exceed the threshold needed for local faulting to occur, causing movement along pre-existing faults previously considered inactive.[57,1327] Because the stresses caused by the water load are small in comparison with the energy released in the earthquakes, it is assumed that the rock masses were close to failure prior to reservoir filling.[1288] This mechanism is most attractive at sites where the seismic activity takes place directly beneath a reservoir. A second and more likely mechanism in most instances is associated with increases in pore pressure (i.e., *the pore pressure effect*).[1330] In this example, reservoir water under pressure would percolate through fractures in the reservoir rock increasing the pore pressure by a few tens of bars. This condition would reduce the effective normal stress (frictional resistance) across lines of weakness allowing sudden slippage to take place.[57,1288] The pore pressure mechanism (1) would not require the presence of pre-existing faults at a reservoir site and (2) is identical to that proposed for induced earthquakes associated with high-pressure fluid injection projects in the U.S. and Japan (discussed in Volume II, Chapter 2). Schleicher[1280] suggests that loading may trigger normal faulting for a short period, perhaps several years, after reservoir filling. If the activity continues, it is possible that the stresses are being modified by changes in pore water pressure.[1280] Klein[1331] further suggests that rapidly fluctuating tidal stresses may influence the time of occurrence (hour time scale) of reservoir-associated earthquakes.

It must be emphasized that the impoundment of a large reservoir does not mean that induced seismic activity will follow automatically. In fact, the filling of more than 90% of the world's large reservoirs has not been accompanied by induced or enhanced seismicity. One or more of the following factors may be responsible for the aseismic behavior of most large reservoirs:[1288,1301,1316,1326,1332,1333]

1. The presence of flexible strata in the reservoir basement which would deform gradually to accommodate any applied load
2. The absence of diaclastic formations (favor large water losses), fractures, and rock heterogeneity, all of which facilitate the movement of water under pressure
3. The absence of large, ambient, effective stress differences in regions that are not critically stressed
4. In areas of high natural seismicity, the stress fluctuations may be so high that the triggering effect of reservoirs is ineffective by comparison.

Until recently, it was thought that only small earthquakes could be associated with reservoir filling. However, during the 1960s four damaging earthquakes of magnitude >6 occurred at reservoir sites in the People's Republic of China, India, Greece, and along the border of Zambia and Rhodesia. These events focused worldwide attention on the fact that reservoir-induced earthquakes could cause deaths and injuries, widespread damage to local communities, and damage to the dam structure itself. The most significant earthquake occurred on December 10, 1967 (M_L = 6.4) in association with the Shivaji Sagar Reservoir and Koyna Dam in the western Peninsular Shield of India, an area considered to be aseismic prior to impoundment. The earthquake claimed 177 lives, injured 2300, and left thousands homeless.[1288,1327] More than 80% of the houses in Koyna Nager Township were destroyed or severely damaged. The 85-m-high concrete gravity dam moved violently during the earthquake and suffered structural damage. The dam was repaired and strengthened substantially during subsequent years and is now thought to be strong enough to withstand similar sized earthquakes in the future.[1288]

Along somewhat different lines, Schleicher[1280] states that, analogous to potentially controlling earthquakes by high-pressure fluid injection programs (Volume II, Chapter 2), fluctuating reservoirs, "under the right circumstances, may provide a convenient way to trigger small quakes, thereby preventing tectonic stress differences from increasing to dangerous levels." He warns, however, that the "repeated triggering of movement along one part of a fault might increase the seismic danger elsewhere along the fault." Similar to the first concept, Bufe[1334] reports that a persistent 10-km seismicity gap along the Calaveras fault coincides with the position of the Leroy Anderson Reservoir southeast of San Jose, California. It is suggested that a small increase in pore pressure provided by the reservoir could "tip the scales" toward increased stable sliding (aseismic creep) and away from stick-slip events. A bridge crossing the reservoir has been severely deformed, apparently by aseismic creep on the Calaveras fault.[1334]

Because of the potential destructiveness of reservoir-induced earthquakes, it is imperative that detailed monitoring and survey programs be carried out for an entire reservoir area. At the first meeting (1970) of the *UNESCO Working Group on Seismic Phenomena Associated with Large Reservoirs*, it was recommended that site investigations be planned in two phases.[1288,1322]

Phase 1—Preliminary State of a Project
1. Study the historical seismicity of a reservoir site and adjacent area
2. Conduct field surveys of the site and adjacent area to identify potentially active faults
3. Monitor preimpounding seismicity and establish minimum noise levels with portable seismographs prior to the start of construction

Phase 2—To Be Initiated 1 or 2 Years Before Impounding if Phase 1 Data Indicate That Site Studies Should Be Intensified
1. More detailed geologic surveys of the reservoir site and adjacent area

2.   Installation of a permanent seismograph network
3.   Possible other work to include precise leveling surveys and studies to evaluate the stability of reservoir slopes

Gupta and Rastogi[1288] suggest that, in addition to the above investigations, the *in-situ* stresses should be estimated by the *hydraulic fracturing technique* whenever possible. Any sites showing rocks to be near-critically stressed should be avoided for dam construction.[1288] Lomnitz[1333] suggests that a fairly simple way of detecting a seismic hazard at a reservoir site would be to drill a borehole in the center of the prospective site and pump water to depths on the order of 300 m or more, gradually increasing pressure until earthquake activity commences. The threshold of induced activity would render a measurement of relative earthquake risk.[1335]

Gupta and Rastogi[1288] describe in detail a number of procedures for establishing and operating seismic stations at reservoir sites. Tsung-ho,[1292] Gupta and Combs,[1300] Kanasewich,[1328] Yaralioglu,[1335] Brown,[1336] Lane,[1337] and Blume and Fuchs[1338] describe seismic investigation schemes for several reservoir sites. The book *Dams and Earthquakes* by Gupta and Rastogi[1288] is highly recommended for those desiring detailed information on reservoir-induced earthquakes.

E. Land Use Problem Areas

Mader[1117] acknowledges that, while there have been advances in recognizing seismic hazards in land use planning, there are a number of areas of concern to which planners and others must address themselves to further improve planning for seismic safety. These areas of concern have been summarized by Mader.[1117]

1. *Education of planners, politicians, and others:* One of the main reasons that seismic problems are ignored in planning and decision-making is that the players simply don't know the score. Until all persons involved have a reasonable understanding of the subject, progress in seismic safety will be slow. Developing information that is meaningful and understandable is crucial. Unfortunately, the scientists have gone for too many years speaking to themselves with pure research, rather than to a larger audience through applied research. It is the application of the research that will educate the players. All avenues, such as regular professional meetings, special educational programs, public schools, and daily contacts, should be used in furthering education with respect to seismic problems and solutions.

2. *Availability of useful data:* Many geologic maps and studies, while well suited to the needs of the geologists and other scientists or engineers, may be entirely useless to the planner. The data need to be presented in readily understandable form. Interpretive maps which describe the probable impact of seismic conditions on land use, for instance, can be very useful. This is a difficult subject area requiring innovation from the producers and users of the data. The on-going San Francisco Bay Region Study sponsored by USGS and HUD is an example of an attempt to provide region-wide geologic data that will be useful to planners and others. While a pioneering effort, the maps and reports from this project are finding their way into planners' offices and are being used. Experience with this and other geologic information programs leads one to believe in the strength of good information itself. It appears that when maps, which clearly describe geologic problems, are publicly available, people buying houses, developers, and others naturally avoid starting purchases or projects which have significant potential hazards unless they are confident that solutions can be found. This tends to avoid the problem of the unsuspecting buyer who tries to complete an ill-advised project because to do otherwise would, he believes, cause him greater financial loss than abandoning the project.

3. *The state's role:* Because local government often seems lethargic in coming to grips with seismic problems, it is probably necessary for states, in many instances, to lead the way. The type of leadership, however, can vary considerably. In California, the newly formed Seismic Safety Commission is attempting to define the state's role. Questions that arise are varied. For instance, should the state attempt to set state-wide land use policy that recognized major areas of potential seismic hazard and divert potential development? Should the state enact primarily enabling legislation and allow local government to go ahead and adopt local regulations and plans if they choose? Should the state attempt to regulate land use directly in order to protect the public against the most obvious hazards? These are difficult and complex questions which will take some time to answer as different approaches are attempted.

4. *Strict regulations versus review procedures:* A problem often faced is whether to adopt strict and specific regulations to govern land use or whether to emphasize the need to have good information and qualified

professionals involved in making land use decisions on a case-by-case basis. This question is particularly apparent in planning related to geology because of the variations in geologic conditions in relatively small areas and the range of design solutions possible to solve problems. The strict and specific regulations are easier to administer as fewer questions have to be answered. The review approach, on the other hand, requires a high level of sophistication on the part of the reviewers. The San Francisco Bay Conservation and Development Commission, which issues permits for all development projects in the Bay, uses a highly qualified panel of experts in matters related to soils engineering, geology, engineering geology, architecture, and structural engineering in reviewing individual permits for bay fill. This type of review approach, however, requires a considerable professional input.

5. *Local geologic advice:* While the employment of a city or county geologist as a staff member or as a consultant has seemed a luxury to many people, it is becoming increasingly clear that in areas with significant seismic or geologic problems it is a must. Just as city engineers review the work of applicants' engineers and as city attorneys review legal matters for a city, it is now increasingly obvious that a city geologist is needed to review and advise on geologic matters. Competent geologic review can help prevent ill-advised development and protect the public. Also such review can help reduce potential adverse legal decisions against the city in the event of seismically induced damage. Nonetheless, there are many who would still hold such advice to be an extravagance.

6. *Use of geologic data at all stages of the planning-regulation-development process:* It is important that geologic information be considered at each step in the planning-regulation-development process, from the preparation of a general plan to the preparation of regulations to the review of projects and finally the construction of buildings. This sequence of decisions flows from general decisions for large areas to detailed decisions for specific sites. Thus, the general type of information that is adequate for the early decisions will usually not be sufficiently accurate for later, more detailed decisions. Accordingly, one should, at each step along the way, consider the geologic data available and determine if more information is needed. To neglect needed information at a critical point can lead to improper decisions.

7. *Definition of acceptable risk:* Actions to minimize exposure of people and improvements to seismic hazards are really actions to minimize risk. This immediately brings one to the question of what degree of risk is acceptable. Civil engineers, for instance, have generally adopted the 100-year flood as the level of acceptable risk for much for their design work related to flood problems. Can some similar standard be established for seismic hazards? Efforts, to date, have not been extremely encouraging. The variables in defining risk are many, and the concept of acceptable risk varies by individuals. Yet, rationally, it appears that a jurisdiction can come to a common definition of what the public will and will not tolerate. Certainly, building codes include risk factors, whether expressed or not. Whether or not acceptable risk is defined, actions taken with respect to seismic safety do imply a level of acceptable risk.

8. *Existing hazardous areas:* As we identify seismic hazards in already developed urban areas, the possibility of decreasing total exposure to hazards by modifying occupancies in structures seems worthy of consideration. Realistically, it is very difficult, if not impossible, to remove structures and uses in even very hazardous areas. If this is the case, changes in land use plans and zoning for an area might attempt to reduce occupancies to a more acceptable level.

Another approach, of course, is to attempt to increase the structural quality of the buildings. The City of Long Beach, California, is often held up as an outstanding example of this approach.

There may be instances where the problem should be addressed as a matter of urban renewal. If a developed area is too hazardous, perhaps it should be cleared of buildings and left open. Then, rebuilding should only take place if new and better designed structures would provide an acceptable level of safety.

In any event, the already built-up portions of our cities are areas of major concern in the event of future earthquakes. A great deal of attention must be focused on this problem.

9. *Planning for land use in seismic areas prior to an earthquake:* The possibility exists that in areas with known seismic dangers, it is possible to predict the types of failures that will occur as a result of the earthquake. If this is true, might there not be advantages to preparing plans or alternate plans for rebuilding following the event? Such plans would likely call for new land use patterns and certainly for new structural requirements. If thought is not given to this matter prior to the earthquake, pressures to rebuild after the catastrophe are usually so intense that there is little time to even think of new land use patterns. In areas that may be obsolete for other reasons as well, such as deteriorated buildings, it would probably be even more possible to consider new land use patterns. Such studies would be expensive and require substantial funding.

10. *Rebuilding following an earthquake:* The pressures to rebuild after an earthquake are very intense. The economic losses to individuals and the city are so great that safety tends to take a back seat to expediency. Yet, in the long run, such an approach may well prove folly. There should be some mechanism by which a city damaged by an earthquake has an opportunity to make the most urgent repairs and then take sufficient time to prepare new land use plans which are responsive to the seismic problems before undertaking the rebuilding process. This probably requires emergency governmental powers, as well as considerable financial aid.

II. EARTHQUAKE DISASTER PREPAREDNESS AND RECOVERY

Disaster preparedness has as its purpose the safeguarding of people in extraordinary emergencies. In an optimum framework, disaster preparedness means that all levels of government and private groups with emergency missions are prepared "to respond promptly to save life and protect property if it is threatened or hit by an emergency of any type, utilizing all available resources."[1339] Disaster preparedness, to be effective, requires planning and preparedness actions to be formulated *before* an emergency arises. Earthquakes are the most difficult natural hazard to prepare for because they can strike suddenly without warning, any time of day or during any day of the year, and create multiple primary and secondary hazards.[509] Added to these difficulties, the need for effective earthquake preparedness programs continues to assume greater importance because of society's increasing vulnerability to seismic hazards. The severity of the problem in the U.S. was described by C. R. Allen in testimony before the Committee on Public Works, U.S. Senate, in 1971.[509]

. . . the pressures of population growth are causing expansion into areas that are more difficult to develop safely than those of past decades — often into mountainous areas, active fault zones, or areas of artificial fill that necessarily have earthquake-related problems associated with them.

Society is rapidly becoming more complex and interdependent; so that we are becoming increasingly reliant on critical facilities whose loss can create major disasters.

The increasing population density in some of our cities creates problems such as a very localized earthquake causing a major catastrophe, such as was not possible some years ago.

And, rightly or wrongly, I think people increasingly look to their governmental agencies to help them or protect them from natural disasters such as earthquakes, even when they individually have demonstrated little foresight — or even negligence — in preparing for such events.

A. The Case of Managua, Nicaragua and San Fernando, California

The events of recent history have taught us repeatedly that communities in numerous countries have been ill-prepared to deal with earthquake disasters. The problem of preparing for, and coping with, earthquake disasters is most acute in the developing countries. For example, Kates et al.[1340] report that very little was done to improve disaster preparedness (e.g., emergency planning, earthquake-resistant construction, redundancy and decentralization of emergency services) in Managua, Nicaragua between the March 31, 1931 and December 23, 1972 earthquakes. More specifically, the only major pre-earthquake prevention, planning, and preparedness measures in effect at the time of the 1972 earthquakes (largest of 3 events, $M_L = 6.2$) were as follows: (1) a few buildings were designed and constructed to be earthquake-resistant, (2) earthquake insurance was in force for only upper-income housing because it was required by mortgage lenders, and (3) a radio frequency had been set aside for emergency broadcasts.[1340]

The following shortcomings characterized the *emergency period** in Managua following the 1972 earthquakes.[627,1340,1341]

1. Emergency facilities performed poorly. All fire fighting equipment was destroyed or trapped in collapsed or damaged buildings. Managua's four major hospitals (1650 beds) were heavily damaged and forced out of operation, and the 5-story Red Cross building collapsed, destroying three ambulances. Destruction of these facilities emphasized the need for special design and/or siting criteria to insure their availability following a destructive earthquake.

2. The water and electrical power distribution systems were severely impaired. For

---

* Period commencing with the earthquake and continuing until the effects of the disaster no longer pose a hazard to life and property.

example, only 10% of the city's population had water service 1 week following the earthquakes.

3. The provision of work space and utilities for emergency organizations was a very difficult task because most buildings were either destroyed or severely damaged (Figures 1 and 2 in Volume I, Chapter 1). Tents and the homes of agency heads had to be used as temporary offices. The homes were very valuable resources because they had certain utilities available by the end of the first week.

4. The mobilization of emergency organizations started late and proceeded slowly; the citizens had little support or direction from public and private organizations for approximately 2 days following the earthquake.

5. Organized search and rescue operations with careful record keeping never took place. Burials took place without body identification efforts.

6. Food distribution was poorly organized and delayed because the highly centralized national government (three-man junta) declared itself the sole distributor of foodstuffs.

7. Efforts by government officials to boost the morale of Managua's citizens and keep the populace informed of what actually was taking place were usually late and meager.

8. The *extended family system** was both a blessing and a hindrance following the earthquakes. For example, approximately 75% of the homeless found shelter in and around the homes of relatives in Managua and more distant communities, and food stored in these host homes supported at least 200,000 people for several days. However, most emergency organizations only had skeleton staffs available during the emergency period because the members' first responsibility was to their extended family, with organizational responsibility ranking a distant second.

Although emergency procedures were generally swift and operated without complications following the February 9, 1971 San Fernando, California earthquake (M_L = 6.4), several important lessons were learned for improving disaster preparedness programs. The following is a summary appraisal of the emergency operations of hospitals, fire departments, law enforcement agencies, and the American National Red Cross. Most of the material for this section was obtained from a detailed and highly recommended report by R. A. Olson.[1342]

For hospitals (17 were damaged or destroyed), the earthquake presented immediate problems of caring for the injured, search and rescue, and transferring patients. Valuable hospital units such as emergency rooms and associated medical resources, including ambulances, were rendered inoperative at certain facilities (Figure 17). One week following the earthquake, 1147 hospital beds were unusable. At the Olive View Community Hospital (Figure 15 in Volume II, Chapter 3), 600 patients had to be removed through two exterior stairways because no elevators were operating, and the main stairwells at the end of each wing separated from the main building (Figure 15 in Volume II, Chapter 3). Flashlights had to be used for the evacuation because the regular and emergency power systems failed.[1342]

The largest search and rescue effort took place at the San Fernando Veterans Administration Hospital (Figure 16 in Volume II, Chapter 3). Organizations involved in the effort included local hospital staffs, fire departments from the City and County of Los Angeles, Los Angeles County Sheriff's Department, Los Angeles County prisoner volunteers, U.S. Forest Service, Salvation Army, and a private contractor. The search

---

* System wherein an individual can expect many types of assistance when needed not only from members of the immediate household, but also from other relatives.[1340]

FIGURE 17.   Damaged ambulance port, Olive View Community Hospital following the February 9, 1971 San Fernando, California earthquake (M_L = 6.4). The shelter had a reinforced-concrete roof slab supported by columns with no walls. The columns failed, usually at the top, and the roof fell on several vehicles.[957] (Courtesy of James L. Ruhle and Associates, Fullerton, Calif.)

for victims lasted for approximately 5 days. The following quotation describes communication and access problems encountered in this operation.[1342]

The hospital was connected to the hospital radio system* . . ., but the equipment was destroyed. A staff member drove into San Fernando and contacted a policeman who was asked to relay the request for assistance. Others were sent to contact the closest fire company, and guards with walkie-talkies were sent to the Sepulveda Veterans Administration Hospital. Hospital equipment was dispatched to the scene immediately, and the first city fire engine arrived at 7:40 a.m., delayed enroute because of congested streets. As a result of a communication problem, the county fire department . . . did not become heavily committed until after 9:00 a.m. The prisoner camp crew had to be flown in because landslides closed the roads. Thus, it was several hours before substantial amounts of assistance were available on-scene to assist in the search and rescue operations. Testimony revealed that search and rescue problems could have been more serious had the hospital been larger and had the earthquake occurred later in the day.

The *Hospital Emergency Assistance Radio Network (HEAR)* "played the key role in minimizing immediate postearthquake medical problems" even though the equipment was destroyed at the Olive View Community and San Fernando Veterans Administration hospitals, and not all institutions in the Los Angeles area were in the system.[1342] Because of HEAR, patients were transferred smoothly from damaged hospitals to ones outside the stricken area. Hospitals closest to the hardest hit area were not saturated with victims because a central command post directed ambulance drivers

* The Hospital Emergency Radio Network (HEAR) was established as a program of the Hospital Council of Southern California for interhospital communications. There were 118 hospitals in the network at the time of the earthquake.[1342]

where to take individuals, depending upon the nature and seriousness of the injuries.[509] The role of HEAR during the emergency period was presented in a detailed staff report of the Hospital Council of Southern California.[1343]

A short time after the earthquake, the Hospital Council queried hospital emergency departments in the affected area with the following question: "What areas of disaster activity caused problems for your hospital during the crisis?" The replies, based upon the frequency of response, were[1342]

Communications
Patient emotion
Water shortage or contamination
Power and/or fuel
Patient transportation
Personnel identification
Personnel and/or supply transportation
Physican coverage
Nursing coverage
Shortage of medical supplies
Clerical assistance
Shortage of blood type
Food
Medical records on transferred or treated patients
Linen shortage

Olson notes that even with HEAR, communication was the most commonly experienced problem, but the specifics were not described by the respondents. Patient emotion was a serious problem in emergency rooms; this problem was not anticipated, and treatment was unsystematic or totally missing.[1342]

The City of San Fernando used two Department of Health, Education, and Welfare Packaged Disaster Hospitals (PDH), each with a 200-bed capacity, in its response to the earthquake. One unit was used only for supplies to treat patients evacuated from the San Fernando Veterans Administration Hospital. The second PDH was partially set up as an outpatient facility, and approximately 300 people were treated for minor injuries over a 5-day period. However, problems affected its operation; these included: "inadequate staffing, lack of direction, unplanned reliance upon volunteers, inadequate consumable resources such as water and cleaning supplies, a lack of knowledge about what the PDH contained and how to set it up, no recordkeeping for either patients or volunteer workers, and no provisions to house and care for the volunteer staff."[1342]

Several operations of the 17 fire departments that responded to earthquake-related incidents have been summarized from the report of Steinbrugge et al.[1344]

1. Manpower shortages did not occur because off-duty personnel from several departments automatically responded to predetermined locations, many oncoming shift personnel were on their way to or about to leave their homes for assigned duty stations when the earthquake struck at 6:01 a.m., and departments needing extra strength retained the off-going shift. The timing of the earthquake resulted in a favorable manpower situation that might not have occurred at less opportune times.

2. Los Angeles City and County Fire Departments had to place personnel and equipment into roving patrol units because of a lack of telephone service. These units encountered emergencies ranging from fires to trapped individuals. There were difficulties responding to several emergencies because of damaged and blocked roadways.

3. Motorized equipment at two stations in the Sylmar area of Los Angeles was not

available immediately because of damaged fire stations. Because of power failures at certain locations, electric doors had to be manually operated, resulting in some minor delays.

4.   In general, all fire sites received departmental service, although sometimes delayed. Because of the large number of emergencies and the resulting reduced apparatus availability, certain departments were able to send only a reduced assignment. In all, there were 1392 fire department incidents on February 9, 1971; this compares with 427 incidents on February 9, 1970. On the day of the earthquake, the greatest number of responses (570) was for first aid and rescue, followed by fire (244).

Law enforcement personnel from San Fernando and Los Angeles, the Los Angeles County Sheriff's Department, and private and institutional security forces were involved in evacuations, providing security for damaged facilities, and traffic control. The largest operation was the evacuation of 80,000 people from the area below Lower San Fernando Dam from February 9 to 12 (Figure 43 in Volume II, Chapter 4). The success of the evacuation was aided (1) by the provisions of the California Penal Code (Section 409.5) which gives law enforcement personnel the authority to arrest any person failing to evacuate a given area and (2) because maps showing potential inundation zones downstream from all dams in Los Angeles had been prepared following the failure of the Baldwin Hills Reservoir in 1963. Maps of the reservoir area were in the custody of the police department at the time of the earthquake. However, conflicting estimates of damage to the dam and potential danger areas downstream delayed the evacuation. It took approximately 4.5 hr following the earthquake "to decide the ultimate scope of the evacuation."[1342] A section of Olson's report describes the problems encountered during and following the evacuation.

Apparently, there was some initial confusion about the directions given to the field forces. This centered on whether residents were being "ordered" or "advised" to evacuate the area. This was clarified by confirming the original instruction ordering evacuation. One observer noted that this confusion about directions and the areas to be included resulted from conflicting instructions being given by responsible officials. It created some anxiety in the area and people were observed driving around not knowing what to do.

The evacuation operation, in general, proceeded very smoothly. Police forces, aided by the radio and television media, notified the residents, and by about 4 p.m., all but an estimated 1 percent had left the area. This 1 percent included people who failed to acknowledge that they were home plus others who may have returned and stayed. Nevertheless, should the dam have failed, this created a potential casualty rate of 700 to 800.

Initially, to keep unnecessary incoming traffic out of the area, a perimeter patrol of motorcycle officers was used. Transportation out of the area was by privately owned vehicle, fortunately in plentiful supply. Reluctance to leave became greatest as one moved farther south from the dam.

The only major occupancy facility within the evacuated area was the Veterans Administration Hospital . . . Some of the patients from the heavily damaged Veteran Administration Hospital in Sylmar were relocated here. This hospital sits on relatively high ground, and evacuation was not required. This decision prevented a transportation and logistics problem of major proportions.

After the residents were moved, the role of the police force changed to a security and antilooting mission. There were very few reports of looting, burglaries, and other crimes against property. Later the same day, February 9, people tried to reenter the area, and until a pass system was established this presented some control problems. Eventually, 1-hour passes were issued at checkpoints around the perimeter, but with people now lawfully in the area, the antilooting force's task became even more difficult. There were cases of people remaining after the pass expired, and others managed to evade the patrols. On the whole, there was no loss of confidence by the residents in the safety of their property, and the area had the lowest crime rate it had experienced in months.

The National Guard was not activated for use in this disaster, but one request was made by the Los Angeles Police Department for National Guard support. As required by California law, the request had to be channeled through the county sheriff's office. The motivation for this request was that, although the police department had not fully exhausted its own forces, the remaining reserve would be insufficient to cope with the problems following the failure of the dam or damage produced by a major aftershock should one occur. Under these conditions there would have been no time to request the Guard, and time is required for its mobilization. In retrospect, one could question the procedural need for going through the local sheriff's office and the requirement that the capabilities of the local law enforcement agency be exhausted before a request for National Guard support can be honored.

Sealing off the evacuated area severed several main traffic arteries in the process. Because of this and damage to other road and freeway structures in the vicinity, congestion became a serious problem on Wednesday, February 10. In response, a few streets farthest from the dam were opened to traffic. There would have been time to clear them had the dam failed, but this traffic added problems to the security patrol's operations.

Several problems affected the emergency response of the police department in the City of San Fernando, one of the communities hardest hit by the earthquake.

1. No incoming calls could be made for police assistance because there was no telephone service for a portion of the emergency period.
2. Only a portion of the department's patrol cars had loud speakers for disseminating information to the community.
3. Because San Fernando had only one radio frequency to handle all city business, it became overloaded immediately after the earthquake. Radio communications were soon restricted to police and fire operations.
4. The city had no *Mutual Aid Agreement* with surrounding communities. A Mutual Aid Agreement creates a formal structure through which a local government in time of disaster relies first on its own resources, then, if necessary, calls for assistance from neighboring cities, counties, the state, and if needed, the federal government.
5. Security at department headquarters became a problem because many organizations were using it as a base of operations.

Unexpected activities occurred in San Fernando that required a police response. Several narcotics arrests were made at a local hospital where people were discovered looking for drugs. Because two 3-day weekends followed the earthquake (Lincoln's and Washington's birthdays), the city was inundated with sightseers which led to excessive congestion and traffic control problems. One of the major tasks was directed at maintaining surveillance of portable toilets that had been set up in the city because of severe damage to the sewage system. According to Olson, there were reports of these being stolen and moved into areas of Los Angeles where coverage was less adequate.

Several chapters of the American National Red Cross conducted or participated in a wide range of activities during the emergency period. For instance, the Red Cross operated 10 emergency shelter facilities, two mass feeding centers, and eight mobile feeding units. Within a 5-day period, 17,000 people were housed and fed at the emergency shelters, and approximately 109,000 meals were served in a 10-day period at the mass feeding centers. The mobile canteens were used for 4 days (12 hr/day). "They dispensed 7,200 gallons (27,692 ℓ) of water, 12,000 cartons each of milk and orange juice, 70,000 sandwiches, and supplied information about assistance available to residents of the area."[1342] A variety of services were available at 10 family service centers, including payments for rent, clothing, medical, and minor household items. The Red Cross disaster welfare inquiry system* was set up to help determine the location and status of people in the affected area. Approximately 20,000 welfare inquiries were processed as a result of the earthquake.[1342]

Although the Red Cross played a major role during the emergency period, several

* A Red Cross service initiated "in response to inquiries from relatives, friends, and others concerned about the safety of those possibly affected by the disaster."[1342]

operational problems were encountered. The problems and their possible solutions have been summarized by Olson.[1342]

. . .First, problems were encountered in effectively reaching minority groups in the damaged area, particularly the Mexican-American community. For example, there were too few Spanish-speaking caseworkers, and nuns from the Santa Rosa Catholic Church were pressed into service as assistants. Also, inadequate knowledge of the minority community hindered operations. As a result, the earthquake added further impetus to the development of "human relations teams" within the Red Cross. These teams will work with minority groups in the affected communities and carry out several important functions. As background, each chapter is studying its jurisdiction and compiling information on the ethnic, linguistic, and cultural characteristics of the areas for use by the human relations team in future disasters. Second, inventories of disaster supplies used by the Red Cross are being increased as a need for greater quantities was recognized. Third, it was found that people in the Mexican-American area would not go to the feeding center in the Anglo area. Thus, one was established at the Santa Rosa Catholic Church to serve this area. Awareness of problems of this nature is increasing. Last, volunteers and local residents hired during emergencies are being asked to perform more sophisticated tasks, such as casework. Thus, training is being increased for staff and volunteers who might have to work on Red Cross disaster assistance programs.

B. Earthquake Disaster Planning — Tokyo, Japan

Tokyo, Japan was severely damaged by the Great Kanto earthquake (M_s = 8.3) of September 1, 1923. The ensuing fire was responsible for at least 60,000 deaths and the destruction of 370,000 buildings. In the following 50 years, little was done to improve Tokyo's vulnerability to earthquake hazards. The city became overgrown and overcrowded, with wood continuing to be the most common building material. To reduce the potential toll, the Tokyo Metropolitan Assembly recently implemented a far-reaching preparedness program.[1345] One important tool for the program is a map that estimates the overall degree of danger from earthquakes and associated hazards for every section of the city (Figure 18).[1346,1347] The map is used as a planning tool for (1) minimizing the potential damage from earthquake and associated hazards, (2) encouraging citizen awareness of earthquakes and ways to minimize deaths, injuries, and property damage, and (3) selecting areas of greatest potential danger for initiating earthquake and fire countermeasures.[1346]

The Great Kanto earthquake, striking on a winter evening during cooking hours, served as the basis for estimating potential dangers. The city was then divided into 500-m² grid squares, and each cell was assessed in terms of ground stability, population density, ratio of population to roads and open space, potential fire sources such as fuel storage facilities and stoves, distance to designated disaster shelters, and the availability of fire extinguishing equipment. These data were then used to establish the degree of danger for each cell. Map updating is required every 5 years, as specified by the *Tokyo Earthquake and Fire Prevention Ordinance of 1971*.[1346]

Upon examination of Figure 18 it is noted that most of the more dangerous areas are located in the eastern part of the city (Koto area) which has been ravaged by floods and fires in the past. More than 1.4 million people currently live below the height of the mean spring tide, and if the dikes or seawalls were damaged, the area would be flooded, and escape routes would be cut off. Other dangerous areas in the city are characterized by dense settlement, limited vacant land, and narrow streets.[1346]

The primary long-term countermeasure for the Koto area is to develop six *anti-disaster complexes*. Each will have a centralized public park of 50 to 100 hectares to serve as a refuge from fire. The open space will be surrounded by fireproof high-rise buildings to shield the area from fire, and these, in turn, will be ringed by a series of canals. It is anticipated that it will take more than 10 years and approximately one third of the total annual Tokyo municipal budget to complete the project.[1345] Other mitigation measures center on "reinforcing structures such as bridges, embankments and dikes; rigidly inspecting outdoor storage facilities and increasing the personnel of fire bri-

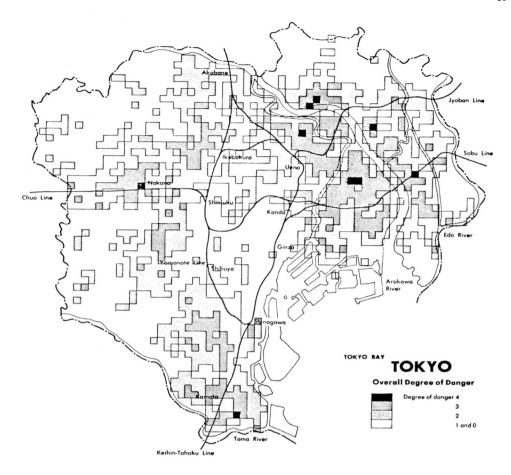

FIGURE 18.   Map estimating the relative degree of danger from earthquakes and associated hazards for Tokyo, Japan. Specific aspects of the rating system are described in the text. (From Anon., *Nat. Hazards Observer,* 1, 7, 1976.)

gades; constructing more emergency water tanks; and improving shelters and access roads to them.''[1346]

Large scale earthquake preparedness training exercises for emergency agencies are held from time to time. One was an evacuation drill in July 1976 that mobilized 58,000 policemen, 2200 emergency vehicles, and a squad of helicopters to simulate a response to an earthquake comparable in size to the 1923 event. The annual *Earthquake Preparedness Week* is designed to encourage citizens to stock food, water, and fire extinguishers, and to be reminded of escape routes.[1346] Earthquake fire preparedness is an important component of *Fire Preparedness Month* in November to early December.[1347]

C. Earthquake Disaster Planning — California

The *Office of Emergency Services (OES)* is the lead agency for California's disaster planning system. OES was established as a part of the Governor's Office in 1950, and its director serves a dual capacity as State Director of Civil Defense and Emergency Planning. OES performs a number of emergency functions.[1347]

1.   In an emergency, OES ''functions as the immediate staff and coordinating organization of the Governor to carry out the state's responsibilities under the Emer-

gency Services Act and applicable federal statutes.'' The *Emergency Services Act* stipulates that the Director of OES will coordinate the emergency assistance activities of all state agencies for war, state, and local emergencies.

2. OES does not divest local government of its responsibility during emergencies, believing that it is the ''day-to-day'' or local government which is best equipped to act effectively in an emergency by making use of its trained personnel, facilities, and equipment needed for routine operations. It is for this reason that disaster planning is envisioned to be a part of everyday activities at the local level.

3. The foundation of disaster planning is a statewide system of *mutual aid* wherein ''each local jurisdiction relies first on its own resources, then calls for assistance from its neighbors — city to city, city to county, county to county, and finally through one of the regional offices of the Office of Emergency Services, to the State.'' This enables each jurisdiction to retain control over its personnel and facilities and to give and receive assistance whenever it is needed. Currently, 398 of the state's 411 incorporated cities and all 58 counties have adopted the *1950 Master Mutual Aid Agreement.* The state is a signatory to the agreement and provides resources to local jurisdictions in emergencies. It will be remembered that the City of San Fernando did not have a mutual aid agreement with neighboring communities at the time of the February 9, 1971 San Fernando earthquake.

4. When the Governor issues an emergency proclamation, an OES task force moves into the disaster area to work with federal and local counterparts to evaluate the extent and nature of damage, to coordinate mutual aid assistance, and to coordinate disaster recovery operations. OES has been called upon 125 times in the past 25 years to coordinate the state's emergency responsibilities.

5. OES staff members do not personally accomplish disaster relief. There are only 100 OES personnel statewide. Approximately one third of the staff work on federally funded projects, while the remainder support mutual aid and emergency assistance activities. The following are OES services available to local California jurisdictions:

ON-SITE ASSISTANCE — a survey of a jurisdiction's current emergency response capability; recommendations and assistance for upgrading that ability.

DISASTER PLANNING GUIDANCE — on-site assistance from federal and state representatives, tailored to local needs.

DISASTER RECOVERY SERVICES (FINANCIAL).

FEDERAL MATCHING FUNDS — for personnel and administrative maintenance of emergency services program.

SURPLUS AND EXCESS FEDERAL GOVERNMENT PROPERTY ACQUISITION.

COMMUNICATIONS AND WARNING INTERTIES.

SUBSIDIZED TRAINING PROGRAMS:
  Simulations of emergency operations in natural and man-made disasters.
  Radiological Hazards — Peacetime and war-related
  Government and Industrial Conferences
  Disaster Recovery Administration

EDUCATION:
  University of Southern California Graduate Program in Disaster Administration
  State Department of Education Teacher Workshops

COMMUNITY EMERGENCY PLANNING — shelter planning and emergency public information.

MUTUAL AID AND DISASTER SERVICES:
Agricultural Assistance
Engineering
Fire and Rescue
Housing and Reconstruction
Law Enforcement
Medical and Health
Radiological
Utilities
Communications
Welfare (Mass Care)
Public Information

PUBLIC INFORMATION & EDUCATION — pamphlets, films, television and radio public service announcements concerning earthquake and other disaster safety.

The Office of Emergency Services recently implemented the *Earthquake Response Plan* in compliance with the *California Emergency Plan,* which identifies the earthquake as one type of peacetime emergency for which contingency plans are required. The purpose of the Earthquake Response Plan is to (1) enhance the overall capability of state and local government to respond to major earthquakes, (2) coordinate the efforts of all governmental agencies and the private sector to insure that essential and available resources can be provided efficiently to stricken areas, and (3) provide suggested operational concepts to be followed by each level of the *Emergency Organization.** Available resources or key facilities, for procurement or use, are to be maintained by appropriate agencies at all levels of the Emergency Organization.[1348]

The general *operational concept* of the Earthquake Response Plan is as follows.[1348]

When a major earthquake occurs, local authorities within damaged areas will be expected to use available resources to protect life and property and alleviate suffering and hardship on individuals. If local resources prove to be inadequate, or are exhausted, assistance should be requested from other areas in accordance with Mutual Aid Plans or Agreements. Jurisdictions in areas sustaining little or no damage will be expected to provide support to affected areas. Where situations are beyond the capability of local governments, requests for resources and other support will be made to the state government. When such support requirements cannot be met by state resources, the state will request federal assistance in accordance with applicable federal laws, policies and procedures.

Local level operations are based on five emergency conditions. The first is a preparedness or standby condition, whereas the remaining four conditions prevail after a damaging earthquake has occurred (Figure 19). Participants for *intermediate level operations* are *Operational Areas* (i.e., county and its political subdivisions) and *Regional Coordinators* from the six Mutual Aid Regions of the Office of Emergency Services (Figure 20). Data supplied by local level operations (Conditions C and D, Figure 19) and supplemented by more detailed situation reports are to be used "by Operational Areas and Regional Coordinators to establish priorities regarding allocation of resources available to support the affected area and to coordinate mutual aid . . ."[1348]

State government operations are handled primarily by the Office of Emergency Services (Figure 21). The following are OES responsibilities.[1348]

1. The occurrence of an earthquake within the State of California will be reported to the State Headquarters of the Office of Emergency Services (OES) by one or more of the following means:

* Includes "civil government including all public employees; employees of special districts (school, utility, transit, etc.); augmented by personnel of private agencies and organizations which control vital emergency resources; and auxiliaries, volunteers and persons impressed into service."[1348]

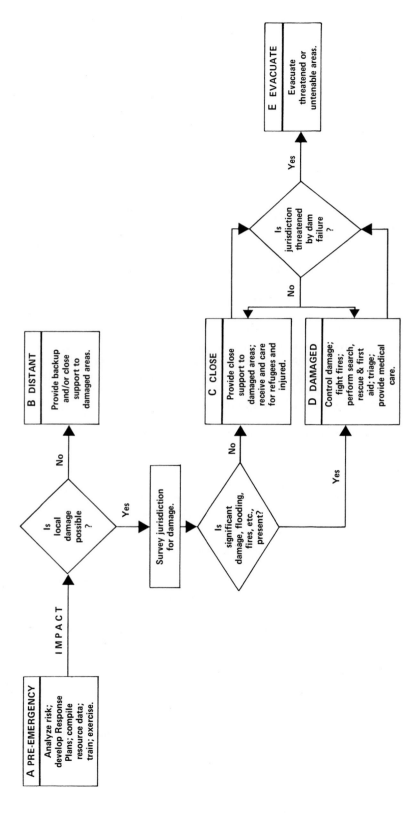

FIGURE 19.   Actions to be undertaken by local government as a part of the California Earthquake Response Plan.   (From California Office of Emergency Services, Earthquake Response Plan, Office of Emergency Services, Sacramento, 1977, II-3.)

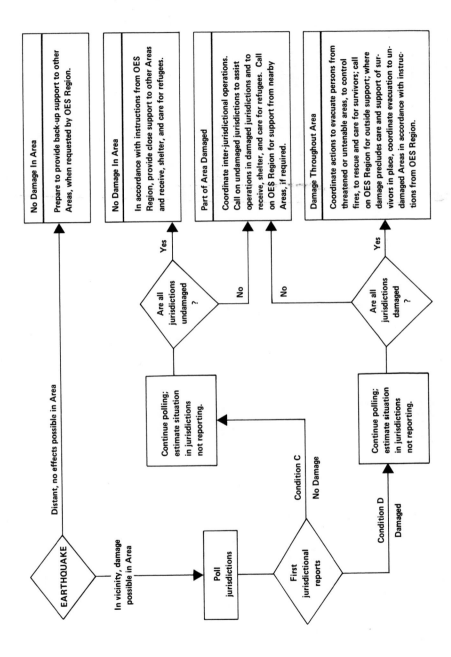

FIGURE 20. Actions to be undertaken by Operational Area and Regional Coordinators as a part of the California Earthquake Response Plan. (From California Office of Emergency Services, Earthquake Response Plan, Office of Emergency Services, Sacramento, 1977, II-7.)

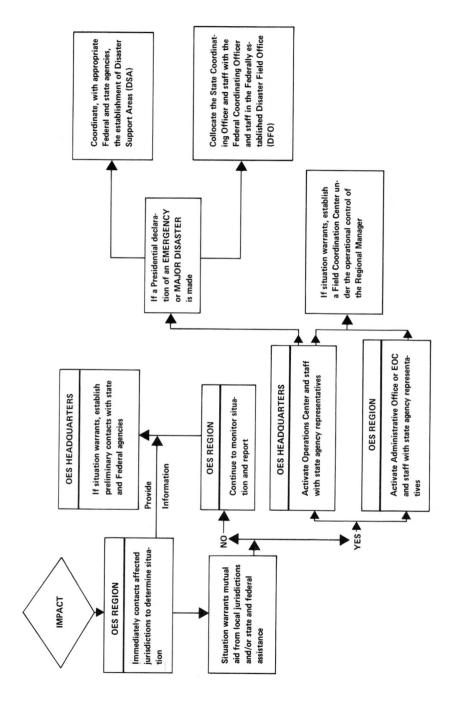

FIGURE 21.   Actions to be undertaken by state government as a part of the California Earthquake Response Plan. (From California Office of Emergency Services, Earthquake Response Plan, Office of Emergency Services, Sacramento, 1977, II-9.)

a. The National Earthquake Information Service, which is operated by the United States Geological Survey.

b. Local authorities in jurisdictions that observe substantial earthquake effects.

c. State agencies whose field units or installations observe substantial effects and report their observations to their headquarters.

2. As soon as a major earthquake in an urbanized area has been confirmed, the Director, OES, will recommend to the Governor that a STATE OF EMERGENCY* be proclaimed in the affected areas and, as required, in areas from which initial mutual aid assistance might be needed. During this time, state agencies will, commensurate with their capabilities, be expected to immediately respond to requests from affected areas for assistance. These activities will be coordinated with the Director, OES.

3. The Operations Center at the Sacramento Headquarters, OES, will be activated. At the request of the Director, OES, authorized representatives from predesignated state agencies will report at the Operations Center to coordinate their departmental response activities. The appropriate OES Mutual Aid Regional Office(s) will also be activated and manned as necessary by OES and other state agency personnel, and representatives of the private sector.

4. If the situation requires, a Command Center will be established in the Governor's Office. The Emergency Planning Council, as described in Part Two of the State Emergency Plan, and the California Emergency Council, as described in the Emergency Services Act, will be convened as necessary.

5. If the situation further dictates, OES will activate a Field Coordination Center within, or adjacent to, affected areas, which will function under the operational control of the appropriate OES Mutual Aid Regional Manager. This facility will be supported by mobile communications from OES and by field, or headquarters staff, personnel from predesignated state agencies.

6. If the Governor requests and receives a Presidential declaration of an EMERGENCY or a MAJOR DISASTER he will appoint a State Coordinating Officer (SCO) and a Governor's Authorized Representative (GAR). If the duly appointed Federal Coordinating Officer (FCO) establishes one or more Disaster Field Offices (DFO), the FCO and SCO will colocate and operate from the DFO(s) along with their counterpart staffs.

7. If the situation dictates and where conditions permit, OES will coordinate the establishment of one, or more, Disaster Support Areas (DSA).

8. After the immediate needs (rescue, medical care, emergency shelter, food, and clothing) of people have been met, governmental actions to fulfill their rehabilitation needs will be taken. Through coordination between the FCO, SCO, and local government representatives, an adequate number of Disaster Assistance Centers will be appropriately staffed by representatives of federal, state, and local governmental agencies, private service organizations, and specific representatives of the private sector.

9. The Director, OES, will assist the Governor with the direction and coordination of the activities of the several departments and agencies of state government, and will coordinate and support the emergency operations conducted by, and under the leadership of, local governments.

10. Each state agency will:

a. Interrogate subordinate field units and facilities to determine the extent of damage and whether they are able to function.

b. As need is evidenced, alert and mobilize personnel and resources.

c. Develop damage assessment data to enable the agency and OES to evaluate the nature and extent of the emergency situation.

d. Establish liaison with allied local agencies in and adjacent to areas that have significant damage or ensuing hazards.

e. Consistent with capability, respond to requests for emergency assistance from other levels of government. All such assistance shall be coordinated with the appropriate Agency Field Office or OES Office.

f. Provide disaster related cost data for the Governor's certification of state expenditures.

Regarding *federal government operations*, agencies whose statutory authorities permit immediate disaster responses will support state and local emergency actions when requested. If there is a Presidential declaration of an EMERGENCY/MAJOR DISASTER, all federal disaster assistance will be coordinated by the Federal Coordinating

---

* Such a proclamation is normally preceded by proclamations of a LOCAL EMERGENCY made by affected jurisdictions. However, due to possible local circumstances which result in the incapability of local officials to make proclamations accordingly, the Governor may, under the authority given him by the California Emergency Services Act, proclaim a STATE OF EMERGENCY.

Officer. The following describes the concept of military operations and the utilization of special federal facilities.[1348]

Concept of Military Operations
1. In all cases, military support should be considered as a complement of, and not a substitute for, local efforts.
2. Local requests for military support are made through the appropriate OES Mutual Aid Regional Office to the OES Headquarters. If communications are not possible, local requests may be directed to the nearest military installation.
3. The State Military Department will respond to requests from OES for National Guard support to state and local agencies.
4. All military groups providing support will be subject only to their immediate military superiors, but will cooperate with local authorities and accept missions commensurate with their capabilities.

Utilization of Special Facilities
1. As required, *Mass Care Facilities* will be established in suitable locations to provide immediate emergency care for displaced and homeless people. Each such facility should be capable of providing temporary shelter and preparing food. Such facilities will be augmented and staffed to provide emergency clothing, limited medical aid, public information, central registry services, and religious counselling. The location of activated facilities will be given wide publicity through the various news media. They will remain open only so long as there is a public need. Responsibility for their establishment and operation is normally a joint County Welfare/American Red Cross function, assisted, and required, by appropriate state and Federal agencies.
2. Where required, *Multipurpose Staging Areas (MSA)* will be activated, in locations selected by local governments, to provide a destination or rallying point for incoming mutual aid, and a staging area for support and recovery activities. The following criteria will be used as a basis for selected MSAs:
   a. A low-risk fire and flood area.
   b. Adequate areas, both paved and unpaved, for all intended uses including assembly, maneuvering and parking areas, and a helicopter landing area.
   c. Adequate facilities, both covered and open, for supplying and maintaining vehicles and heavy construction equipment.
   d. Facilities for first aid and medical treatment sufficient for use as a medical evacuation point for disaster victims.
   e. Feeding, sleeping, and sanitation facilities sufficient for mutual aid and outside resources personnel.
   f. Provision for security and safety, such as a fence or natural barriers.
   g. Ready access to major transportation routes.
   h. Adequate communications.
   i. If required, and feasible, auxiliary power.
3. Where required, State OES will coordinate the establishment of *Disaster Support Areas (DSA)* on the periphery of the disaster area(s), usually on, or in close proximity to, operable airports with runways capable of accommodating heavy aircraft. They will be used to receive, stockpile, allocate and dispatch disaster relief supplies (manpower and material), and may be used for the receipt and emergency treatment of casualty evacuees arriving via short-range modes (air and surface) of transportation and the subsequent movement of a select number of heavy, long-range aircraft to adequate medical care facilities.

Because state government will have a major role in supporting local emergency operations following major earthquakes, the requirements for providing such support will be as follows: (1) all appropriate state agencies will include provisions in their plans for accomplishing assigned tasks, (2) emergency activities of state agencies will be coordinated by the Director of OES, as defined by the California Emergency Services Act, and (3) upon order from the Governor or the Director of OES, each agency will provide personnel to OES at all appropriate levels "for liaison, intelligence and operational matters related to the agency's emergency functions."[1348] The following lists the specific agencies that are responsible for various emergency functions.

1. *Direction and Control*
   The Office of Emergency Services, assisted by representatives of other State agencies, will:
   a. Exercise overall coordinating responsibilities for those activities involving emergency operations of state agencies.
   b. Coordinate and cooperate with appropriate local and federal agencies.

    c. Establish priorities for the utilization of resources to support state and local disaster relief operations.

    d. Coordinate emergency public information, damage assessment, emergency communications, and other essential activities.

2. *Fire and Rescue*

The Office of Emergency Services will coordinate the emergency operations of state agencies necessary to support local fire fighting, rescuing, evacuating and related functions. (Note: All fire fighting and rescue mutual aid operations will be conducted under the California Fire and Rescue Emergency Plan.) The following state agencies will provide support as required:

    California Conservation Corps
    Department of Corrections
    Department of Fish and Game
    Department of Forestry
    Department of Youth Authority
    State Fire Marshal
    Military Department

3. *Law Enforcement and Traffic Control*

The Office of Emergency Services will coordinate those efforts of state agencies supporting local emergency actions in maintaining order and traffic control, to include evacuation. This responsibility will include the implementation of the California Law Enforcement Mutual Aid Plan, which provides for the coordination of mutual aid assistance for law enforcement operations. The following state agencies will provide support as required:

    California Highway Patrol
    Department of Corrections
    Department of Fish and Game
    Department of Forestry
    Department of General Services (California State Police)
    Department of Justice
    Military Department
    Department of Transportation

4. *Disaster Public Health*

The Department of Health will be responsible for the statewide coordination of disaster public health programs, activities and functions, to include the allocation of health resources to support emergency operations. The following state agencies will provide support as required:

    Air Resources Board
    Department of Fish and Game
    Department of Food and Agriculture

5. *Disaster Medical Care*

The Department of Health will be responsible for the statewide coordination of disaster medical care programs, activities and functions, to include the allocation of medical resources to support emergency operations. The following state agencies will provide support as required:

    Department of Corrections
    Department of Veterans Affairs
    Military Department

6. *Emergency Welfare*

The Department of Health will be responsible for coordinating support requirements for local operations to meet the mass care needs of disaster victims. The following state agencies will provide support as required:

    Department of Aging
    Department of Corrections
    Department of Forestry
    Department of Motor Vehicles
    Department of Parks and Recreation
    Department of Rehabilitation

Department of Veterans Affairs
Department of Youth Authority
Military Department

7. *Resources and Support*

The Office of Emergency Services will coordinate those activities relative to the procurement and allocation of those essential resources (engineering, manpower, supplies, transportation, utilities, etc.) required to support state and local governmental operations. The following state agencies will provide support as required:

Air Resources Board
California Highway Patrol
Department of Conservation
Department of Corrections
Department of Education
Department of Finance
Department of Fish and Game
Department of Food and Agriculture
Department of Forestry
Department of General Services
Department of Industrial Relations
Department of Motor Vehicles
Department of Navigation and Ocean Development
Department of Parks and Recreation
Department of Transportation
Department of Water Resources
Department of Youth Authority
Employment Development Department
Energy Resources Conservation and Development Commission
Military Department
Public Utilities Commission
State Personnel Board

Despite the uncertainties involved with earthquake prediction, a mechanism for official response was recently established in California when the Office of Emergency Services formed the *California Earthquake Evaluation Council* on March 26, 1974. The council's purpose is to judge the scientific merit of all earthquake predictions and to advise OES of its findings. If a prediction is judged to be scientifically valid, it is then the responsibility of OES to take appropriate actions.[1350] The latest guidelines for evaluating predictions were adopted by the council on February 22, 1977;[1351] these are presented in Appendix H.

The first council evaluation of an earthquake forecast occurred on April 14, 1976 in connection with the "Palmdale Bulge" in southern California that had been discovered by scientists of the U.S. Geological Survey (Figure 11 in Volume II, Chapter 1 and attendant discussion). According to Manfred,[1352] the council determined that the scientific data were valid, but that the estimate of when the predicted earthquake would take place was not supported with sufficient evidence. The council issued the following statement:[1352]

The council has concluded that the area of concern, the so-called Palmdale "Bulge," definitely warrants further detailed study but that there is no reason at this time, on the basis of the data presented, to conclude whether or not a major earthquake will occur at any specific time in the future. However, in our judgment, the uplift is probably a manifestation of the gradual buildup of earthquake producing stresses, and it should serve to give us a renewed sense of urgency in preparing for the large earthquake that some day inevitably will occur in this region.

The Office of Emergency Services passed this information on to local governments in the region and asked them to:

1. Review ordinances and emergency plans to insure public safety responsibilities are assigned.
2. Convene emergency or disaster councils to review the status of emergency preparedness programs.
3. Determine the readiness of agencies which have life-saving missions, such as fire, medical, and public works.
4. Review, update, and/or prepare instrumental material for release by the news media to citizens.
5. Coordinate plans with officials of school districts, Red Cross, hospitals, and news media.
6. Be prepared to speak directly to the people through local radio broadcasting systems, outlining precautionary measures to be taken before an earthquake. Also be prepared to use these systems to inform and instruct citizens immediately after an earthquake.
7. Have each first response service prepare procedures to insure emergency equipment will not be damaged or trapped inside its normal storage location (fire, law, communications).
8. Identify hazardous areas and structures which may have to be evacuated prior to or immediately after an earthquake.
9. Consider ordinances and enforcement to reduce hazardous structural features which are nonfunctional to a facility, such as parapets or cornices.
10. Conduct training for emergency service personnel to insure a reliable reporting system on the nature and extent of damage in the community and to provide direction and control of life-saving operations.
11. Inventory essential resources, anticipate deficiencies, and review procedures for requesting and implementing mutual aid.

A second OES advisory committee, the *Advisory Panel on State Government Response to Earthquake Prediction,* is composed of members representing social, economic, and government disciplines. Its purpose is to make recommendations to OES concerning government actions to be taken in response to earthquake predictions.[1352]

The *Contra Costa County Office of Emergency Services* is one of the nation's most progressive disaster organizations. The county OES has the following responsibilities: (1) preparation and coordination of emergency functions relating to disasters such as "floods, large fires, massive civil disturbances, earthquakes, severe storms, landslides, epidemics, ruptured petroleum pipelines, transportation accidents involving people or hazardous materials, aircraft accidents, and war related problems," (2) preparing or revising plans on dam-failure evacuation, earthquake response, oil and hazardous material spills, and air pollution, (3) conducting simulated earthquake exercises and emergency-procedure workshops for county emergency services sections, (4) providing disaster planning guidance and assistance to cities, schools, and hospitals, and (5) distributing federally supplied literature on emergency preparedness and county OES literature on earthquakes, floods, fires, emergency supplies, etc.[1339]

Following an earthquake, the county OES staff will establish an operations center that will include representation from all utility companies and essential county services divisions. The primary function of OES will be to assist local cities through its coordination of county-wide disaster services* and the distribution of vital information. It is not intended for OES "to relieve cities of their own disaster preparedness responsibilities."[1339]

The U.S. Defense Civil Preparedness Agency developed a checklist to assist local communities in determining their readiness to cope with a disaster. The Tri-Cities Advisory Committee on Seismic Safety to the Cities of El Cerrito, Richmond, and San Pablo has modified the checklist to apply directly to natural disasters. The committee suggests that the following questions be used by a city's safety commission or other similar board and by each city council when a community's disaster preparedness program is reviewed.[1339]

1. Is the city served by a full-time professional disaster preparedness coordinator who fits together the planning and preparation by all local public agencies for dealing with disasters?

* County resources are integrated into one of several services comprising the *Disaster Organization:* (1) public safety services, (2) operational services, (3) public assistance services, (4) supply, logistic, and support services, and (5) continuity in government.[1339]

2. If the city does not have a full-time disaster preparedness coordinator, does it have a paid coordinator who also serves local government in some other paid capacity?

3. Is the disaster preparedness coordinator encouraged to attend courses available in the state through the Civil Defense University Extension Program or at the Defense Civil Preparedness Staff College at Battle Creek, Michigan?

4. Does the city's disaster preparedness coordinator report directly to the City Manager?

5. Do the City Manager and the heads of operating departments understand that the role of the disaster preparedness coordinator is to coordinate emergency planning and operations by all local departments, on behalf of the City Manager, not to act as an operating department head or to "take over" in case of an emergency?

6. Has a recent "hazard analysis" been made of the major dangers, such as earthquakes and industrial or transportation accidents, that the city could face?

7. Does the city have a local warning system which can alert its citizens of impending hazards?

8. Does the city have an Emergency Operating Center, a protected control site with communications to all city forces, where key executives can receive information on a disaster and make coordinated decisions on operations required by the emergency?

9. Is the Emergency Operating Center providing day-to-day benefits to the community? Is it being used for police, fire or ambulance dispatching?

10. Does the city have an up-to-date emergency plan covering all likely disasters?

11. Was the emergency plan developed by a team, including planners, from all major operating departments, rather than solely by the disaster preparedness coordinator?

12. Have the City Manager and department heads participated in emergency exercises, based on simulated disasters affecting the jurisdiction, to check disaster plans and decision making in major emergencies?

13. Does the city's emergency organizations include means to advise and inform the citizens in case of disaster via local news media?

14. Are the emergency plans for hospitals, schools, industries and utilities tied in with the overall plan?

15. Has the community's level of readiness of major emergencies been recently evaluated by the City Manager and department heads? Has an "action program," which covers specific methods to improve emergency readiness been prepared?

D. Disaster Planning — U.S. Government

The *Disaster Relief Act of 1974* (Public Law 93-288)[1353] established a broad-based program of assistance to help states and local governments (1) in relief and rehabilitation support following a major disaster and (2) develop disaster preparedness programs. The *Federal Disaster Assistance Administration (FDAA)* of the Department of Housing and Urban Development administers most of the programs.

1. Disaster Assistance[1353]

When the President determines that an emergency or major disaster exists in any state or territory, FDAA administers both public and private assistance of up to 25 federal agencies and voluntary organizations. Requests for a declaration by the President are made by the governor of the affected state. The governor's request is based upon the finding that effective responses to cope with the disaster are beyond the state and affected local governments and that federal assistance is necessary. Immediately upon the declaration of a major disaster, a *Federal Coordinating Officer* from FDAA is sent to the affected area to (1) make an initial determination of the types of relief most urgently needed, (2) establish field offices, (3) coordinate the administration of relief, including the activities of state and local governments, the American National Red Cross, the Salvation Army, and the Mennonite Disaster Service, and (4) establish actions that are deemed necessary "to assist local citizens and public officials in promptly obtaining assistance to which they are entitled." *Emergency Support Teams* comprised of federal personnel are deployed in the affected area to assist the Federal Coordinating Officer in carrying out his/her responsibilities. The state in question then is required to designate a *State Coordinating Officer* for coordinating state and local assistance programs with those of the federal government.

The Disaster Relief Act provides for supplemental assistance to federal agencies, state and local governments, and individuals and families in the affected area. These assistance programs are presented in Appendix I. Following the period of emergency aid and replacement of essential facilities and services, a number of programs are initiated in a disaster area that "has suffered a dislocation of its economy of sufficient severity to require (1) assistance in planning for developments to replace those lost in the major disaster; (2) continued coordination of assistance available under federal aid programs; and (3) continued assistance toward the restoration of the employment base." These programs are described in Appendix I. According to Dunne,[1354] the disaster assistance authorities of the Disaster Relief Act of 1974 "have been adequate to cope with 74 major disasters declared by the President since enactment."

2. Preparedness Programs

The Disaster Relief Act also allows for the allocation of $250,000 development grants to states "for the development of plans, programs, and capabilities for disaster preparedness and prevention."[1353] The state is required to create or designate an agency to plan and administer the disaster-preparedness program and submit the plan to the FDAA which will "(1) set forth comprehensive and detailed provisions for assistance to individuals, businesses, and local governments; and (2) include provisions for appointment and training of appropriate staffs, formulation of necessary regulations and procedures and conduct of required exercises."[1353] All states are participating in this program, and several states are using a portion of the grant for earthquake response and mitigation studies.[1354] FDAA is authorized to award up to $25,000/year in matching funds for improving or maintaining state disaster plans. Technical assistance is also available from FDAA to supplement a state's resources in planning to meet natural disasters.[1355]

FDAA has funded four studies to estimate the damage and casualties likely to result from large earthquakes that could strike the San Francisco,[1356] Los Angeles,[1357] Puget Sound,[1358] and Salt Lake City[1359] metropolitan regions. The studies have been and are being used as a basis to assist all levels of government for planning earthquake disaster relief and recovery plans. The following discussion is concerned with several of the findings for the nine-county San Francisco Bay region study.[1356]

Six earthquakes were simulated — shocks with local magnitudes (M_L) of 6.0, 7.0, and 8.3 with epicenters on the San Andreas fault near San Francisco and on the Hayward fault near Hayward at three times of day, 2:30 a.m., 2:00 p.m., and 4:30 p.m., in both the wet and dry seasons. Different times of occurrences were used because of the strong influence of time on determining the number of casualties (e.g., 2:30 a.m., the greatest proportion of the population would be at home in bed; 2:00 p.m., the greatest proportion of the population would be away from home; 4:30 p.m, the beginning of the rush hour). The season would strongly affect the potential for fire and landslides.

Casualties would vary dramatically not only with earthquake magnitude, but also with the time of occurrence (Table 1). The largest number of deaths and hospitalized injuries in the bay area would occur at 4:30 p.m. for an $M_L = 8.3$ earthquake on the San Adreas fault because a large portion of the population would be concentrated in or near sources with the greatest hazards. Excluding casualties from dam failures, the number of deaths could approach 11,000, and hospitalized injuries could exceed 40,000 for the largest earthquake (Table 1). Estimated probable deaths from the rapid failure of any one of 14 important dams (important with respect to fault location and downstream populations) could range from 3000 to 35,000. A credible assumption is that a magnitude 8.3 earthquake on either the San Andreas or Hayward fault would cause the failure of six or more dams with a potential life loss in excess of 100,000.

The number of long-term homeless is estimated at 19,500 if a large event was to be followed by fire (Table 2). In the study area, the prevalent type of construction is the

Table 1
DEATHS AND HOSPITALIZED INJURIES EXCLUSIVE OF DAM FAILURES

Fault	Magnitude	Time of day	Deaths				Hospitalized injuries			
			Schools	Hospitals	Other sources	Total	Schools	Hospitals	Other sources	Total
San Andreas	8.3	2:30 a.m.	0	550	2,300	2,850	0	1,600	9,200	10,800
		2:00 p.m.	2,100	1,320	6,040	9,460	5,800	4,000	24,600	34,400
		4:30 p.m.	200	1,320	8,840	10,360	600	4,000	35,760	40,360
	7	2:30 a.m.	0	100	400	500	0	300	1,600	1,900
		2:00 p.m.	350	230	1,060	1,640	1,300	700	4,200	6,200
		4:30 p.m.	40	230	1,720	1,990	100	700	10,880	11,680
	6	2:30 a.m.	0	5	20	25	0	20	80	100
		2:00 p.m.	10	10	60	80	50	30	240	320
		4:30 p.m.	0	10	90	100	0	30	360	390
Hayward	8.3	2:30 a.m.	0	820	2,300	3,120	0	2,400	9,200	11,600
		2:00 p.m.	1,900	1,950	3,350	7,200	8,700	6,000	13,800	28,500
		4:30 p.m.	200	1,950	4,500	6,650	900	6,000	18,000	24,900
	7	2:30 a.m.	0	300	740	1,040	0	900	2,960	3,860
		2:30 p.m.	1,400	700	1,100	3,200	3,400	2,100	4,400	9,900
		4:30 p.m.	100	700	1,440	2,240	300	2,100	5,760	8,160
	6	2:30 a.m.	0	100	230	330	0	300	920	1,220
		2:00 p.m.	140	240	350	730	560	700	1,400	2,600
		4:30 p.m.	10	240	450	700	50	700	1,800	2,550

Note: The ratio of nonhospitalized injuries to deaths is 30:1 in all cases.

From Algermissen, S. T., Rinehart, W. A., and Dewey, J., A Study of Earthquake Losses in the San Francisco Bay Area, U.S. Government Printing Office, Washington, D.C., 1972, 121.

TABLE 2

Long-Term Homeless

County	San Andreas Fault			Hayward Fault		
	M = 8.3	M = 7	M = 6	M = 8.3	M = 7	M = 6
Alameda						
Wet season	2,100	1,000	N[a]	16,600	16,600	3,000
Dry season	2,100	1,000	N	25,900	25,900	2,600
Contra Costa						
Wet season	300	N	N	8,400	8,400	2,100
Dry season	300	N	N	10,500	10,500	1,400
Marin						
Wet season	1,400	N	N	3,800	3,500	N
Dry season	600	N	N	1,700	900	N
Napa						
Wet season	N	N	N	300	N	N
Dry season	N	N	N	300	N	N
San Francisco						
Wet season	16,500	6,100	900	7,100	4,700	N
Dry season	19,500	6,100	900	7,100	4,700	N
San Mateo						
Wet season	9,100	9,100	1,400	700	700	N
Dry season	9,600	9,100	1,400	700	700	N
Santa Clara						
Wet season	3,400	1,500	N	10,600	10,600	2,700
Dry season	3,900	1,500	N	10,600	10,600	2,700
Solano						
Wet season	N	N	N	300	N	N
Dry season	N	N	N	300	N	N
Sonoma						
Wet season	400	N	N	400	N	N
Dry season	400	N	N	400	N	N

Note: Homeless due to potential dam failure must be *added* to these figures.

[a] Negligible (less than 100).

From Algermissen, S. T., Rinehart, W. A., and Dewey, J., A Study of Earthquake Losses in the San Francisco Bay Area, U.S. Government Printing Office, Washington, D.C., 1972, 206.

single-family wood-frame dwelling. Dollar losses in terms of repair costs to this type of structure alone is estimated to be approximately $1.2 billion for a magnitude 8.3 earthquake on the San Andreas fault and $1.4 billion for a magnitude 8.3 earthquake on the Hayward fault. Total earthquake damages could exceed $20 billion.

There would be serious effects on several important resources.

Medical facilities — The tactical and logistical problems to be faced by hospitals and related health facilities will be considerable, including many that will not be expected. Caring for the injured will be one of the greatest area-wide problems. Based upon the experience in the 1971 San Fernando earthquake, it is highly possible that many more medical facilities will be severely damaged than originally suspected. Many hospitals surviving an earthquake may have restricted functions because of damage to service facilities and access problems caused by the collapse of freeway overpasses. The problem of hospital survival is especially acute on the east side of San Francisco Bay. There, seven general hospitals and one military hospital are located on or immediately adjacent to the Hayward fault; five additional hospitals are located close-by.

Dams — The rapid failure of any one of 14 dams would lead to catastrophic results in terms of deaths and injuries. The failure or significant leakage from a dam also would cause secondary effects such as the closing of roadways, rapid transit systems, airports, schools, and factories.

Transportation systems — There will be severe damage to the railroad, highway, freeway, bridge, and airport systems. The largest postulated earthquake would stop rail service for an indefinite period. Damage is expected to be heaviest in the structurally poor ground areas and where rail lines cross the Hayward fault. Principal damage to roadways will be due to landslides, structurally poor ground movements, overpass damage, and damage to bridges crossing San Francisco Bay. Should a large earthquake occur during the wet season, landslides will be common in high intensity areas. Runways constructed on poor ground at four airports, including San Francisco International, could be badly broken up by a larger earthquake. The nonoperational period could vary from a few days to several weeks. Overall, port facilities, especially those that are pile supported, are not expected to be severely damaged. However, cranes will likely be derailed, and pipelines from storage tanks may be severed where they cross structurally poor ground areas near the docks. Access to the docks may be restricted because of the failure of approach fills and nearby freeways.

Public utilities — Experience from the moderate 1971 San Fernando event and other recent earthquakes "has shown that the superior earthquake-resistive precautions which are practiced by the public utilities do substantially reduce the earthquake hazard, but certainly do not eliminate it." For the study area, utility systems are expected to be damaged because earthquake forces and their effects are still not precisely known, and certain geologic hazards, at best, can be only minimized (e.g., structurally poor ground, liquefiable sands, landslide zones). The water supply system for Oakland, Berkeley, Richmond, El Cerrito, and San Leandro will be especially vulnerable to larger earthquakes. Water is stored in reservoirs located east of the Hayward fault and is then transmitted to the distribution system through four aqueducts that cross the fault. If any significant movement occurs on the Hayward fault, it is reasonable to assume that all aqueducts will fail or be severely damaged for magnitude 7.0 and 8.3 earthquakes, thereby isolating the storage areas from the main distribution system. It could reasonably take 6 months to fully restore water service if the East Bay area was hit by the largest earthquake. Numerous natural gas transmission lines also cross the Hayward fault. All lines are expected to fail in 7.0 and 8.3 magnitude earthquakes, and two lines are expected to fail in a magnitude 6.0 earthquake.

Telephone system — In general, central and exchange buildings will probably survive a large earthquake because they are designed more conservatively than conventional structures. A possible exception is a station located within the Hayward fault zone. Local service may be seriously affected by underground cable breaks, and electrical problems may put some stations out of operation that use electronic switching components. Even without damage to the system, circuit overloads will be a serious problem, as it has been in many previous earthquakes.

Radio and television facilities — In a magnitude 8.3 earthquake on the San Andreas fault, all radio and television facilities would likely be out of operation for 1 day in San Francisco and San Mateo counties because of in-house, power supply, or transmission line problems. Approximately one third of facilities elsewhere in the nine-county area would also be inoperative for one day. For the same size event on the Hayward fault, all facilities in East Bay cities west of the Berkeley Hills, plus those in Santa Clara County, would be out of service for 1 day. On the San Francisco Peninsula, two thirds of the facilities are expected to be out of service for 24 hr.

On May 3, 1976 the *Director of the U.S. Geological Survey (USGS)* of the Department of the Interior assumed the authority for issuing warnings of impending earthquakes, volcanic eruptions, landslides (including mudflows and mudslides), glacier-related phenomena, and subsidence as specified in Section 202 of the Disaster Relief Act of 1974. This function was originally delegated to the Federal Disaster Assistance Administration by the President. In the April 12, 1977 issue of the *Federal Register,* the USGS presented a position statement describing its responsibilities for issuing warnings[1360] (Appendix J). The statement gives a brief description of the geographic distribution and effects of the hazards and reviews the present capabilities of scientists to predict the location, time, and magnitude of hazardous events. For example, the statement stresses that it is now possible to predict impending geologic events in only certain cases for landslides, volcanic eruptions, and subsidence. These predictions require the detailed monitoring of specific locations with various types of instrumentation and systematic ground measurements.

Information acquired by Geological Survey scientists that indicates a locality may be susceptible to a geologic condition that could pose "a significant potential hazard to life or property" will be conveyed immediately to the Director with supporting evidence and documentation. Selected *scientific evaluation panels* will review the evidence and transmit their findings and recommendations to the Director. Identifications or predictions of potentially hazardous events made by non-USGS scientists will also be evaluated if requested by an appropriate state or federal official.

The *Earthquake Prediction Panel* is the first evaluation body to be established by the Director. It is charged with reviewing scientific evidence that could warn of an impending earthquake and, if warranted, recommending to the Director that a prediction be issued. An authorized prediction could only then be issued by the Director. Emphasis will be placed on evaluating evidence for potentially destructive earthquakes (i.e., $M_L \geqslant 5.5$), but data will also be evaluated for predicted smaller events to establish a "track record" for various prediction techniques. Predictions made by nonscientists are to be considered "in a manner appropriate to the nature of the evidence."[1361]

The following material from the *Federal Register* describes the three types of statements that might be issued by the Director following an evaluation panel review and the communication procedures for releasing the statements.[1360]

Notice of Potential Hazard.

(1) Where the Director has authenticated identification of an area as susceptible to a potentially hazardous condition, but available evidence is insufficient to suggest that a hazardous event is imminent or evidence has not been developed to determine the time of occurrence, the information will be prepared for normal publication.

(2) The Director or his designee will transmit such information, as soon as possible, as a Notice of Potential Hazard to appropriate Federal, State, and local officials responsible for the public safety and welfare and to the public by a press release. The reports and maps cited earlier that show the distribution of earthquake, volcanic, landslide, and subsidence hazards are examples of identifications of potentially hazardous conditions that will form the basis for notices of potential hazards.

(3) Notices of Potential Hazard will be accompanied by a description of the geologic and hydrologic conditions that exist, the factors that suggest that such conditions constitute a potential hazard, and the location or area they may affect. In most instances, it will not be possible to estimate the severity of the hazard or the time it might occur. Information such as possible earthquake recurrence intervals will be given, however, if justified by available information.

(4) Where available evidence suggests that a hazardous event could occur and that precursory phenomena exists that will better define the time, location, and magnitude of the event, the geologic conditions or processes likely to trigger a hazardous event will be monitored by the U.S. Geological Survey within the limits of available funds and manpower.

Hazard Watch.

(1) If existing or new information indicates that a region, area, locality, or geologic condition is undergoing change that may be interpreted as a precursor to a potentially hazardous event within an unspecified period of time (possibly months or years), such information will be evaluated and, if authenticated, the Director will assure that such information is transmitted promptly to civil authorities and the public as a Hazards Watch.

(2) Federal, State, and local officials responsible for public safety will be notified in advance of the intent to issue a Hazard Watch to enable them to invoke emergency preparedness plans for an orderly public response.

(3) Hazard Watches will be accompanied, to the extent possible, by a definition of the parameters of the expected event, including, in addition to the place, magnitude, and general time, the possible geologic or hydrologic effects and the uncertainties associated with each.

Hazard Warning.

(1) When developing information from precursory phenomena, which have been monitored through an experimental or operational hazard assessment program, appears to signal a potentially hazardous event within a specific period of time (possibly days or hours), the information will be conveyed promptly to the Director for evaluation and consideration as a hazard prediction.

(2) The Director or his designee will determine whether or not the prediction has a sound scientific basis and is authenticated by a comprehensive evaluation. If a prediction is issued as a result of this review and authentication process, any uncertainties that may exist will be evaluated and stated.

(3) The Director, upon authentication of a prediction of an event of possible catastrophic proportions, will assure that such information is promptly transmitted as a Hazard Warning, first to Federal, State, and local officials responsible for public safety, to enable them to invoke emergency preparedness plans for an orderly public response, and then to the news media.

(4) Hazard Warnings will be accompanied, to the extent possible, by a definition of the parameters of the expected event including, in addition to the time, place, and magnitude, the possible geologic or hydrologic effects, and the uncertainties associated with each.

Communication of Notices of Potential Hazard, Hazard Watches, and Hazard Warnings.

(1) Information leading to a Notice of Potential Hazard or a Hazard Watch will generally be obtained well in advance of an event and can be transmitted directly to concerned officials by letters and to the public by press releases to the news media.

(2) Where potentially hazardous conditions are monitored, local, State, and Federal authorities will be informed periodically of the results of such investigations and technical assistance, to the extent possible, will be extended as requested by these officials to assist in developing possible mitigation measures.

(3) At the present time, a capability to predict a geologic event of possible catastrophic proportions within days or hours does not exist except in rare cases. In such cases, where the information becomes available that suggests a potentially disastrous event may be imminent, public officials will be notified by telephone and such information will be transmitted directly to the public as a Hazard Warning. Public and existing Federal communication facilities, such as the Department of Commerce's Weather Radio System and the Department of Defense's National Warning System will be utilized whenever possible and appropriate.

(4) The Geological Survey will also communicate to responsible Federal Agencies and State and local governments, as soon as practicable, all available new knowledge as to geologic conditions or processes that may affect or alter public response to Notices of Potential Hazard, monitoring programs, Hazard Watches, or Hazard Warnings. This may result in the cancellation of the notice, watch, or warning, or a change in the hazard classification to better reflect an increased degree of uncertainty as to the time of occurrence of the event or a lessened sense of urgency.

(5) Notices of Potential Hazard, Hazard Watches and Hazard Warnings to governmental agencies will also include:

(a) A statement of the authority of the U.S. Geological Survey for issuing the notice, watch, or warning;

(b) Copies of scientific papers or authentication reports that form the basis of the notice, watch, or warning;

(c) An offer to consult with any reviewers that the Governor or Governors of affected States may wish to appoint;

(d) An offer to provide appropriate technical assistance within areas of expertise in the Geological Survey in evaluating possible geologic hazards, as they may affect people and property;

(e) A statement of what additional steps, if any, the U.S. Geological Survey proposes to take to better define the degree or area of hazard; and

(f) A list of all parties to whom the notice, watch, or warning is being transmitted.

E. Earthquake Disaster Relief — The International Role

Although it is the responsibility of governmental agencies in a country ravaged by a natural disaster to assist the victims, it is unrealistic to expect most countries to supply all the help required alone.[1362] The response to a major disaster is usually global, with assistance provided by *national governments, nongovernmental agencies,* and *international level groups* such as the United Nations and the International Red Cross. For example, within a 6-month period following the Managua, Nicaragua earthquakes of December 23, 1972, some 51 nations had sent aid through governmental agencies, and 23 additional countries sent aid through nongovernmental organizations. In all, more than $65 million in cash or services was contributed, and more than $180 million in loans and credits was pledged for reconstruction projects.[1345]

A recently published report describes the *bilateral assistance program* established by the U.S. Government to help the victims of the May 6, 1976 earthquake (M_s = 6.5) that devastated the Friuli region of northeastern Italy.*[1363] One day after the earthquake, Senator Edward M. Kennedy, Chairman of the Subcommittee to Investigate Problems Connected with Refugees and Escapees, urged in the U.S. Senate that:[1363]

. . . while the immediate and longer effects of the earthquake are not fully known, all sources confirm that a close friend of the United States has suffered a great tragedy, and the emergency relief needs, let alone rehabilitation and reconstruction, are urgent and massive.

As chairman of the Subcommittee on Refugees, I want to express my deep personal sympathy and concern to the people and Government of Italy. And I am hopeful that our Government in concert with . . . the international community will spare no effort in responding to any appeals for help from the Italian Government . . .

Officials in the Office of Disaster Relief in AID (Agency for International Development) informed me this morning that our Government has already moved . . . Private American relief organizations are also lending their support. And I am confident that our country's response to the human tragedy in Italy will fully reflect our traditional concern for people in need.

Emergency legislation calling for $25 million in humanitarian assistance was speedily enacted into law. The contribution for the emergency phase of the relief operation totaled close to $1 million.[1363]

Ambassador's declaration (helicopter support)	$25,000
283 tents (40th Tactical Command)	$180,000
1550 tents (AID stockpile-Camp Darby)	$424,316
Foodstuffs (military field rations)	$50,000
Earth-moving equipment	$70,000
Blankets, sheets, stretchers, and medical supplies	$13,000
TDY for 4 U.S. foreign-disaster-relief specialists	$13,755
Communications support	$4,000
Engineer unit (68 men)	$25,000
Italy field task force	$37,835
Engineering assistance contract	$69,600
Total	$912,506

Approximately $24 million of the American contribution was available for rehabilitation and reconstruction purposes. After consultation with the Italian Government at all levels, most of the money was allocated for the building of facilities for the care of the elderly, many of whom were displaced by the earthquake, and for rebuilding schools in several communities.[1363]

Besides the U.S., contributions for emergency relief operations came from Austria, Switzerland, Sweden, Yugoslavia, Soviet Union, Canada, Australia, several NATO members, and the European Economic Community. Contributions also came from private organizations such as the League of Red Cross Societies in Geneva. American citizens contributed through established voluntary relief organizations, including Cath-

---

* In cold, statistical terms, nearly 1000 people were killed, more than 3700 were injured, and up to 100,000 people were left homeless.[1363]

olic Relief Services, and special relief organizations established specifically to assist the victims in the Friuli region. Many of the latter group were created by Americans of Italian descent. In addition to the U.S., the European Economic Community and Saudi Arabia contributed a total of $54 million for rehabilitation and reconstruction assistance. The total cost for the people and Government of Italy could run as high as $3 billion.[1363]

A number of specialized disaster relief programs are available through the United Nations (UN) for the international response to natural disasters. The following briefly describes several UN agencies whose services can be requested by the governments of countries affected by natural disasters.[1345,1362]

1. The *United Nations Disaster Relief Office (UNDRO),* established in 1972 by the General Assembly, has the responsibility of coordinating assistance from the UN relief system and ensuing cooperation with governments and voluntary agencies such as the League of Red Cross Societies at the time of natural disasters. UNDRO services were first implemented following the December 23, 1972 Managua earthquakes; coordination was urgently needed because of the large amounts of bilateral aid pledged for the disaster victims.

2. The *United Nations Development Program (UNDP)* with its *Resident Representative,* serves as the contact for governments requesting UN assistance and as the representative of the *World Food Program (WFP).* UNDP provides assistance in economic and social development.

3. The *United Nations Children's Fund (UNICEF)* was established in 1946 to provide food, housing, medical, and sanitation services for children in developing countries. In emergency situations, aid is given when help from other sources is not immediately available. Supplies usually consist of children's food, vitamin capsules, and medical supplies, and at times, food grains, blankets, clothing, motor vehicles, and well-drilling equipment have been supplied. Approximately 5% of UNICEF's expenditures are for emergency assistance.

4. The *Office of the United Nations High Commissioner for Refugees (UNHCR)* can make an allocation available to a country affected by a disaster where the inhabitants include refugees. The allocation is to meet the immediate needs of the refugees.

5,6. The *Director General* of the *Food and Agricultural Organization (FAO)* decides on the amount of food commodities to be given to a stricken country. Once this decision is made, the *World Food Program (WFP)* has the responsibility for locating the stocks of food in donor countries and arranging for external transportation. Emergency food aid is usually provided in approximately 3 months unless there is WFP food already stored in the stricken country, in a neighboring country, aboard a ship near the affected area, or when it is possible to borrow from local stocks. A 3-month period is normally required because the WFP has to "shop around" with donor countries to ascertain the availability of commodities from pledged stocks and because of the time needed to transport the materials.

7. The *International Labor Organization (ILO)* can assist in the rehabilitation phase by training construction workers.

8. The *World Health Organization (WHO)* can supply technical personnel and urgently needed supplies and equipment until other outside assistance arrives.

9. The *International Civil Aviation Organization (ICAO)* can supervise and operate ground-based aeronautical facilities, and in certain instances, it can provide equipment, but with some delay.

10. The *Inter-Governmental Maritime Consultative Organization (IMCO)* can immediately supply experts for emergency situations that involve mass evacuations by ship, the unloading of equipment and commodities at port facilities, and search and rescue operations.

F. Long-Term Post-Earthquake Planning

Following the emergency period of a damaging earthquake when the immediate needs of the victims have been satisfied and the community is beginning to function once again, there will be a strong desire to rehabilitate and rebuild the community as quickly as possible. According to Duchek,[1364] the findings from several recent earthquakes indicate that pressures to rebuild communities will commence approximately 45 to 60 days following the earthquake. It is at the end of the emergency period that constraints on manpower, equipment, materials, supplies, and funding will fade, and massive reconstruction efforts will be possible. Duchek further observes that, unless these pressures are anticipated, it is unlikely that local government can manage reconstruction in a rational manner.

It is during the pre-earthquake period that government has the best opportunity to thoroughly analyze potential actions that would assure reconstruction would be to the benefit of the future community.[1364] Duchek[1364] describes plans and strategies for recovery operations that should be considered during the pre-earthquake period.

1. Procedures to quickly evaluate structural damage and performance of structural types and identify actual impacts from the kinds of ground movement which occurred.
2. Evaluation of land-use, zoning, and building standards to consider possible revision or amendments that might be warranted after a major damaging earthquake. These might prevent rebuilding in areas of proven hazard or assure that re-use or re-building will minimize hazards in the future.
3. Preparation and adoption of contingency legislation to become effective immediately upon the occurrence of a major damaging earthquake. This could effect changes in existing codes, institute interim recovery-period controls, or establish procedures for rapidly drafting and adopting new controls or standards based upon the findings of damage evaluation.
4. Evaluation of projected potential major damage areas and consideration of ways to capitalize on the damage to effect positive or desired changes in development patterns that might not be feasible prior to extensive major destruction.
5. Strategies to assure the normal procedures for development review and approval are not circumvented due to extreme pressures for rapid reconstruction.
6. Thorough evaluation of funding sources — the financial institutions, including government — and incorporation of them into development review and approval processes to assure funds for development will not be available without the required governmental approvals for construction.
7. Strategies to establish priorities for use of labor, materials, and other resources to control reconstruction and assure priority is given to projects of greatest public need and benefit.
8. Consideration of creation of a special task force to prepare a plan for recovery. It could be composed of a cross-section of community leaders from all fields in both public and private sectors. An effort of this kind could assure concensus on objectives prior to a major earthquake and establish the framework for efficient, rational re-building to the benefit of the entire community.

Unfortunately, predisaster planning, if present, is almost always directed towards emergency response planning and not reconstruction planning. However, as noted by Mader,[1118] the *City of San Francisco's Community Safety Plan* does specifically address the issue of post-earthquake reconstruction. The plan stresses the numerous opportunities that are present for effecting needed improvements during the reconstruction period.[1118,1365]

In a positive sense, post-earthquake reconstruction presents opportunities to affect actions and changes not possible prior to extensive damage; these opportunities would be related to transportation systems, land uses, building sizes and heights, location and connection of open space systems, and other factors. Properly directed reconstruction can provide the means for long needed improvements while correcting or eliminating past mistakes.

Three important policies were adopted by the city to guide the reconstruction planning:[1118,1365]

- Policy 1: Maintain the sound and rational redevelopment of San Francisco, following a major disaster, by rebuilding in accordance with established comprehensive plan objectives and policies, appropriate city codes, and other community concerns and needs.
- Policy 2: Adopt contingency legislation to provide for anticipated needs following a disaster and to reduce pressures for unnecessarily rapid reconstruction.
- Policy 3: Create a reconstruction planning committee to insure that development following a major disaster takes place in a timely fashion according to established objectives and policies.

Following a destructive earthquake, the *Reconstruction Planning Committee*, described above, would have the following duties:[1118,1365]

1. Insuring that post-earthquake building code and design standards are as advanced in terms of seismic safety as possible.
2. Implementing objectives, policies, and criteria of the Comprehensive Plan.
3. Recommending contingency legislation to be enacted now, but taking effect after an earthquake to authorize such actions as provision of temporary housing.
4. Determining priorities for allocating resources, particularly building materials.
5. Seeking joint agreements with lending institutions, insurance companies, and Federal disaster assistance agencies to require a valid building permit before money for new construction is released.
6. Developing an information booklet setting forth all requirements pertinent to reconstruction and sources of financial assistance.

A recently published and highly recommended book, *Reconstruction Following Disaster*,[1366] describes in detail how four cities recovered from natural disasters. Three recovered from earthquakes: San Francisco, California; Anchorage, Alaska; and Managua, Nicaragua; one recovered from a flood, Rapid City, South Dakota. An analysis of reconstruction experiences for the above cities reveals a number of major insights that are of particular value to the policy maker.[1367,1368]

1. Cities struck by disasters will be rebuilt, and except for small communities like Valdez, Alaska (Figure 22), they will be rebuilt on the identical predisaster site. Several factors working against relocation include: (a) the social problems in uprooting the inhabitants, (b) the lack of nearby risk-free sites, (c) the enormous capital investments in the fixed stock of the existing community, (d) the time and money involved in preparing a new site (Figure 22), (e) the necessity to restore the economic base of a region as soon as possible, and (f) the feeling that if seismic regulations are either adopted or improved, the next earthquake will have less serious consequences at the original site. The authors believe that only minimal attention should be directed towards the relocation issue with the major focus directed towards questions such as how fast will the recovery be and how much safer will the reconstructed city be?
2. Reconstruction efforts for the functional replacement of urban systems will be about 100 times the length of the emergency period. For most disasters, the time frame is between 2 and 8 years. The rate of recovery is directly related to the magnitude of the damages suffered, resources available for recovery operations, prevailing predisaster trends (i.e., rapidly growing cities will recover rapidly, whereas stable or declining communities will recover slowly), and the qualities of leadership and degree and comprehensiveness of planning available for the reconstruction effort.
3. In a disaster-stricken city, the primary objective is to return to normalcy as quickly as possible. Pressure for this action is usually strongest among displaced families and the business community. This is countered by the bane of planners

FIGURE 22. New site of Valdez, Alaska and dock facilities. Old Valdez, devastated by the March 27, 1964 Alaskan earthquake (M_s = 8.5) is at center-left. Loss of the Valdez waterfront by submarine landslides and extensive earthquake damage throughout the community led to recommendations by U.S. Geological Survey scientists to abandon the town and to rebuild Valdez on the flat near Mineral Creek, 5.6 km northeast of the old site. Bedrock ridges along the shore protect the new townsite and also mean that there is no danger of offshore slides. It took the federal government approximately 3 years to complete the project at a cost of $37.5 million, or a per capita cost of $50,000. (From George, W. and Lyle, R. E., The Alaska Earthquake, March 27, 1964: Field Investigations and Reconstruction Effort, U.S. Geological Survey Professional Paper 541, U.S. Government Printing Office, Washington, D.C., 1966, 87.)

and others who would like to rebuild the community only after a careful study of reconstruction issues and their long-term consequences have been evaluated to insure that the catastrophe will not be repeated. Therefore, grandiose plans for reconstruction are usually counterproductive because they will lead to uncertainty about the future of the city by those who live and work in it, conflicts between various elements of the community, and timely delays. Ultimately, such plans are doomed to failure. The reconstructed city will be less changed than inferred originally from the destruction, and it will be less vulnerable to a recurring hazard, but the achievement will be less than the opportunity for change that was afforded by the disaster.

4.   It cannot be assumed that decisions makers in the private sector will delay their decisions about reconstruction until the most important policy decisions have been made by governmental agencies. For example in Anchorage, the decision by a hotel builder to proceed with his reconstruction plans in the central business district ''short-circuited'' a government reconstruction plan fraught with indecision and delays.

5.   Although a stricken city is initially perceived as a community of common suffer-

ing, recovery is not uniform for each group of victims. For example, those with greater knowledge and access to resources prior to the disaster continue to have such access after the disaster, and those "with a high position in the hierarchy of commerce, industry, and residence are favored by the heirarchy of return." Simultaneously, marginal industries, businesses, and low-income families are usually hard put to reestablish themselves because the cost of available buildings and vacant land is inflated. Marginal industries and businesses may vanish or move to outlying areas of the city, which in turn, can eliminate jobs or force employees to travel longer distances to and from work. Low-income families may have to move frequently within the city or move from it forever.

Chapter 2

SOCIAL ASPECTS

I. INTRODUCTION

Social aspects associated with the earthquake hazard have only recently been analyzed to any great extent. For example, the first concerted effort to focus attention on seismic-related social problems in the U.S. did not come until the March 27, 1964 Alaskan earthquake. Following this disaster, the National Academy of Sciences (NAS) established a *human ecology group* and a comprehensive report was completed,[1369] even though the President of the U.S., in charging NAS with the responsibility for compiling and preserving the findings of various governmental agencies, "made no specific mention of social sciences, of humans, or of ecology."[1370] This chapter is concerned with the following social topics: earthquake insurance, public response to the earthquake hazard, human response to earthquake and tsunami warnings, and emotional problems experienced during the post-earthquake period.

II. EARTHQUAKE INSURANCE

One type of long-term earthquake protection measure is *earthquake insurance*, which is available throughout most of the U.S. at a reasonable cost from a number of private companies. The most common type of earthquake insurance is an addendum to the standard fire policy. Rates vary with the risk zone in which the property is located, coverage desired, and the type of property to be insured. Two basic systems, the *Western Method* and the *Eastern Method,* are used for writing earthquake insurance. Each is subject to the rules and regulations of rating bureaus in the individual states. The primary difference between the two systems is that the Western Method includes a mandatory deductible of a standard sum (usually $500 for a wood-frame dwelling) to 5 to 15% of the building's value at the time of loss, while the Eastern Method does not, but a 25% credit is given to the policyholder if a 2% deductible clause is elected.[1371] Policies including low-standard deductibles have somewhat higher annual rates than the percentage-of-value deductibles. Most small residential dwellings qualify for the 5% deductible or the standard deductible of $500.[1021]

Rates under the Western Method are also higher than under the Eastern Method because of the greater risk of exposure in the more earthquake-prone states of Alaska, Arizona, California, Hawaii, Idaho, Montana, Nevada, Oregon, Utah, and Washington.[1371] Each of these states is divided into risk zones reflecting the likelihood of damaging earthquakes. California, for example, is divided into three zones:[57]

- *Zone 1 —* Coastal counties and counties east of the Sierra Nevada
- *Zone 2 —* Central section of the state from Kings and Tulare counties north to the Oregon border
- *Zone 3 —* Imperial County

Zone 2 carries the lowest rates and Zone 3 the highest. For a single-family wood-frame dwelling in Zone 2, the annual rate is a about $1.50 per $1,000 of coverage. The rate for the same structure is $3.00 per $1,000 in Imperial County.[57] In Los Angeles and San Francisco (Zone 1), the annual rate for a single-family wood-frame house varies from \$1.50 to \$2.50 per \$1,000 of coverage.*[1021] Assuming a \$2.00 rate per \$1,000 value of dwelling and a 5% deductible clause, the annual premium for a \$35,000 house (exclusive of land) would be \$70, and the owner would pay the first \$1,750 of any earthquake damage.[1372] Personal property is not covered by the typical earthquake policy and must be insured in a separate addendum with another deductible clause ranging from a standard sum to 5 to 15% of the value of the personal property. The amount of the deductible is determined by the type of building and the damageability of the personal property.[1021] The *Pacific Fire Rating Bureau* has a special rate category for buildings incorporating earthquake-resistant designs. Rates are determined by the bureau after a building's plans and specifications have been examined.[1371] In areas where the Eastern Method is used, rates for a single-family wood-frame dwelling vary from \$0.45 to \$1.50 per \$1,000 coverage with a 2% deductible clause.[1371]

Even with reasonable rates, relatively few policies have been purchased for commercial and private properties exposed to the earthquake hazard in this country. For example, in a 1972 Office of Emergency Preparedness report prepared for the U.S. Congress,[1371] it was estimated that the aggregate value of property exposed to potential earthquake damage was in the tens of hundreds of billions of dollars, but only about \$3 billion in property was covered by earthquake insurance. Only about 200,000 homes in the entire State of California have earthquake coverage.[1372] In 1975, one major insurer in California experimentally dropped its rates to \$1.50 per \$1,000 to determine if price was a barrier for the purchasing of earthquake insurance by the homeowner. Although the lower rate was promoted throughout the state, only 18 additional policies were sold.[1372]

The obvious question suggested by the foregoing discussion is why so little property is covered by earthquake insurance in the U.S., especially in high risk areas. The Office of Emergency Preparedness suggests several factors that could account for the low level of public participation.[1371]

. . . First, there is *ignorance about the need* for earthquake coverage in the more earthquake-prone areas. From the standpoint of disaster preparedness, widespread earthquake coverage is desired insofar as it contributes to hazard reduction and avoidance, and provides a more equitable arrangement for disaster relief. Many citizens in high-risk areas, however, are apparently unaware of or unconcerned about their exposure to risk. Public officials have contributed to this problem insofar as they have not made sufficient efforts to inform the public of the hazards and have allowed unsafe and unwise building practices to take place.

. . . In addition to lack of information concerning [the] earthquake hazard, *complacency* — particularly in California — about the threat of a serious earthquake is probably due to the infrequency of high-intensity earthquakes that have affected major population centers. In the last 50 years, only three earthquakes with intensities greater than VIII on the Modified Mercalli Scale . . . have occurred in heavily populated areas of the United States, all in California — Long Beach in 1933, Bakersfield-Kern in 1952, and San Fernando in February 1971 . . . It is probable therefore, that many property owners have been lulled into a false sense of security. Furthermore, there is no evidence indicating a substantial increase in the purchase of earthquake insurance in the affected area following the San Fernando quake.

. . . The *inability to purchase earthquake insurance conveniently* also has an impact on availability. For example, earthquake insurance in California is normally written as an endorsement to the homeowner's policy for fire and extended coverage. Since not all property insurers write earthquake insurance, or some restrict the areas or the types of dwellings which are eligible for coverage, the property owners must find a company willing to sell him a policy. Although most leading insurers issuing homeowner's policies in California will write earthquake insurance . . ., the property owner is nevertheless placed in a position of "shopping around" for a company willing to insure him.

. . . Availability of earthquake insurance also is affected by the *reluctance of the companies to write this coverage in large amounts.* Accordingly, they do not sell it aggressively . . . As a result, property owners may simply be unaware that it can be purchased. For a large extent, the reluctance is due to the unpredictable, singular nature of the earthquake itself. Risk cannot be spread easily, and companies fear they will not

---

*   Rates vary with a building's class of risk. The annual cost per \$1,000 coverage for the highest risk building in Zone 2 (e.g., unreinforced masonry, masonry veneer, or adobe) is \$25.00.[1021]

be able to build and maintain reserves adequate to meet the claims resulting from a single, sizable earthquake in a populated area. Concomitantly, the growing commercial concentration in high-risk areas adds to the financial burden the companies would have to assume if they agreed to provide full earthquake coverage for commercial as well as residential structures. It is noted that the "growing values of commercial and industrial properties have been concentrated in certain high-risk areas and have increased to the extent that the potential liability on a single building may exceed the area-underwriting limits of a reasonably large insurance company" . . . Related to this is the question of whether adequate reinsurance — or some form of financial backup — would be available to the primary insurer if earthquake underwriting were greatly expanded.

    . . . Finally, *little positive institutional action* has been taken to encourage property owners to insure against the earthquake hazard. Lending institutions in the private sector, as well as such Federal agencies as VA and FHA, do not require earthquake insurance as a prerequisite for home loans . . .

An important study conducted by scientists at the Wharton School, University of Pennsylvania, explored the reasons why some people in flood- and earthquake-susceptible areas purchase insurance and others do not.[1373-1375] A total of 2055 homeowners residing in 43 flood-susceptible areas in the U.S. and 1066 homeowners in 18 earthquake-prone areas in California were interviewed. Approximately half of each sampled population had purchased insurance. The following summary describes several principal findings for the earthquake segment of the study.

1.   Only 62% of those interviewed were aware that earthquake insurance was available in their areas. The news media was the primary initial source of insurance availability. More than 85% of the insured individuals contacted their agents first rather than being contacted by the agent.

2.   In reference to awareness of the terms of the insurance policy, 76% of the uninsured homeowners were unable to provide any estimate of the premium or the deductible. Quite surprisingly, 25% of the insured population did not know the deductible on their policy, while another 9% assumed there was no deductible clause. All policies had, in fact, a 5% deductible clause.

3.   For the uninsured who provided an estimate of the maximum amount of money they would be willing to pay for coverage, 34% were willing to pay more than the cost of the actual policy. Had this group been aware of the actual rates, they would have found earthquake insurance to be an attractive buy.

4.   A relatively large percentage (12%) of the uninsured did not expect any damage from an earthquake even though they were residing in seismic-prone areas. Quite interestingly, 2% of the insured did not anticipate property damage from an earthquake. It is difficult to understand why homeowners who estimated zero damage would voluntarily purchase a policy.

5.   An argument raised against a system of liberal disaster relief (e.g., forgiveness grants, low interest loans) is that it discourages property owners from purchasing insurance during the pre-disaster period. In order to determine what role the respondents expected the federal government to play should they suffer losses from an earthquake, each person was asked to enumerate the sources and amounts of funds that would be expected for restoring losses. A substantial majority of both insured and uninsured homeowners expected to receive *no* aid from the federal government. A surprising finding was that two thirds of the uninsured expected no federal assistance. This factor appears to be relatively unimportant in explaining a homeowner's decision not to buy earthquake insurance.

6.   Two factors, when taken together, carry the most weight in the decision process to purchase insurance. These are whether the individual considered the earthquake to be a serious hazard and whether a person knew someone who had already purchased insurance. The latter factor is apparently important for interpersonal exchanges on insurance availability, although it could not be determined

specifically from information supplied by the respondents. Socioeconomic variables such as income and education had a relatively unimportant impact on the purchase decision or how people viewed the seriousness of the hazard.

One key finding of this investigation was that most property owners refused to be concerned about future losses from floods and earthquakes because they are perceived as having little chance of occurring. If insurance is brought to their attention, it is apt to be viewed as a poor investment because of the unlikelihood of receiving a return on their cash investment. To make voluntary insurance more attractive, the researchers make several recommendations for improving the present lack of awareness. For example, much more attention must be given to disseminating information about potential problems so that the traditional view of seeing floods and earthquakes as low probability events is translated into a potential loss perspective that is probable enough to warrant the purchase of insurance. The news media, vivid graphics such as films and easily read maps, and talks to community groups by government officials should be widely used to increase the individual's awareness of the consequences of a flood or earthquake disaster.

The awareness of hazard insurance, in terms of availability, rate schedules, and deductibles, can be improved by the agent contacting clients who have purchased other policies from him/her. However, small commissions have not encouraged such personal contact. One of the principal functions of the agent should be to educate the property owner that the insurance policy is not an investment, but that the biggest return on a policy is no return at all. If property owners do not adopt this view, they are apt to purchase insurance only after suffering losses from a flood or earthquake and then cancel the policy after several years because they have not received a return on their investment. This education process will continue to be slow unless insurance agents play an active role.[1373-1375]

As opposed to the U.S. system, New Zealand offers a unique national insurance program to cover damages incurred from extraordinary and unforeseen disasters. All property covered by a private fire insurance policy is *automatically* assessed an addition levy of 50c per $1,000 of coverage which is paid by the private insurance companies to the *Earthquake and War Damage Commission.* The annual revenue is deposited into two accounts: 90% goes to the *Earthquake and War Damage Fund* from which the commission pays for the damage caused by earthquakes or enemy action in time of war; the remaining 10% goes to the *Extraordinary Disaster Fund* for repairing or replacing property damaged from natural hazards including volcanic eruptions, floods, storms, landslides, and tsunamis.[1376] Property owners bear 1% of each loss with a minimum of $10 (New Zealand) and a maximum not to exceed $100.[57]

Although in theory, at least, the New Zealand scheme appears to approach the optimum in hazard management, O'Riordan[1376] notes that the program is beset with several problems which largely center on the property owner's selfish desire to make use of the communal fund. For example, the Earthquake and War Damage Commission is seen as a claim-processing governmental agency sitting on vast sums of public money which should be spent and not saved. Consequently, the commission has been forced by public pressure to extend its coverage to events which are by no means unforeseen (e.g., landslide damage in areas of known slope instability, flood damage in areas with a record of past floods) and to pay damage claims regardless of the prior condition of buildings insured under the program. Spurious claims are frequent because the participants feel they have paid their fair share, the funds are enormous, and they personally feel they have suffered from a "natural" disaster. As a consequence, claim payments are increasing, although the incidence of hazards is apparently not increasing.[1376]

III. HUMAN RESPONSE TO THE EARTHQUAKE HAZARD

Several important studies have assessed the *human response to the earthquake hazard* in North America. Jackson[1377,1378] conducted a response survey in 1972 among 302 residents of the seismically hazardous areas of Los Angeles, Anchorage, and Vancouver. The survey assessed how residents perceive the earthquake risk to which they are exposed and the extent to which they adopt precautionary measures or adjustments to reduce the possibility of injury, loss of life, and property damage from a future seismic event. Precautionary measures include the purchase of earthquake insurance, modifications to improve the structural integrity of a dwelling, selection of a house away from hazardous ground, family earthquake drills, and storage of emergency supplies such as water, canned food, blankets, flashlight, and a medical kit.

Survey results demonstrate a response to the earthquake hazard that closely approximates the response patterns identified by Slovic et al.[1379] for other natural hazards. Three primary features characterize the response: "(1) a narrow range of adjustments is adopted; (2) the possibility of future occurrence of earthquakes tends to be downgraded; and (3) there is a strong preference for crisis-response rather than precaution."[1378] Regarding the adoption of precautionary measures for the earthquake hazard, 209 of the 302 respondents (69.2%) had not adopted any kind of adjustment, 72 (23.8%) had adopted only one adjustment, and only 21 of the respondents (7.0%) had adopted two or more precautionary measures. The most common precautionary measure was earthquake insurance; 52 of the respondents (17.2%) claimed to have insurance. However, Jackson[1378] cautions that the percentage may be inflated because of the "respondents inaccurate knowledge of their insurance policies." Other adoptions were as follows: storage of emergency supplies, 26 (8.6%); family earthquake drills, 8 (2.6%); structural modifications, 4 (1.3%); and house selection away from hazardous ground, 3 (1.0%).[1378]

According to Jackson,[1378] the response to the future occurrence of earthquakes showed a greater degree of denial than was established by Kates[1380] in studies on flood and coastal storm hazards. However, the common reasons for denial did appear in the earthquake investigation; these included the "perception of the event as cyclic, belief that future events would be minor, and an absolute unwillingness to consider the possibility."[1378] More specifically, 177 respondents (58.6%) either were uncertain that they would ever be in an earthquake or denied the possibility outright; 98 respondents (32.5%) accepted the possibility of future earthquake occurrence, but they believed their property would not sustain any damage; only 27 respondents (8.9%) accepted the possibility of being affected personally by a future seismic event.[1378]

Jackson[1378] established a preference for crisis-response in two ways. First, while only 93 respondents (30.8%) had adopted precautionary measures, 138 (45.7%) had taken avoidance action solely at the time of an earthquake. Second, the adoption of precautionary measures was much more common among residents whose property had suffered severe damage in a previous earthquake. For example, 29 respondents (20.1%) who had not experienced an earthquake had adopted precautions, whereas 39 (52.0%) who suffered property losses in excess of $1,000 from a previous seismic event had adopted precautionary measures (Table 1).

One surprising finding of the Jackson study not encountered in other types of hazard research was that 270 respondents (89.4%) believe it is government's responsibility to prepare for future earthquakes; 163 identified the federal government; 70 identified local government; and 57 identified state or provincial government. Jackson believes that certain government strategies discourage citizens from initiating precautionary measures on their own. The negative impact of governmental policies on citizen behavior likely occurs for two primary reasons: "(1) the provision of funds for reconstruc-

TABLE 1

Relationship Between Adoption of Precautionary Measures and Intensity of Past Experience

Precaution (action taken)	Intensity of experience							
	Losses over $1,000		Losses under $1,000		No past experience		Total	
	No.	%	No.	%	No.	%	No.	%
No precautions	36	48.0	58	69.9	115	79.9	209	69.2
Precautions taken	39	52.0	25	30.1	29	20.1	93	30.8
Total	75	100.0	83	100.0	144	100.0	302	100.0

From Jackson, E. L., *Calif. Geol.*, 30, 279, 1977.

tion after disaster makes it cheaper for the individual to rely on government assistance than to invest his own funds in precautionary measures, and (2) other governmental strategies (seismic risk zoning, land use planning, building codes and structural measures, disaster planning and relief, and earthquake prediction and control research) generate a widespread public sense of security related to faith in technology, assuming that no further precautions are necessary."[1378]

Jackson[1378] stresses that the public must be made aware of the limitations of current governmental policies and actions that the individual can carry out to reduce injury and damage from future earthquakes. This can best be accomplished by a detailed and widely disseminated information program.

Jackson and Mukerjee[1381] analyzed the human adjustment of 120 San Franciscans to the earthquake hazard in their city. Analogous to the previous study, the respondents gave a low priority to the seismic hazard. The following listing describes several of their findings.[1381]

1.   The authors experienced a 78% refusal rate of citizens in nine areas of the city to participate in the questionnaire survey. Possible reasons for the unusually high refusal rate include an unwillingness to spend approximately 30 min in answering the questionnaire, an unwillingness to participate in surveys of any kind, an unwillingness to allow strangers in their homes, and a denial of the earthquake problem. The latter factor may be the most important reason. According to Jackson and Mukerjee, the tone of most refusals suggested that many of the citizens are worried about the earthquake hazard, but they do not allow themselves to think about it. This could represent a minimization of the seismic threat through the *dissonance reduction process.*

2.   When asked about disadvantages of their place of residence and of the city, 44.2% could think of no problems. The problems that were mentioned by the other respondents were usually social in nature, but a few mentioned natural phenomena such as cold temperature and fog. The earthquake hazard was *never* mentioned as a problem.

3.   Although the earthquake was not mentioned as a problem, 75% of the respondents were willing to admit that there would be earthquake occurrences at some period in the future; 20.8% did not know; and only 4.2% thought that an earthquake would not occur again. Even though most respondents expect future events, there was a tendency to minimize the damage expected from such activity.

For example, 20% of the respondents thought their property would escape damage. For those that did anticipate damage, structure and contents were specified by 36.6 and 13.3% of the respondents, respectively. Regarding the extent of these damages, however, only 6.7% thought the damages would be total; 20% thought they would be substantial; 25.8% believed the damages would be either slight or nonexistent; and 22.5% did not know. This tendency to minimize personal vulnerability probably relates to the recent lack of damaging earthquakes in the region.

4. Some 55.9% of the respondents thought precautions could be taken to prevent earthquake-related damage. Of the remainder, 33.3% stated that nothing could be done, and 10.8% did not know of any precautionary measures. However, the next question revealed that only 26.7% of the respondents did not practice any of the adjustments contained in the questionnaire, suggesting that certain precautions are not perceived as loss-reducing. Adjustments contained in the questionnaire were "do nothing; pray; evacuate; protect home against fire and looters; structural changes to home; earthquake insurance; and impact and postimpact emergency action, for example run outside to an open space or take shelter in a safe place." The two most practiced adjustments (37.6% each) were *do nothing* and *pray!* Adjustments that required investment and preparation to modify the loss potential or prepare for losses had been adopted by only a few respondents. For example, earthquake insurance and structural changes to the home were each reported by only 7.5% of the respondents. The authors emphasize that the most popular adjustments represent a "poor adaptation" to the potential for damages.

5. Suggestions have been made that scientifically based earthquake warnings might encourage people to adopt optimal, preparatory adjustments. However, in this study, 49.2% would do nothing if a warning of up to one year was given. Many respondents did not believe that earthquake warnings would be reliable. Of the remaining respondents, action to be taken by most was evacuation (24.2%). Preparatory adjustments such as structural modifications to a dwelling would be adopted by only 4.2% of the respondents.

In 1970 and again in 1976, Sullivan et al.[1382] conducted a survey of residents in San Mateo County, California who resided in a narrow strip of land (0.8 × 10.5 km) straddling the San Andreas fault. The boundaries of the study area, although defined in 1970, closely approximate the limits of the Special Study Zone for fault-related hazards established after the passage of the 1972 Alquist-Priolo Special Studies Zones Act (described in Volume III, Chapter 1). The study area contains the most extensive residential development to straddle the 968-km-long San Andreas fault and is depicted in Figure 3 in Volume I, Chapter 1. The purpose of the 1970 survey "was to evaluate the awareness of, and attitudes toward, earthquake hazards in a densely populated area close to the San Andreas fault." Because government agencies and the news media began to focus increasing attention on the potential destructiveness of a major seismic event after the 1971 San Fernando earthquake, a second objective "became the evaluation of any effect that this publicity had in changing public awareness and attitudes."[1382]

Responses for the two periods are summarized below and summarized in Table 2.[1382]

1. *Awareness and attitudes toward the earthquake hazard.* Although almost all of the residents had heard of the San Andreas fault in both of the survey years, only about one third were aware of the direct relationship between faulting and earthquakes. Residents were generally aware that their homes were located within 1.6 km of the San Andreas fault, but a smaller percentage knew of the fault's

TABLE 2

Summary of Earthquake Hazards Survey, 1970 and 1976, San Mateo County, Calif.

Respondents	Percentage of respondents	
	1970[a]	1976[b]
1. Age and sex		
Female adults		52
Male adults		42
Female minors		4
Male minors		2
2. Residents		
Own residence	93	82
Rent or lease	7	18
3. Lived in the San Francisco Bay area		
For more than 20 years	51	48
For more than 10 years	64	71
4. Lived in their current residence for less than 4 years	67	50
5. Had heard of the San Andreas fault	96	98
6. Were unaware of the relationship between fault movement and earthquakes	31	38
7. Knew that the San Andreas fault was within 1 mile (1.6 km) of their residence	77	77
8. Knew of the proximity of the San Andreas fault before taking up residence	72	63
9. Would feel no safer from fault-related hazards if they lived 5 miles (8.1 km) further from the San Andreas fault	82	74
10. Think the building industry is doing all it can to make residences safe in earthquakes	20	26
11. Were informed of potential earthquake hazards by the previous owner, developer, or landlord	8	16
12. Would inform future residents of potential earthquake hazards	22	51
13. Have felt earthquakes while living in their present residence	52	44
14. Would react to an earthquake occurring now by		
Standing under a doorway	43	43
Getting under a desk, table, or bed	8	15
Staying away from windows, mirrors, and chimneys	5	3
Remaining stationary and calm	20	16
Running outside	12	14
Unsure	10	5
Other responses	2	4
15. Have earthquake insurance	5	22
16. Reasons for not obtaining earthquake insurance		
Too expensive	59	42
Not needed	29	28
Unaware of availability	9	14
Other responses	3	16
17. Feel that earthquakes can be predicted		61
18. Feel that geologists and/or other earth scientists will make the most reliable earthquake predictions		94
19. Reactions to prediction of a major earthquake with 20 years		
Would do nothing	94	85
Would move	2	5
Other responses	4	10
20. Reaction to prediction of a major earthquake within one year		
Do nothing	77	61
Leave the area temporarily		3
Move		16
Move or leave the area temporarily	11	
Unsure	9	8

TABLE 2 (continued)

Summary of Earthquake Hazards Survey, 1970 and 1976, San Mateo County, Calif.

	Percentage of respondents	
Respondents	1970[a]	1976[b]
Reinforce the residence	1	4
Purchase insurance	1	1
Other responses	1	7
21. Reaction to prediction of a major earthquake within 1 week		
Do nothing	52	38
Leave the area temporarily		35
Move		5
Move or leave the area temporarily	36	
Unsure	9	8
Reinforce the residence	1	3
Purchase insurance	1	1
Other responses	1	10
22. Had received earthquake hazard information directly from their local government	22	28
23. Support legislation requiring that all prospective occupants be informed of potential earthquake hazards before buying or signing a lease	81	77
24. Were aware that a law (Alquist-Priolo Special Studies Zones Act) dealing with fault-related hazards was passed in 1972		16
25. Were in favor of federal and/or state government funding of earthquake research	75	88

[a] Results of 1970 survey.
[b] Results of 1976 survey.

From Sullivan, R., Mustart, D. A., and Galehouse, J. S., *Calif. Geol.*, 30, 8, 1977.

location before moving into the area. A common attitude was that the entire San Francisco Bay region was equally vulnerable to earthquake damage. Most respondents would feel no safer if they lived 8 km further away from the fault. Relatively few respondents in both surveys claimed that they were informed of possible hazards by the previous owner, developer, or landlord. However, the percentage of respondents who were informed has doubled from the number who knew of the hazard in 1970. This suggests that recent legislation and media publicity might have had a small effect on the public awareness of earthquake hazards. A strongly defined trend was the increased number of residents who had purchased earthquake insurance. The percentage more than quadrupled between 1970 and 1976. The primary reasons, in declining order of importance, for not purchasing earthquake insurance were (1) too expensive, (2) not needed, and (3) unaware of its availability.

2. *Attitudes toward earthquake prediction.* A significant percentage believe earthquakes can be predicted on a scientific basis. This is most likely attributable to the attention given the topic in the news media. In the questionnaire, respondents were invited to give their reactions to a major earthquake predicted to occur in 20 years, 1 year, and 1 week. Surprisingly, a large number of people would do nothing in the advent of a prediction for any of the time windows. Most surprising was the fact that for a 1-week prediction, 38% of the respondents indicated they would do nothing. The most encouraging aspect of the earthquake prediction part of the survey was that the percentage of respondents who would do nothing to reduce the impact of a seismic event decreased from 52 to 38% over the 6-year study period.

3.   *Awareness of public agency information policies.* This part of the study assessed the role played by government agencies in informing residents of earthquake hazards and the activities of government in earthquake research. Generally, a relatively low percentage of respondents had received earthquake hazard information directly from their local government. A large majority would favor legislation that required all prospective occupants to be informed of potential seismic hazards at a particular site. However, only a small percentage of the 1976 respondents were aware of the enactment of the 1972 Alquist-Priolo Special Studies Zones Act which is designed to encourage such disclosures and limit construction on seismically active faults. The vast majority of residents were in favor of using federal and state funds for earthquake research.

Sullivan et al.[1382] believe that a program of public education that is well planned and imaginative should be initiated immediately because in their study, for instance, most respondents were not concerned about the potential destructiveness of a major earthquake and were generally uninformed as to the proper actions to take in the advent a scientifically based earthquake warning is issued. The authors recommend that an appropriate agency of California state government organize and coordinate this program even though there may be concerns about continually alarming the public about earthquake hazards while sustaining their attention over a long period of time. The public should be informed on disaster preparedness measures, hazard abatement programs, and long-term social and economic disruptions that would be likely to follow a catastrophic earthquake. Such topics could be most logically discussed at community-centered meetings because many of the hazards are unique to each region.

In 1975, the Field Research Corporation conducted in-home interviews with 1004 residents in northern and southern California. Survey responses were cross-tabulated along lines of sex, age, income, education, and area of residence.* The study revealed several interesting facts about the public's knowledge and attitudes towards earthquakes and their potential prediction.[925]

1.   At least 75% of the respondents expressed some concern about a damaging earthquake occurring at their location. Expectation of a major earthquake diminished as age increased. Los Angeles respondents thought there was a strong likelihood of a major seismic event affecting them, while those living in northern California outside the San Francisco Bay region were not at all concerned.
2.   Most respondents were aware that there are more earthquake hazards than just ground shaking. Older people, the less educated, and respondents with lower incomes described fewer effects that could happen during and after an earthquake. Groups naming the most effects were between 30 and 39 years of age, members of households with incomes in excess of $15,000, and those with some college education.
3.   Some 72% of the respondents had previously heard or seen earthquake information; 54% named the broadcast media as the source of this information. Los Angeles area respondents depend far more heavily on the broadcast media for earthquake-related information than Bay area repondents. Moreover, 10% fewer respondents in the Bay area mentioned any public source for earthquake information when compared with the Los Angeles group.
4.   Most respondents (72%) thought it was a good idea to give people several months

---

*   Sex: male, female. Age: 18—29, 30—39, 40—49, 50—59, 60 and over. Income: under $7000, $7000—$9999, $10,000—$14,999, $15,000 and over. Area: southern California (Los Angeles and Orange counties), other southern California, northern California (Bay area), other northern California.[925]

warning when scientists believe a major earthquake is likely to occur. Younger age groups, the college educated, and those with higher incomes were more favorably inclined toward earthquake prediction than other groups. From 10 to 20% of respondents in most groups thought prediction to be an unsound concept; the most negative segment (34%) was the age group over 60 years of age.

5. Three times as many respondents thought that a prediction with an 18-month lead time would have a negative effect on the local economy than felt it would have a positive effect. Some 20% thought there would be no effect on the economy. Negative effects were mentioned most often by respondents in the 40 to 49 age group and those with higher incomes and educational levels. Positive economic effects of prediction were most popular with respondents aged 18 to 29 and those with the lowest incomes.

6. More than half of the respondents indicated they would either do nothing or leave the area in response to an open-ended question that asked what precautionary measures they were likely to take in the event an earthquake prediction was was made by government scientists. More specifically, younger people would be more apt to leave the area, whereas older persons feel they would be less able to take action.

In the past, it has been suggested that the public does not want to be reminded of the dangers associated with future earthquake activity and would prefer less media coverage of earthquake-related matters.[1383] To determine if 2 years of news coverage of the Palmdale bulge in southern California (Figure 11 in Volume II, Chapter 1) had produced a saturation effect, R. Turner of UCLA initiated a project in 1978 to study the community response to the potential earthquake threat. It was believed that the results would be of practical importance for media representatives and officials involved in earthquake safety education programs. A telephone survey of 500 adults in Los Angeles County was conducted to elicit opinions about the amount of media coverage for the most recent 6-month period.[1283]

The findings were overwhelmingly one-sided. More than two thirds of the respondents wanted *more* media coverage on matters relating to the Palmdale bulge, scientific earthquake prediction research, earthquake preparedness measures, and what to do in the event an earthquake strikes. Only on the subject of predictions by nonscientists did a substantial number of respondents (43%) believe that the coverage had been excessive. The evidence from this study clearly indicates that well-conceived earthquake news and feature studies will not be rejected by a "saturated" public; in fact, most of the public is ready for a more extensive treatment of discussions concerning the earthquake threat in southern California.[1383]

Slosson[1384] analyzed earthquake-related bills introduced in the California Legislature and suggests that a strong emotional reaction prevails. The study evaluated the impact of the 1971 San Fernando earthquake upon legislation and a review of legislation trends for 3 years before and after 1971 (Figure 1). Several of Slosson's conclusions are described below.

The pre-1971 legislation was in part stimulated by the engineering and scientific professions' continuing effort to improve earthquake science and technology and by the 1964 Alaskan earthquake and other damaging earthquakes outside the U.S. (e.g., 1967 Caracas, Venezuela earthquake). The 1969 Santa Rosa earthquake stimulated a doubling of proposed earthquake-related bills and an increase in the failure record of legislation. Immediately following this earthquake, 18 bills were introduced, but according to Slosson, only six were enacted because "losses were not great enough to stimulate concern." The lull in seismic activity in late 1969 and the 1970 legislative

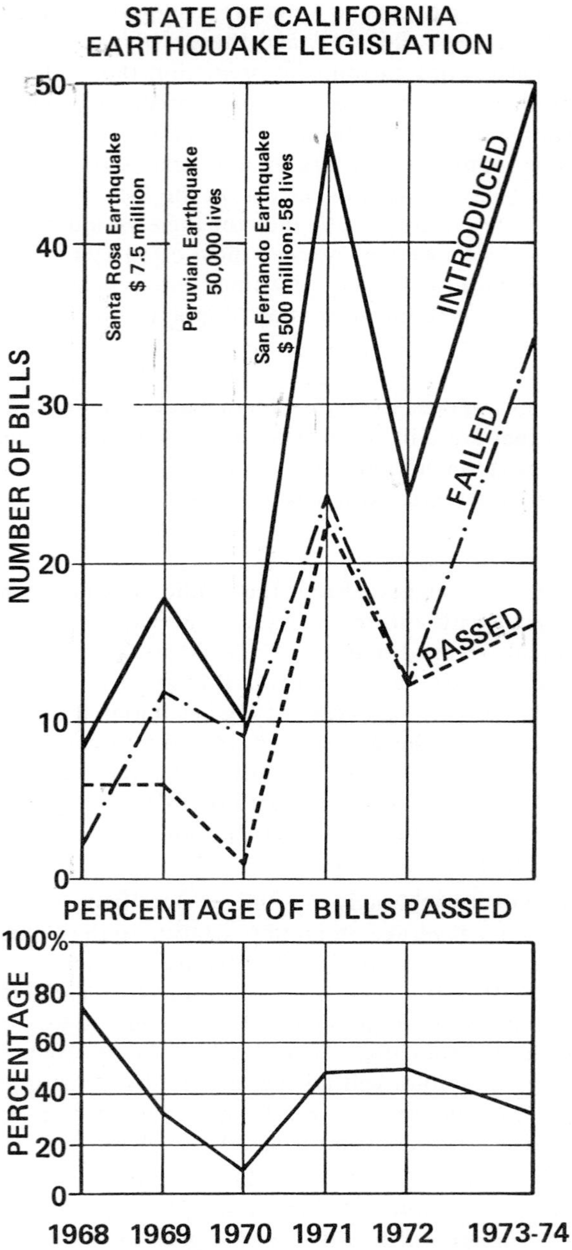

FIGURE 1.   California earthquake legislation and percentage of bills passed into law for the period 1968 to 1974. (From Slosson, J. E., *Calif. Geol.*, 28, 37, 1975.)

session produced a decline in earthquake concern as only one of 10 bills was passed (Figure 1).

The human and property losses caused by the 1971 San Fernando earthquake stimulated great legislative interest. According to Slosson, the "public and press were both very emotional as a result of this earthquake, and, in turn, this emotion was transmitted to the elected officials." Of 47 earthquake-related bills introduced in the legislature, 23 were enacted into law. The interest was still at a peak in 1973, and technical advisory committees worked closely in recommending legislation. This resulted in the introduction of 24 bills with 12 passing (Figure 1).

An analysis of bills introduced in 1973 to 1974 "was complicated by the initiation of a new 2-year legislative program which allowed 2 years for processing of bills with introduction and passage being possible in either 1973 or 1974." During this period, 50 bills were introduced with only 16 passing (Figure 1). Most of the bills enacted "were amendments or corrections to bills that were passed in 1971 and 1972 when the emotion was still high."[1383]

Slosson's[1383] interpretative results for the legislative period 1968 to 1974 are as follows:

1. Legislation related to hazards or disasters is greatly affected by major catastrophes.
2. Legislators appear to react to the *emotional* desires of the public.
3. There is generally a lack of action during the lulls between disasters, accompanied by strong overreaction immediately following disasters.
4. Many of the legislative bills passed during the height of the emotion are hurriedly prepared and ill-conceived requiring corrective legislation during the ensuing 1 to 3 years.
5. As a result of waning emotions and interest coupled with the examples of confusion caused by the hastily prepared legislation, it is noted that good, well prepared, and technically sound legislation generally fails as indicated by the 1973—74 legislative results.
6. This sequence pattern strongly suggests that it is the responsibility of concerned people in science and technology to have recommendable legislation prepared prior to the disaster which is bound to occur and then be willing to volunteer time and effort to assist the legislators (i.e. the Governor's Earthquake Council and the Joint Committee on Seismic Safety).

IV. POSSIBLE PUBLIC RESPONSES TO EARTHQUAKE WARNINGS IN THE U.S.

Because it may someday be possible to specify the location, time, magnitude, and probability or certainty of occurrence of anticipated earthquakes in the U.S., scientists are beginning to examine the potential socioeconomic effects that an earthquake warning might have on target regions. Most agree that if potential socioeconomic problems are not seriously examined now, earthquake prediction technology could come back to haunt society.

J. E. Haas and D. S. Mileti[1385, 1386] of the University of Colorado completed one of the first studies to treat the anticipated impacts of scientifically based earthquake warnings on various segments of a target region. Basic data for the California study were obtained by interviews and discussions with 35 business and government executives throughout California, 22 San Fernando business and community leaders, executives of eight large news organizations (newspapers, television, and radio), 31 federal and state agencies which could have involvement in making earthquake predictions, 37 firms representing the largest and most influential operating in the state, 41 local units of government in urban areas of northern and southern California that would be heavily involved in the event a credible prediction was made, 38 local businesses from the same urban areas, and 246 families in the earthquake-prone areas of northern and southern California. Several principal findings of the study are described below.[1385, 1386]

A credible earthquake prediction would result in a drastic reduction in deaths and injuries and property damage, but at the same time it is highly likely that the prediction will cause a local economic depression and accompanying social disruption. Several reasons would account for a reduction in casualties. There would be, for example, a considerable reduction (up to 50%) of people in the area of the expected earthquake. Physical and socioeconomic adjustments would further reduce the risk. Physical adjustments would include securing bookcases to walls, storing items to prevent toppling, making major structural modifications to buildings, and lowering water levels behind dams. Up to 75% of businesses, industries, and government agencies contacted in this study would (1) make an effort to assess the vulnerability of their buildings, equipment, and inventories in the event a prediction was made and (2) initiate appropriate precautions. Approximately 50% of the families interviewed would also initiate precautionary measures, with emphasis on nonstructural measures to reduce injury or death. The cessation of new construction in a target area would further reduce casualties because the fewer the number of buildings there are in a harder-hit area, the fewer the people at potential risk.

However, the above gains will not come lightly, as a target community will suffer significant social disruptions and a decline in the local economy from the time the warning is issued until the expected date of the event, especially if the lead time for the prediction is a year or longer. For example, the nonavailability of earthquake insurance following an official prediction would initiate changes in "mortgage availability, property values, investment patterns, and employment opportunities." These changes would then lead to a reduction in tax revenues for local governments, which in turn would lead to a reduction in public services.

Businesses, governmental agencies, and families will respond in a manner most favorable to their own interests. For example, investors are likely to place their money outside the target area, construction in the private sector will eventually stop, and local government will drastically reduce or stop capital construction projects. Influenced by the construction industry, business will slow, and unemployment will rise sharply. Local residents will reduce their level of purchases, and tourism will fall off to nothing. If governments urge evacuation of high-risk areas, local businesses will be further affected. Approximately 10% of all families will leave the area permanently. The percentage will be much higher if many businesses and industries permanently relocate outside the target region.

According to Haas and Mileti,[1385,1386] there is a sizeable price to pay in social and economic terms for the lowered risk of casualties, but the disruptions do not have to be so massive "if wise action is taken in advance of the prediction." In order to stimulate consideration of new or revised policies to improve the positive consequences of earthquake predictions while counteracting negative impacts, Haas and Mileti[1386] have outlined a series of broad and specific issues in question form that deserve serious attention before the first prediction for a damaging earthquake in the U.S. is released.

Broad Issues
- What are the most effective and feasible measures for insuring public safety?
- Should the trend toward population decline in the area be encouraged or discouraged? How?
- How shall the tendency toward a local economic downturn be combated?
- Should special outside financial assistance be provided: for local government? for local business? for area families?
- What changes, if any, are needed in liability legislation?
- How should the demands for increased local government expenditures be balanced against declining revenues?
- How should property values be stabilized?

Specific Issues
- Is an adequate assessment of potential damage feasible? Should damage assessment maps be made public? What, if anything, can or should be done to prevent the publication of damage assessment maps which are based on inadequate data?
- Who should take the responsibility for helping citizens to understand the "true" nature of the threat from the predicted earthquake? Some information about the threat may hurt the local economy. Should that factor be taken into account in informing the public?
- Should buildings with anticipated low earthquake-resistance be ordered evacuated? Should public and

private buildings be treated alike in this regard? Should different criteria be applied to buildings with different uses — should hospitals, stores, and apartment houses be treated alike?

- Should evacuation from higher-risk areas be encouraged? urged? ordered?
- Should the owner of a building which meets current building code standards be required to evacuate the building? If so, should such an owner receive compensation for lost income?
- Should government — local, state, and/or federal — act to offer property-owners in the threatened area access to earthquake insurance and its equivalent? If so, who will bear the cost?
- How does the liability of a private employer change, if at all, in the face of the "yet to be proven" science of earthquake prediction? Are the liability considerations different for the government as an employer? How may these issues be clairified prior to an earthquake prediction?
- Under what conditions, if any, should a seismologist developing an earthquake prediction attempt to withhold its release? What are the liability considerations of the scientist who prepares the earthquake prediction? of the person(s) who releases it to the public?
- If the temporary relocation of the operation of schools, hospitals, jails, etc. is necessary, how will the extra cost be financed?
- If businesses need to relocate temporarily, will loans be available to cover the extra costs?
- If property values fall significantly in the predicted earthquake, should a reassessment of valuation for tax purposes by carried out uniformly within all affected areas?
- Should the number of weeks during which unemployment compensation is available be lengthened?
- Should disclosure of earthquake damage risk be required in all real estate transactions?
- Should the news media "play down" news which is likely to have a negative impact on the community? Can the news media carry out their obligations to the public without threatening their own financial status?
- How can vacated dwellings and commercial properties be protected from a possible increasing threat of vandalism?
- Shall special, perhaps heroic, efforts be made to minimize the slowdown in local construction? Shall special employment programs be instituted to employ laid-off construction workers? other hard-hit occupational groups?
- Shall special incentives be offered to attract new employers to the area and/or to retain current employers?
- How shall the competency of engineering evaluations of structures be assured?
- Will there be an adequate number of engineering specialists available to meet the demand?
- Following disaster impact, a state must certify that the relief and recovery costs are beyond the capacity of the state to handle. Only then will the White House consider a Presidential Disaster Declaration. Should the same approach apply during the period of ANTICIPATED disaster impact? Does the "target" community have to be on its "economic knees" (as a consequence of an earthquake prediction) before special assistance will even be considered?
- Above and beyond "normal" assistance, what is or should be the responsibility of the state to the "target" community? special financial assistance? to local government only? special technical assistance? to local government only?
- Should regulations governing the operation of financial institutions be altered for firms involved in the target community?
- Should regulations governing insurance carriers be altered?
- If there is evidence of "scare tactics" being used by speculators to drive down property values, what can or should be done?
- In making mortgage and other loan decisions, shall "redlining" be permitted?

The National Academy of Sciences published a report in 1975 that identifies and clarifies several important public policy implications (e.g., social, economic, legal, and political issues) that could arise as the U.S. attempts "to make constructive use of the new prediction capability."[5] The report was prepared by the Panel on Public Policy Implications of Earthquake Prediction, established in April 1974 and mandated to[5]

. . . provide advice to the Federal Disaster Assistance Administration, Department of Housing and Urban Development, that will serve as a basis for the formulation of public policy relating to an expected earthquake prediction capability. The types of governmental response with which the Panel will be concerned include warning of public officials and of the general public; governmental actions to mitigate the loss of life and property and the need for further studies and research.

Panel members believe that experience with actual predictions will improve our comprehension of the social and economic dynamics of prediction, and hence, there will be important differences between social responses to the first and later predictions. However, in either case, the panel stresses that "preparation *now* for future predictive capability should greatly improve the chances for responding constructively to save lives and property and to maintain public order."[5] The following listing describes several of the panel's recommendations.

1. The highest priority should be directed towards saving lives with secondary attention to social and economic disruptions and property losses.

2. Prediction should supplement, *not* replace, other earthquake hazard mitigation measures such as seismic engineering practices and land use planning. The report stresses that we should not wait until a prediction is released to start practicing land use management and adopting seismic building codes to encourage the construction of earthquake-resistant buildings.

3. Socioeconomic monitoring should be established concurrently with geophysical monitoring to develop baseline data and methodologies to serve as standards "for measuring the social, political, and economic impacts of earthquake prediction, and to refine techniques that can be applied to other regions as the geophysical monitoring networks are expanded." Geophysical networks now being established or expanded in a number of high-risk areas by the U.S. Geological Survey "provide an excellent opportunity for coordinating the observation of physical and socioeconomic data."

4. Understanding the public policy implications of earthquake prediction will be a slow process because of the infrequency of damaging earthquakes in the U.S. Therefore, investigations should take advantage of the prediction experiences in countries such as Japan, the Soviet Union, and the People's Republic of China.

5. The prime responsibility for planning and responding to earthquake predictions should be given to agencies of government that have a broad-based concern for community and economic planning and with disaster preparedness planning and response, rather than to new agencies established just to handle earthquake predictions. This is because an "agency established exclusively to cope with earthquake prediction would surely stagnate and suffer reduced funding during an interval of years without a substantial earthquake crisis."

6. Legal determinations and clarifying legislation should be established as quickly as possible to minimize legal ambiguities that could hamper governmental officials in making constructive responses to predictions made by the scientific community. More specifically, it is imperative that a legal inquiry be initiated to clarify what powers now exist for responding to earthquake predictions under the provisions of the Disaster Relief Act of 1974 (see Volume III, Chapter 1 and Appendix I) and what additional powers may be necessary.

7. The interests of various segments of a community are likely to be unequally served in planning and responding to earthquake predictions. Less powerful and less well organized groups are likely to be overlooked. Consequently, some public agency should bear the responsibility of "(a) identifying groups of people likely to need special assistance in the event of an earthquake or to suffer disproportionate loss and disruption when an earthquake is predicted, (b) developing a plan to offset, insofar as is practicable, the inequitable costs and suffering attendant on both the quake and the prediction, (c) monitoring events after the prediction from the point of view of equity, and (d) helping unorganized population segments to recognize how the earthquake prediction affects their interests."

8. Earthquake predictions* should be developed and issued to the public by scientists. In addition, procedures must be available that guarantee prediction data reach all segments of the population. Special legislation may be needed to assure that prediction information "will not be withheld from general knowledge to the advantage of special groups."

9. An earthquake warning** should be issued by elected officials after the prediction of a potentially damaging earthquake has been authenticated. The warning should contain "a frank assessment of the prediction, noting the possibilities for error, information on the types and extent of damage that the earthquake could cause, a statement concerning plans being developed to prepare for the quake, and advice concerning appropriate action to be taken by individuals and organizations." Although public officials may be deterred from issuing warnings "by the unjustified fear of panic, by the fear that a prediction judged to be false will cause people to disregard the next warning altogether, by pressure from local interests who fear economic loss from the disruption of normal community life, and from fear of the political consequence of the warning," they will eventually have to be released. Delays can only diminish the public's trust in public officials that is so essential in preparing for a potential earthquake.

10. At least for the immediate future, predictions are apt to originate from a variety of sources and range in validity from scientifically based to irresponsible. It is, therefore, imperative that the federal government establish a panel of scientists to determine the validity of all predictions. (The U.S. Geologcal Survey has essentially adopted this recommendation with the establishment of the Earthquake Prediction Panel. See Volume III, Chapter 1.)

11. A federal agency should establish mechanisms for monitoring the attitudes and actions of the public at every stage of the "prediction-warning-earthquake sequence." This information should be made available immediately to responsible public officials. Special attention should be directed at the problem of communicating warnings and collateral information to special segments of the population such as "foreign-speaking minorities, the physically handicapped, tourists, and the socially isolated."

12. The future success of earthquake prediction and warning programs will depend on whether public officials, the business community, and the populace will accept their validity. Unlike other types of disaster warnings (for tornadoes, hurricanes, and floods), we really do not know how the public will accept the several unique features of an earthquake prediction such as "potentially long lead times (months or years) before the predicted occurrence of a destructive quake; the public's lack of experience with severe quakes because of their infrequent occurrence; the absence of external signs by which the public can confirm that an earthquake will shortly occur, could have occurred, or still may occur; and the likelihood of false alarms and unpredicted quakes." Therefore, the panel recommends several lines of research.

---

* To be useful in developing public policy, the panel made a clear distinction between a prediction and a warning. "Prediction is a statement indicating that an earthquake of a specified magnitude will probably occur at a specified location and time, based on scientific analysis of observed facts. A prediction is strictly information; it says nothing about how people should respond and takes no account of the consequences that may follow from the issuance of the prediction. Issuance and assessment of predictions are strictly technical matters to be debated only on technical grounds."[5]

** A warning "is a declaration that normal life routines should be revised for a time. Warnings are issued because of a judgment that public welfare will be served thereby. Warnings are normally based on predictions or other types of technical information, but not all predictions will be followed by warnings."[5]

. . . Circumstances influencing the credibility of earthquake predictions and warnings, and techniques for improving their credibility, need more careful study.

. . . Research is needed on how people process information regarding low-probability disasters and how this processing changes when a prediction alters the probability. It is important to gain more understanding of how people establish acceptable levels of risk in such instances.

. . . Popular perceptions and understandings of earthquakes and earthquake prediction should be investigated comparing populations in different earthquake-prone regions of the United States and also comparing people who have had no previous experience with earthquakes with those who have experienced severe or minor quakes.

13. Because advance warnings may be measured in weeks, months, or years, hazard reduction programs, involving both public and private agencies, can be put into effect or accelerated during the period between warning and earthquake. Every threatened community should determine the applicability of the following hazard reduction measures: "(a) evacuating limited areas and vacating dangerous structures; (b) accelerating structural design and maintenance programs; (c) employing land-use planning and management powers in relation to the predicted locale of the quake; (d) protecting essential natural gas lines and other community lifelines; (e) dealing with such possible hazards as nuclear plants, vulnerable dams, highly flammable structures and natural cover, and facilities involving the risk of explosion or the release of dangerous chemicals." Because all hazard reduction programs are expensive, various agencies should adopt a policy that a considerable amount of financial assistance, usually available to a community after an earthquake strikes, should be made available for hazard reduction measures in response to a scientific prediction of a potentially destructive earthquake. Legislation should be enacted to achieve this end.

14. Preparation for post-disaster emergency response and recovery is easily understood by the layman, and accordingly, all appropriate programs (e.g., widespread training in first aid) should provide for active and widespread citizen involvement. Such involvement in preparing for the crisis would serve three important purposes: "to upgrade the effectiveness of the community's response when the quake occurs; to enhance the credibility of the prediction by involving people in readily understandable action; and to augment public support for some of the less popular but essential measures in preparing the community for the earthquake."

15. Because a target community's economy may be negatively affected by the issuance of a warning, a joint commission with members from government and private sectors should be established after issuance "to monitor the economy in the threatened area to ensure early detection of changes and to make recommendations to government, business, and labor organizations as needed." Policy makers most likely will have to determine the merits of sustaining the economy at its prewarning level or of encouraging the outflow of capital in an orderly fashion. Subsidies may be needed to sustain the economy or to protect certain segments of the population who might suffer undue hardships as a result of economic dislocations.

Weisbecker et al.,[925] under sponsorship of the National Science Foundation, conducted a *technology assessment* (i.e., a process in which scientifically based information is translated into public action) of earthquake prediction. The primary goal was to identify potential impacts of the impending prediction technology and suggest procedures to deal with critical issues before they arise. The following discussion is a brief summary of the principal conclusions of the study.[925]

1. If people were warned that an earthquake would strike from the developing technology of earthquake prediction, they could take actions to save lives and reduce property damage. However, for a scientific prediction to be taken seriously by the public, it "must take full account of the complex social processes by which words are translated into action." If practical benefits are to be realized, a number of stringent conditions must be satisfied.

   The prediction must in fact be a warning; it must convey a sense of danger, not just neutral "scientific" information.

   The warning must be specific as to time, place, and intensity and must accurately identify the areas, people, and structures at risk.

   The warning must contain prescriptions or at least strong suggestions for action.

   The warning must be disseminated through society's existing communication system, using both formal and informal networks of communication.

   The warning system as a whole should adhere to the "principle of redundancy"; that is, it should provide alternative independent sources of communication that are mutually confirming and consistent when cross-checked.

2. Earthquake prediction in the U.S. could take centuries to reach maturation. During the transition or development period, scientists will develop hypotheses that are based on probabilities rather than on any historical track record. Consequently, many predictions made during this period could have uncertain parameters that will present difficulties to the public "as to whether and how to respond to these transition predictions." To respond effectively, the public will have to establish tactical response plans for dealing with a number of uncertain prediction hypotheses. Although such response plans represent a difficult task, they are feasible "because, within any region that an earthquake might affect and for which an earthquake-prediction system might be established, the range of predictions and the contingent effects can be bounded." In turn, the people of a particular region must then establish the *political will* to cooperate in the appropriate response plan once a prediction is made. Weisbecker et al. developed an *Earthquake Prediction Impact Statement (EPID) process* which entails a number of steps that lead to the development of response tactics for various types of predictions for a given region. Details of the EPID process are presented in Appendix K.

3. The major benefits to be anticipated in any U.S. prediction program are reductions in deaths, injuries, and property losses "if society pursues tactics that remove people and property from the danger zone or increase the likelihood of their survival." Various risk-mitigating tactics are available. Their number and type will be strongly influenced by the prediction lead time. Accurate short-term predictions (e.g., 30 days or less) will most likely be associated with tactics that will have a high potential for saving lives and reducing property damage and causing the least disruption. Risk-mitigating measures for long-term predictions (e.g., 50 to 100 years) are likely to be similar to those used currently in the absence of an earthquake prediction capability — earthquake engineering practices, land use zoning, and disaster preparedness measures. A long-term capability to predict earthquakes should complement and strengthen the use of the above measures for two important reasons.

. . First, a long-lead-time prediction would provide society with better information than it has today. Actions might be rational on the basis of long-lead-time predictions that are not obviously rational on the basis of earthquake probabilities assigned accordingly to recurrence theory. Second, a long-lead-time prediction capability may hasten actions that would be rational, but more difficult to adopt, in the absence of any prediction capability.

Weisbecker et al. suggest that a planning and operations guide can be developed to identify short- and long-term risk-mitigating measures that can be initiated within and outside a region targeted for an earthquake (Appendix L).

As noted by the National Academy of Sciences' (NAS) Panel on Public Policy Implications of Earthquake Prediction,[5] a "crucial question for public policy is how to release earthquake predictions and issue warnings in such a way that the response will be constructive and not counterproductive." The most appropriate information currently available is from three California studies, wherein respondents were asked what they would do if an earthquake prediction was issued at some time in the future. In each study, the largest percentage of the respondents either would *do nothing* or *leave the area.*[*][925,1381,1382] In addition, the NAS panel proposes that an examination of the problems associated with (1) *disaster warnings for floods, hurricanes, and tornadoes,* (2) *long-term disaster preparedness programs,* and (3) *the slowly developing energy crisis* may provide insights as to how people and organizations will respond to actual earthquake predictions and warnings in the future. The following summary, largely from the NAS panel report,[5] examines each of the above sources for possible hints.

Several problems in the disaster warning process start with the individuals or agencies responsible for detecting and releasing the warnings to the public. For example, warning issuances have been delayed until those in authority were reasonably satisfied that a danger would actually materialize. This can occur because of (1) concerns relating to the effects false alarms may have on the credibility and future effectiveness of warnings, (2) legal concerns with erroneous predictions, or (3) the widespread myth that people usually panic when they are warned of an impending danger. For these reasons, warnings have been withheld until it was too late, and evidence from many disaster studies indicates that people *do not* panic if warned.[1387-1394]

People in authority often view the warning "as a direct, stimulus-response type of communication, in which the person issuing the warning gives the signal 'danger' and people automatically respond as though danger were imminent." This perspective fails to consider several important social influences that help to determine the public's interpretation of danger and how they respond to it. Previous research suggests that there likely will be differences in a population's response to earthquake predictions and warnings and that these differences will be based on factors such as previous disaster experiences and social-cultural characteristics. For example, recent research suggests that it is easier to obtain a more positive reaction when people have had a recent disaster experience.[1391] Groups (public officials and citizens alike) living in areas susceptible to recurrent natural disasters "tend to build cultural defenses against them, including organized ways of responding to warnings." Such groups are referred to as *disaster cultures.*[1395,1396] It is, therefore, possible that the response to warnings in California could be more favorable than in Missouri or South Carolina — areas of definite risk, but areas with a much lower level of seismic activity.

Especially during the developmental phase of the prediction-warning system when accuracies could be marginal, a possible citizen response could be derived from "*normalcy bias*",[1397] in which people will readily accept "any information that enables them to disbelieve the prediction, minimize the danger, and view the situation optimistically." The panel believes that normalcy bias will be strengthened by the absence of

---

* Various aspects of these studies are described in the preceding section of this chapter.

observable changes in the environment that reinforce the presence of danger in the case of floods, hurricanes, and tornadoes and by potentially long lead times between the issuing of warnings and earthquake occurrence. The research results of Mileti and Krane[1398] suggest that the longer the lead time, the less likely people will be to take precautionary actions. A long lead time, in essence, imparts a sense of unreality. Keeping a population from becoming complacent during a long lead time will be one of the most difficult tasks of responsible agencies.

Based on the results of existing disaster studies,[1398] the NAS panel anticipates that the most serious problem in securing desirable responses for earthquake predictions and warnings could come from groups "outside the mainstream of society" (e.g., the elderly, the handicapped, the poor, and various ethnic and minority groups). As noted by the panel:

. . . People in these groups are especially likely not to receive, understand, or believe earthquake warnings. Foreign-speaking ethnic groups may not understand earthquake-warning messages given in English. Because of past grievances and hostilities, many members of minority groups may have difficulty believing the disseminators of earthquake warnings. Such grievances and hostilities may undermine or diminish the credibility of the official sources issuing the predictions and warnings and thereby increase the possibility that large numbers of the population at risk will not take appropriate precautionary and protective actions.

In summary, experiences with other natural disasters suggest that official warnings will be discounted by segments of the public, with inaction, rather than evacuation in panic, being a more common response. For example, the pattern of inaction was vividly evident during the official warning period that preceded the arrival of Hurricane Camille which struck the Gulf Coast of the U.S. on August 17, 1969.*[1390,1393] Although warnings were issued over the public media by the U.S. Weather Service and 75,000 people calmly evacuated the worst-hit area in Mississippi, some people decided to totally disregard the warnings that appraised the situation as "extremely dangerous." To determine why some people chose to ignore the official warnings, 384 people from the 100,000 affected by Hurricane Camille were interviewed. From this total, 107 (28%) did not vacate their homes even though most of the respondents had received and understood the seriousness of the Weather Service warnings. Most of the respondents were from the upper-middle class with at least a high school education. Those that remained were generally the best educated in the sampled population.

The responses of why some people did not evacuate coastal areas indicated that the decisions were based upon ignorance and a sense of fatalism. Some, based upon their previous experiences, simply discounted the potential destructiveness of the predicted tides. They completely overlooked the fact that the tides in this case were being driven by one of the strongest wind systems ever recorded. Only 2.8% of those who did not evacuate believed that the predicted tides would damage their homes. Others believed their homes would be out of danger from a 6.1-m predicted tidal surge, even though they did not know the elevation of their homes above sea level. Tragically, a few of the elderly people in the survey expressed the view that if their homes were going to be destroyed, they too might as well be destroyed. Of the 123 people who died in Mississippi, more than 44 were 66 years of age or older.

The most highly organized preparations for long-term disaster preparedness programs usually occur in communities that have experienced repetitive and recent events. For this reason, earthquake preparedness programs are largely confined to just two states, California and Alaska, but even there, many communities have failed to initiate adequate programs. Normally, disaster preparedness programs must compete une-

* Hurricane Camille moved inland over Mississippi producing winds in excess of 300 km/hr and 6.1-m waves along portions of the Gulf Coast. Property damaged exceeded $1 billion and there were 300 deaths.

qually for limited national, state, or local resources with "more immediate, imperative, or well-defined social needs." The NAS panel stresses there "is a real danger that the usual preoccupation with immediate and pressing social and personal concerns will prevent the adoption and implementation of those long-term preparedness plans and programs that will be most effective in saving lives, reducing property losses, and minimizing social disruption."

The efforts to obtain constructive responses to earthquake predictions and warnings during the developmental phase of the program could be analogous to the efforts to mobilize the government and the population to deal with the slowly developing energy crisis. M. K. Hubbert was probably the first scientist (ca. 1956) to predict a domestic shortage of petroleum and natural gas. During the ensuing years, the petroleum industry attempted to disprove his statements,[1399] while various "interest groups entered into the public debate by forecasting damaging consequences to the United States if it did not curb its excessive use of energy resources, increase domestic production of fossil fuels, or develop alternative sources of energy supply." However, the public showed little concern for the growing energy shortage until the Arab oil boycott commenced in October 1973. The NAS panel describes the American response during and after the boycott.

> The public response to the government's call for energy conservation during the oil boycott was generally favorable. Speed limits on the highways were lowered, many industries and businesses voluntarily cut back on their use of energy, and families and individuals attempted to do their part by forming car pools, lowering the temperature in their houses, and purchasing smaller, energy-efficient autos. However, the resulting public awareness and the impetus to develop basic societal changes in solving the nation's long-term energy problems were relatively short-lived. Within a year's time, numerous observers reported that the government had failed to exert adequate leadership in developing a coherent program to solve the country's long-term energy problems. At one point, the President, to the dismay of many, even publicly declared that the energy crisis was over. A year later, automobile manufacturers had grudgingly made limited shifts in production quotas toward smaller cars, and automobile drivers remained ambivalent about reduced speed limits. In a news program aired on July 28, 1974, entitled "Whatever Happened to the Energy Crisis?" CBS [Columbia Broadcasting System] had asked, "How are you going to get action before the day of reckoning?" The same question is relevant for earthquake predictions and warnings.

The following quotation from the NAS report describes analogies between the energy crisis and some expected responses to earthquake predictions and warnings.

> The energy crisis raises serious questions about the length of time required to draw the necessary attention to the earthquake hazard so that response to warnings will be adequate. It also raises doubts about sustaining that awareness once it is developed. Predictions covering long time spans are usually marked by considerable uncertainty and are greeted with skepticism in many segments of society.
>
> A lack of consensus on the validity and value of earthquake prediction may spawn opposing interest groups and countermovements. The merits and credibility of such predictions will be debated in scientific circles and by various interest groups, utilizing the mass media of communication. The opposition of some scientists and engineers to the development of an operational earthquake-prediction system is already beginning to surface. This opposition is based on a variety of doubts and fears: scientific skepticism about the current theoretical underpinnings for earthquake prediction; fears that increased emphasis on prediction will detract from or undermine continuing efforts to achieve improved hazard reduction in earthquake-prone areas; fears that inaccurate predictions will undermine scientific credibility; and fears that predictions may produce "mass panic" among the affected populace . . .
>
> With the advent of specific, credible predictions, various businesses and other groups whose economic, social, and political interests are adversely affected by a prediction may publicly oppose the acceptance of the predictions and urge that no action be taken on the warning . . . Especially during the early developmental phase of earthquake-prediction technology, they are likely to attack the credibility of the predictions and warnings by pointing to the uncertainties in the state of the art, to alternative explanations for the precursory signs used as evidence by the persons and agencies issuing the predictions and warnings, to inaccuracies in previous predictions, to earlier false alarms, and so on. Large-scale businesses will undoubtedly hire their own seismologists to evaluate the evidence, and, in some cases, such seismologists may be utilized to refute the prediction or to cast doubt on its validity and reliability. The resulting differences or

judgment among presumably equally qualified experts are likely to cause the public officials to resist taking positive action on the prediction.

The panel's examination of the three types of problems previously described led it to conclude that "constructive responses to earthquakes will not be an easy task." However, by the same token, the panel believes there is hope if certain items are satisfied.

All this does not mean that there is no hope of achieving adequate, constructive responses to earthquake predictions and warnings. It does mean that these resistances and constraints must be squarely confronted by the relevant federal, state, and local officials in the development of earthquake-prediction and warning-response systems. Officials must take major responsibility for planning and implementing a coherent, continuing program of hazard reduction and disaster preparedness. To the extent that this program requires public participation, the leadership must specify the realistic nature of future danger, the means for dealing effectively with the danger, and the concrete steps needed to secure the required state of preparation prior to the quake's impact. Means for facilitating public compliance with the requirements of the plan must also be incorporated in the program of preparation.

The development of this coherent earthquake prediction and warning system will require cooperation among scientists, public officials, and the communication media to provide understandable and unsensational interpretations of reported predictions. A continuing informational program is needed to ensure that public officials and citizens learn directly from scientists the nature of their thinking about earthquake mechanisms and prediction. Public officials, the media, and the general public will require the advice of a disinterested group of scientists in distinguishing valid from doubtful predictions. Cooperation of the communication media will be important in helping people visualize the laboratories, the seismographic networks, and the panoply of instruments and devices through which predictions are developed. Outlining concrete response plans should help to add a sense of reality to the warnings as well as to forestall some disorganized and disruptive responses. The development of constructive ways in which citizens and groups can participate actively in the preparedness program should also help to bolster public credence. Emergency plans should provide for activation of citizen involvement directly upon issuance of a warning, with intensified and broadened involvement as the predicted time approaches.

Many of the problems that may be expected on the basis of past experience will, of course, tend to diminish if a successful record of accurate predictions of major damaging earthquakes is established. This will be especially true if it can be shown that the hazard-reduction and emergency-preparedness measures taken on the basis of predictions and warnings were highly effective in saving lives, preserving property, and minimizing social disruption.

V. PUBLIC RESPONSE TO EARTHQUAKE WARNINGS IN THE PEOPLE'S REPUBLIC OF CHINA

The most appropriate country in which to examine the public response to earthquake warnings is the People's Republic of China.[1400] The Chinese apparently have predicted 13 earthquakes between 1974 and 1976, and each prediction resulted in the evacuation of people from their homes.[868] The following listing describes the political and social climate in the People's Republic that produces a highly favorable public response to earthquake warnings.[868,1401]

1. Earthquake prediction is not a minor experiment, but rather, it carries the highest government support.
2. The almost total lack of aseismic construction* fosters "community-wide cooperation." Turner[1401] notes that in reference to the warning issued for the 1975 Liaoning Province earthquake (see Volume II, Chapter 1), very few people could feel secure in their homes or places of work. Therefore, the threat of an earthquake was hard to discount, and a firm basis "existed for a development of a sense of common concern."

* The percentage of earthquake-susceptible structures in the People's Republic of China is perhaps as high as 80 to 90%.[868]

3. There is an intensive program for educating the public about the cause of earthquakes, the meaning of earthquake prediction, techniques for recognizing precursory anomalies, and ways to strengthen structures. This information is available via films and pamphlets that are available in bookstores and distributed widely in earthquake-prone areas. Amateur groups commonly distribute the pamphlets door-to-door in a region "targeted" for an earthquake. Representatives from these groups also conduct public meetings to stress the importance of earthquake prediction and favorable community response after a warning has been released.

4. Because *folk wisdom precursors* (e.g., closely spaced foreshocks followed by a period of calm and anomalies in animal behavior and well water composition), officially recognized by the government, can be seen or heard by people, they help to enhance the credibility of an official warning. There are reports that when people detect such precursors they vacate unsafe buildings on their own initiative before a community order has been issued.

5. Large numbers of lay people are a significant part of the prediction program. As the predicted date for an earthquake approaches, all people in target communities are called upon to report any precursors that they may have witnessed. This solicitation program contributes strongly to obtaining community cooperation for initiating short-term mitigation measures, such as evacuation of unsafe buildings.

6. Compliance from the people is sought by various levels of government through the "use of authority, support, and persuasion." After the Liaoning Province warning was issued, the first step was to enlist local people in the task of constructing temporary shelters. Instructions were given on how to warm the shelters. Local militia units ordered the evacuations, and persuasive efforts were directed at the disbelievers by leading cadres. In some cases, people were bodily removed from their homes if the persuasive efforts failed. Outdoor motion pictures are widely used as an inducement to keep people out of doors. The movies are not only popular, but assembling with friends and neighbors undoubtedly helps to produce "group solidarity to counter the otherwise isolating effects of fear and uncertainty."

7. For a warning program in the U.S., there is a possibility of counterproductive responses, the most serious being economic disruption based on the operation of the free marketplace. In the People's Republic, the "substitution of community ownership for private control of capital eliminates the pattern whereby one person's loss is another's gain."

VI. PUBLIC RESPONSE TO TSUNAMI WARNINGS IN THE U.S.

The U.S. Tsunami Warning System (TWS), headquartered in Honolulu, Hawaii, is based on first detecting and locating earthquakes in the Pacific Basin large enough to create a tsunami (seismograph stations), identifying tsunami-type ocean waves (tide stations) produced by the earthquake (Figure 42 in Volume I, Chapter 3), and providing timely information about a tsunami to participating agencies. A *watch* is issued when there is a possibility of tsunami generation, and a *warning* is issued when tide stations indicate that a tsunami has been generated. A warning contains any reported wave heights and arrival time forecasts for populated areas in and around the Pacific Basin (Figure 42 in Volume I, Chapter 3). The TWS has never been responsible for disseminating watches and warnings directly to the general public; rather, messages are transmitted to a single warning point in a country, territory, or state. In addition, the TWS cannot warn participants against locally generated tsunamis. Consequently, regional warning systems have been established in Alaska and Hawaii for providing a measure of protection during the first hour after a local tsunami has been generated.[584,587,607,1402] The following discussion is concerned with the human behavior for two tsunami disasters in the U.S.

On May 23, 1960 a tsunami, caused by a large earthquake off the coast of Chile, struck Hilo, Hawaii. Despite more than 10 hr of potential warning, the tsunami killed 61 people, injured several hundred more, and destroyed more than 500 buildings.[1388] Lachman et al.[1388] examined, via a questionnaire survey, how a segment of the public interpreted and responded to the *siren warnings* in Hilo. The sample consisted of 327 adults and was considered to be fairly representative of the 1000 to 1200 adults displaced by the tsunami. The sample contained 28 people who lost at least one member of their immediate family in the tsunami disaster and 50 who suffered injuries; 48 people from the injured group did not evacuate their homes. Several of the important findings have been summarized from the Lachman et al. report.[1388]

Approximately 4 hr before the tsunami struck Hilo, the public sirens were sounded for 20 min.* Only 127 of the respondents (38.8%) evacuated the low-lying coastal area, even though 309 people (94.5%) had heard the warning sirens. Only 18 respondents from this group indicated that they did not know the meaning of the sirens. However, of the 291 people who stated they knew what the siren signal meant, they had different understandings of its significance (e.g., alert, 4.8%; warning, 4.5%; preliminary signal preceding an evacuation signal, 24.4%; evacuation signal, 28.9%; signal to wait for additional information, 8.9%; signal to make preparations, 6.2%; other, 23.3%). In the Hilo telephone directory, an official medium for disseminating Civil Defense tsunami information, the siren signal was defined as an *alert,* but without indication as to what the population in low-lying areas was expected to do upon hearing it. It is interesting to note that of the 84 respondents who interpreted the sirens as a signal to evacuate, 74 (88.1%) did in fact evacuate their homes before the first wave arrived.

There were several responses to the sirens. A "do-nothing" group (44 people) maintained a normal routine because they believed they were safe. Other respondents in this group were too tired, disabled, or in theaters; 94 people evacuated their homes for safety reasons; and 131 people who heard the sirens waited for another signal, for additional news on radio or television, or for an official notice to evacuate. Because of the multiple responses to the sirens, it was concluded that future disaster warnings must be unambiguous.

Of the people who did not evacuate, 43% were sleeping when the tsunami hit a few minutes past midnight, indicating a feeling of absolute safety. Most of those who were awake (58.1%) were awaiting additional signals. This establishes the fact that this group knew of a possible danger, but because they were not sure of the sirens' meaning, they failed to evacuate. Some 14 people were on the beach waiting to see the tsunami arrive! Of the 197 people who remained, 112 (56.9%) were trapped in debris and 47 (23.9%) suffered injuries. This stressed the need for the establishment of danger or inundation zones that should be completely evacuated during a tsunami warning period.

The respondents were asked if they had received any information about the tsunami other than from the siren alert. The results showed that 66 people (20.2%) said no, and 261 people (79.8%) responded in the affirmative. Of these people, 206 (63%) had received information from radio or television, and 45 (13.8%) from relatives and friends. However, based on the large number of different responses to communications

---

* The sirens were sounded at 8:30 p.m.; the first wave, but not the highest, entered Hilo harbor a few minutes past midnight.

from the above sources, it appears that each person was forced to interpret the communications for information about the tsunami.

The amount of formal education for the respondents who evacuated and those who did not was examined, but the difference was too small to be of any significance (9.2 and 10.1 years, respectively). However, the people who had experienced previous disasters responded somewhat better to evacuation. Of several ethnic groups represented in the sample, a higher proportion of Hawaiians evacuated.

A number of changes were made to tsunami warning procedures and preparedness programs in Hilo and throughout Hawaii following the experiences of the May 23, 1960 disaster. These included: (1) improvements in emergency communication procedures, (2) improvements in the radio broadcasting of emergency information, (3) delineation of low-lying coastal areas to be evacuated during a tsunami emergency and evacuation routes to be followed, (4) maintaining up-to-date lists of disabled people living in evacuation areas who will require assistance if an evacuation is ordered, (5) designation of additional emergency shelter facilities, and (6) publishing and widespread dissemination of warning information and evacuation plans.[1388]

Now when an TWS tsunami warning is received by the Hawaii Civil Defense Agency, an *Attention/Alert Signal* (public sirens) is sounded in communities thought to be in danger. The signal is a steady 1-min siren tone followed by 1 min of silence. The sequence is repeated as necessary. This signal is used to obtain public attention to all major emergencies, including tsunamis. When heard, people are to tune their radio to any station for emergency information and instructions (*Civ-Alert system*). This system is a joint effort of state government and the broadcast industry. When activated, civil defense information and instructions are broadcast simultaneously over all radio stations in the state. The *All Clear Signal* is broadcast over the Civ-Alert System. No siren signal is used. Tsunami inundation maps and civil defense information are printed in the directories of the Hawaiian Telephone Company.[1403] The public reponse to tsunami warnings in Hawaii has been much improved since the 1960 disaster.

Wide variations in community response outside of Alaska were associated with the tsunami caused by the March 27, 1964 Alaskan earthquake.[1391,1404,1405] Weller[1404] believes that "the most crucial variable in determining the nature of the warning received by the public was the judgment and experience of officials responsible for determining the measure of response to be taken within their political subdivisions." In addition, public response appeared to be dependent upon the individual's "experience and understanding of tsunamis and tsunami warnings."[1404] These shortcomings were especially apparent in California.

TWS transmitted three messages to states and participating nations bordering the Pacific Ocean after the earthquake. The first, issued at 5:02 Greenwich Mean Time (GMT), was an "advisory" and indicated that a large earthquake had occurred near Seward, but that it was not known if a tsunami had been generated. The second message, issued at 5:30 GMT as an "information bulletin", reiterated the unknown aspect of tsunami generation, but estimated arrival times for each station were given. The third message, issued at 6:37 GMT, was a "warning." It reported that a tsunami had been generated and estimated arrival times were repeated.[1404,1405] The TWS messages were as follows:

(5:02 GMT) THIS IS A TIDAL WAVE (SEISMIC SEA WAVE) ADVISORY. A SEVERE EARTHQUAKE HAS OCCURRED AT LAT 61 N LONG 147.5 W VICINITY OF SEWARD, ALASKA AT 0336Z 28 MARCH.* IT IS NOT KNOWN REPEAT NOT KNOWN AT THIS TIME THAT A SEA WAVE HAS BEEN GENERATED. YOU WILL BE KEPT INFORMED AS FURTHER INFORMATION BECOMES

---

* The earthquake began at 5:36 p.m., Friday, March 27, 1964, Alaska Standard Time or 0336Z or Zulu Time (same as Greenwich Mean Time), Saturday, March 28, 1964.

AVAILABLE. IF A WAVE HAS BEEN GENERATED ITS ETA FOR THE HAWAIIAN ISLANDS (HONOLULU) IS 0900Z 28 MARCH.

(5:30 GMT) THIS IS A TIDAL WAVE (SEISMIC SEA WAVE) INFORMATION BULLETIN. DAMAGE TO COMMUNICATIONS TO ALASKA MAKES IT IMPOSSIBLE TO CONTACT TIDE OBSERVERS. IF A WAVE HAS BEEN GENERATED THE ETA'S ARE: ATTU 0745Z, ADAK 0700Z, DUTCH HARBOR 0630Z, KODIAK 0530Z, SAMOA 1430Z, CANTON 1530Z, JOHNSTON 1100Z, MIDWAY 0845Z, WAKE 0930Z, KWAJALEIN 1430Z, GUAM 1315Z, TOKYO 1030Z, SITKA 0530Z, SAN PEDRO 0930Z, LA JOLLA 1000Z, BALBOA 2330Z, ACAPULCO 1300Z, CHRISTMAS 1230Z, CRESCENT CITY 0800Z, LEGASPI 1430Z, NEAH BAY 0730Z, SAN FRANCISCO 0915Z, TAHITI 1430Z, TOFINO B.C. 0730Z, VALPARAISO 2200Z, HONOLULU 9000Z, HUALEIN TAIWAN 1430Z, LA PUNTA 1900Z, MARCUS 1100Z, HONG KONG 1530Z, SHIMIZU 1130Z, HACHINOHE 1000Z. ALL TIMES ARE FOR MARCH 28.

(6:37 GMT) THIS IS A TIDAL WAVE (SEISMIC SEA WAVE) WARNING. A SEVERE EARTHQUAKE HAS OCCURRED AT LAT 61 N LONG 147.5 W VICINITY OF SEWARD, ALASKA AT 0336Z 28 MAR. A SEA WAVE HAS BEEN GENERATED WHICH IS SPREADING OVER THE PACIFIC OCEAN. THE ETA OF THE FIRST WAVE AT OAHU IS 0900Z 28 MAR. THE INTENSITY CANNOT REPEAT CANNOT BE PREDICTED. HOWEVER THIS WAVE COULD CAUSE GREAT DAMAGE IN THE HAWAIIAN ISLANDS AND ELSWHERE IN THE PACIFIC AREA. THE DANGER MAY LAST FOR SEVERAL HOURS. OTHER ETA INFORMATION IS AS FOLLOWS: . . . [same ETAs as those listed in previous bulletin].

The California Disaster Office (CDO) received the three TWS messages at 9:36 p.m. Pacific Standard Time (PST), 10:44 p.m., and 11:13 p.m. Apparently because of the tentative nature of the first TWS dispatch, CDO did not disseminate the information to public and emergency agencies in communities vulnerable to tsunamis. The second and third messages were transmitted via the State Department of Justice's (DOJ) Teletype System for CDO as *All Points Bulletins* "to all sheriffs, chiefs of police, and Civil Defense directors of coastal counties and cities."[1404] These two messages were received by local officials at 11:08 and 11:50 p.m.[1404]

Unquestionably, one of the best responses to the tsuanmi danger occurred in Humboldt County. There, the sheriff's office responded to the first CDO message by notifying fire and police departments in coastal communities and appropriate county personnel. These groups were fully mobilized by 11:18. By 11:40 p.m., beaches and homes in low-lying areas were evacuated and roads leading to these dangerous areas were blocked. Fortunately, no tsunami damage was reported.[1404,1405]

At Crescent City, where 11 people were killed and 35 were injured by the tsunami, the response was quite different. There county and civil defense authorities arrived at the Del Norte County Sheriff's Office by 11:20 p.m. in response to the 11:08 CDO message. Although the estimated time of arrival for the potential tsunami was midnight at Crescent City, no action was taken to warn the community until the second CDO bulletin was received at 11:50 p.m. Sheriff's deputies and local police officers were then dispatched to the waterfront area to warn residents in a door-to-door alert. The first wave reached the city at midnight, as predicted, but it was a mild surge and caused only negligible damage in the business district. A second mild surge was experienced at 12:40 a.m.; by then, the evacuation had been completed. However, because of the "mild action of the water," a number of people returned to the waterfront area, believing that the worst was over. Unfortunately, this premature return was responsible for most of the deaths and injuries because the highest wave height above mean tide (6.3 m) was associated with the third (1:20 a.m.) and fourth waves (1:45 a.m.).[602,1405]

Elsewhere in the state, responses ranged from partial evacuations to essentially no response. For example, in San Francisco, where there were no deaths and only slight property damage, 2500 people were evacuated from low-lying areas. However, efforts to protect the population were thwarted by an estimated 10,000 people who jammed beaches to watch the tsunami arrive. In San Diego, unsuccessful attempts also were made to evacuate the curious from several beaches. Spaeth and Berkman[1405] note that the death toll would have been much higher if large-amplitude waves had hit these areas. No attempt was made to evacuate low-lying coastal areas in Los Angeles County.[1405]

According to Weller,[1404] no Alaskan community had sufficient time before the impact of the tsunami to have profited from the TWS messages. For many Alaskans, there was simply no warning; for others more fortunate, the sea provided visual clues that made escape possible (e.g., mild surge for the first wave).[1404] The following quotations describe the tragic effects of the tsunami in two Alaskan communities.[1404]

. . . at Chenega, a native village of 76 residents on Chenega Island in Prince William Sound, there was hardly any opportunity to escape a massive wave that arrived no more than 10 minutes after the onset of the earthquake. There was a slight recession of sea level before the wave, but it was followed so quickly by an inward flood that few could react to the warning. Caught in their homes or along the beach, 23 Chenegans were killed.

On the southeastern shore of Kodiak Island at the head of small Kaguyak Bay, the 45 people of the native village of Kaguyak were more fortunate. One member of the community foresaw the possibility of a tsunami resulting from the earthquake. While there was no immediate attempt to evacuate to higher ground, he did communicate his fears to a few friends. Soon he sighted a seismic swell approaching the village. Warnings were shouted as others became aware of the danger. The height and force of the first wave were not sufficient to reach even those who were slowest to respond. At intervals of about 50 minutes there was subsequent wave action by similar mild effect. Warned, but lulled, by the low power of the waves, some people began moving to their boats, homes, and other low-lying areas. The fourth wave, however, was more powerful. It was estimated to have been between 30 to 50 ft [9.1 and 15.2 m] high and caught many who had left high ground, killing three, while others narrowly escaped.

As noted by Haas,[1406] the lessons learned from the tsunami disaster in Alaska are simple.

1. At the first sign of an earthquake (e.g., ground shaking), move as rapidly as possible away from low lying areas around any sizeable body of water . . Take persons who cannot walk with you but do not pause to take any of your material possessions. Go to high ground away from the water; distance from the normal waterline is not the crucial factor; it is the height of your location that may save you. Stay at your safe location until *official* word is received that all danger is past. Do not allow the receding water to deceive you; additional and larger waves or tides after the first are almost a certainty.

2. If you are in a boat, especially a small boat, and are near shore, you should get to the nearest land as soon as possible and run for high ground. In Kodiak, Kaguyak, and Valdez, persons in boats that were near the shore lost their lives. Those who managed to get to shore and stayed there did not. In a few cases, persons who were in boats that were at sea or who managed to get their boats out a considerable distance away from the shore were safe.

Since the March 27, 1964 Alaskan earthquake, several changes have been made to the Tsunami Warning System.[589,1406,1407]

1.   The *Alaskan Regional Tsunami Warning System* was created with the initial responsibility of detecting and issuing warnings for tsunamis off the coast of Alaska. Subsequently, Washington, Oregon, and California were made a part of the Alaskan system. Following the tsunami caused by the November 29, 1975 earthquake ($M_L = 7.2$) that struck the southeast coast of Hawaii, the *Hawaiian Regional Tsunami Warning System* was established for issuing local warnings.

2.   Deep ocean sensors for evaluating the destructive potential of tsunamis were employed for the first time. Additional seismograph and tide stations were established to improve the detection capability of the system.

3.   Use is now made of the *Geostationary Operational Environmental Satellite (GOES)* as a data collector for seismic and tide stations and communication relay

for transmitting these data and bulletins to users and disseminators. The use of satellite telemetry can improve the warning time by as much as 60%.

4.   New terminology for TWS messages was introduced to reduce misinterpretations. A tsunami "*watch*" is now used for the first alerting message, replacing the term advisory. When it has been confirmed that a tsunami has been generated, a tsunami "*warning*" is issued to system participants that a potentially dangerous tsunami can be expected at their location.

VII. EMOTIONAL EFFECTS

Compared to other natural hazards such as hurricanes, tornadoes, and floods, earthquakes are uniquely different because they usually strike without warning, and their force is often perceived as being capable of destroying the very foundation of the planet. This uncomparable force and its destructive effects can cause emotional stress that envelops all elements of society during the impact and immediate post-disaster periods. This condition is often apparent when earthquake victims are interviewed for mass media reports. However, there is universal agreement that mental health problems associated with earthquake disasters have not been fully explored and analyzed. For example, because there have been virtually no follow-up studies on the survivors of earthquake disasters, it is not known if emotional effects are residual. The following summary describes the effects on mental health reported for several earthquake disasters.

According to Marshall,[1408] two of the first studies dealing with mental health observations in earthquakes were conducted by Komine and Maki[1409] and Brussilowski.[1410] Komine and Maki, while making observations during the September 1, 1923 Kwanto, Japan earthquake ($M_s = 8.3$), found that psychiatric disorders had the following characteristics:[1408,1409]

1.   Acute onset of psychic symptoms
2.   Cloudiness of consciousness
3.   Restlessness, often accompanied by anxiety
4.   Favorable prognosis, recovering mostly within one month
5.   The symptoms encountered resembled those found in the war neuroses

Brussilowski, upon observing human reactions to a series of earthquakes in the Crimea (Soviet Union) in 1927, noted that the most common medical symptom was dizziness, followed by symptoms such as "cardiac palpitations, pallor, irregular pulse, and cold extremities."[1408] Additionally, paralysis and disturbed coordination were observed, and younger children had fewer symptoms of nausea and vomiting than children 13 years and older. Based upon his observations, Brussilowski developed an "*earthquake syndrome*" which consisted of:[1408]

1.   Tachycardia (abnormally fast heartbeat)
2.   Pallor of face
3.   Perceptions of vibrations of the earth
4.   Dizziness
5.   Nausea
6.   General weakness
7.   Restlessness (expectation neurosis)
8.   Disturbances in sleep
9.   Diminished ability to work

The "earthquake syndrome" was substantiated by Marshall[1408] for the March 10, 1933 Long Beach, California earthquake ($M_L = 6.3$).

One of the first systematic attempts to collect and analyze multiple accounts of reactions during and immediately following a damaging earthquake was made by Franz and Norris.[1411] Following the 1933 Long Beach earthquake, 392 students enrolled in psychology courses at UCLA were asked to write brief accounts of their feelings and experiences during and after the main shock and to give their location at the time of occurrence. The authors acknowleged that the students did not represent the general population, but it was their contention that they were "probably more capable of describing their feelings, with less exaggeration and candor."

Upon analyzing the summaries, it was determined that the students generally identified multiple primary effects or reactions and after effects. These are plotted against each other in Table 3. For example, a person who was panicky during the earthquake might also have been nervous and depressed at a later time (Table 3). Note that 132 students exhibited panic, paralysis, and hysteria, termed "great fear" by Franz and Norris, but that 187 "acted socially or with intelligence" (responsible and no effect reactions in Table 3). The former number may be misleading because many of those who exhibited panic reactions at the onset of the earthquake sought to save or comfort others before the quake ceased (i.e., they "became too busy looking after others to pay much attention to themselves.").[1411]

Franz and Norris describe a number of psychological reactions to the earthquake that they categorized as "more interesting." For example, *contagion effects* were mentioned by several students, whereby "some who were not immediately affected became emotionally unbalanced by the terrors of those around them." *Mental conflict* was mentioned by six male students who wanted to run but did not and by five male and female students who knew they should not run but did. Reports of loss of memory were not uncommon, and in a few instances, there were reports of recurrent amnesia. Five male and twenty-two female respondents reported claustrophobia. A few students also mentioned hallucinations of constant earth movements.

Popovic and Petrovic[1412] provide a description of the human reactions to the July 26, 1963 Skopje, Yugoslavia earthquake (M_L = 6.0). Most of the population experienced mental disorders, including confusion, amnesia, stupor, and puerile or immature behavior. In addition, the people tended to congregate in groups, and rumors were spread attributing the earthquake to a punishment for past sins. Moric-Petrovic et al.[1413] collected data on the behavior patterns of 632 Skopje children (6 to 16 years old) who were housed in three nearby communities for several months following this disaster. Almost 70% of the children displayed some type of abnormal behavior (e.g., increased food intake because of the fear of starvation, use of pacifiers by 12- to 16-year-old girls, enuresis or involuntary bed wetting, destructiveness, decline in school performance). It was suggested that the magnitude of the above problems could have been lessened or eliminated if improvements had been made in the scheduling of daily activities, parental communications, and professional counseling services.

Although there was no systematic research completed on the mental effects of the March 27, 1964 Alaskan earthquake (M_s = 8.5), several reports were written that contained information about mental health problems.[1414-1420] A number of these reports were summarized by Lantis[1414] for the National Academy of Sciences' "human ecology" report on the Alaskan earthquake.[1369] Langdon and Parker,[1415] drawing upon medical case records and personal observations of public reaction, noted the following general patterns in Anchorage.

1.   There was no panic during the impact and immediate post-disaster periods.
2.   It was generally 4 days following the disaster that "psychiatric casualties" began to appear for professional treatment. This was because in the days immediately following the earthquake people were vigorously involved, often to the point of

TABLE 3

Relation of Primary and After Effects Reported by 392 UCLA Students for the March 10, 1933 Long Beach, California Earthquake ($M_L = 6.3$)

After effects	Primary effects								
	Panic	Paralysis	Hysteria	Surprise	Annoyance	Thrill	Responsible	No effect	Totals
Physical upsets									
M	9	3	1	3	1	2	14	3	36
F	22	3	11	3	0	4	21	2	66
Nervous									
M	6	3	0	3	0	2	12	5	31
F	17	0	3	4	2	9	18	4	57
Depression									
M	3	0	0	0	0	0	7	0	10
F	1	2	1	2	2	0	7	1	16
Excitement									
M	1	1	1	4	0	1	6	1	15
F	2	0	3	1	0	2	9	0	17
Interest									
M	0	2	0	2	1	9	5	2	21
F	3	0	3	2	1	4	11	4	28
Enjoyment									
M	2	0	0	4	1	3	8	2	20
F	1	0	0	0	0	0	4	1	6
Indifference									
M	6	1	0	1	0	2	4	0	14
F	5	0	0	2	0	1	10	4	22
No report									
M	5	2	0	1	0	0	5	9	22
F	5	2	2	0	0	0	1	7	17
Totals									
M	32	12	2	18	3	19	61	22	169
F	56	7	23	14	5	20	81	23	229

From Franz, S. H. and Norris, A., *Bull. Seismol. Soc. Am.*, 24, 112, 1934. With permission.

exhaustion, in surviving and bringing their living conditions under some degree of control, leaving no or only limited time for depression.

3. The most prominent types of reactions to the disaster are best described as "*separation phenomena*," which indicate "an unconscious psychological separation from the overwhelming part of the disaster as determined by the individual."

4. Humor as a *defense mechanism* (e.g., oral funny stories, comical signs) occurred within hours after the earthquake. The humor "was generally not the projective, destructive type, but regressive in service of the ego — most of the funny stories were about oneself or (one's) possessions." This response played some part in the various phases of public reaction to the disaster.

According to Langdon and Parker, a number of psychological effects were observed during a 2-month period following the earthquake by practicing Anchorage psychiatrists. In approximate order of onset, these included:[1412,1413]

1. Acute feeling of fear (real or potential danger).
2. Anxiety-acute onset including tremor, visceral symptoms, tendency toward flight.

3. Concern for others.
4. Depersonalization (if present).
5. Thirst for information.
6. Seeking human companionship; avoidance of aloneness.
7. Impairment or heightened appreciation of reality.
8. Guilt (particularly at being lucky).
9. Depression.

Additional symptoms that overlapped the above included (1) excess fatigue and (2) changes in eating and drinking patterns.

The earthquake also gave a glimpse of how mental patients responded to the disaster. At the Alaska Psychiatric Institute in Anchorage, Ray[1419] notes that, although several of the patients were quite frightened, there was no panic reaction. In addition, the patients were not concerned about their personal safety, although several were concerned about the safety of their families. According to Landis,[1414] these people "were so self-involved, so withdrawn, that many did not react fully and immediately to the external threat." For example, on the evening following the earthquake, there was a dance as usual at the institute.[1418]

Two organizations, the *San Fernando Valley Child Guidance Clinic* [1342,1421-1424] and the *ENKI Research Institute,*[1342] documented in detail the care and treatment of people with emotional problems caused by the February 9, 1971 San Fernando, California earthquake ($M_L = 6.4$) The guidance clinic undertook an extensive program directed primarily at relieving the anxieties and fears of children, and the research institute studied the reactions of mental patients and professional staff at the Olive View Community Hospital.*[1342]

The clinic's activities in response to the earthquake "began with the supposition on the part of the staff that anxieties were probably generated by the relatively unfamiliar event and that perhaps the clinic could provide a useful community service in relieving them."[1342] The day after the earthquake the clinic, by means of television, radio, and newspaper announcements, offered its services to help parents and children with their fears and anxieties. Although the staff expected, at most, about 20 inquiries, more than 1200 calls were received by the end of the second week.[1342]

Levine[1421] describes the problems that were heard repeatedly, their probable causes, and why parents sought professional assistance.

. . . The children, mostly between three and twelve years of age, had immediate problems of sleeping and increased dependency on adults. Most of the children were afraid to sleep in their own rooms even if an older brother or sister slept in that room with them. When faced to go to bed in their own room . . . they couldn't sleep and would usually get out of bed and stand by their parent's bed. The helpless parents would make room for the child in their bed, or move the child's bed into their own room, and thus create a vicious circle, because after all if the child is getting a reward for sleeping with the parents, he's not about to go back to his own room.

The increased dependency was also indicated by the children's fear of remaining alone in a room of the house during the day. They tended to cling to their mothers, following them around the house, insisting that the mothers accompany them to the bathroom. Many of them refused to go out and play with their friends . . . Children and parents sought out each other (when the earthquake commenced**) and clung together as the earth moved under their feet, but the children we saw were those who continued to seek reassurance and stability by attempting to maintain contact with those persons who represented safety and security, their parents.

The behavior of the parents reflected their own fear of the earthquake, and this had two consequences. Small children take their cues from their parents. It's rather difficult for a frightened parent to convince

---

*   The ENKI Institute was studying the impact of recently enacted legislation on the California Mental
     Health System, and as a part of the Los Angeles County evaluation, the Olive View Community Hospital
     was being monitored when the earthquake occurred.[1342]
** The earthquake struck at 6:01 a.m., a time when most parents and children were still sleeping.

frightened children that they should relax and feel safe. Also the fear, in addition to being experienced as fear for their own children's and their own safety, acted to interfere with their behavior as parents. They expressed much uncertainty and hesitancy in setting limits and in dealing with the child's maladaptive fear, for example the continued demands to sleep with the parents. Now, remember, these were parents who would not normally contact a Child Guidance Clinic, but they seemed to be afraid that if they did not give in to the child's demands, this would have permanent negative psychological effects on the children . . . Relationships between parents and child continued to deteriorate. In effect, most of the parents . . . had become temporarily immobilized in their roles as parents.

The primary mechanisms for treatment were telephone conversations with the parents and group therapy sessions held at the clinic. In most cases, the parents only needed reassurance over the telephone that they were acting in an appropriate manner. It was suggested in the course of the conversation that "they help their children unwind by talking with them, giving them some warm milk or hot chocolate at bedtime and reading to them, using a nightlight, reminding the children that they were safe and the parents would take care of them." The parents were also urged "to begin to have their children return to the usual sleeping, eating, and playing patterns."[1421] For those parents who continued to express a need for assistance, they were invited to attend a group therapy session at the clinic.

Crisis counseling groups for approximately 250 families were provided for a 5-week period following the earthquake. Some 75% of the families needed to attend only one session; the remaining 25% "either returned for an additional meeting, or were referred for immediate short-term individual or family treatment, or for behavior modification groups in which desensitization techniques were used."[1421]

For those families who visited the clinic, the primary function of the group sessions was to reduce the anxiety level of the parents and children and to reestablish the family unit and family roles. Olson[1342] describes the usefulness of the group sessions.

The group setting proved particularly useful in handling the emotional problems. First, the children were greatly relieved when they heard others of about the same age talk about similar fears, feelings, and reactions. This helped to reestablish his normal frame of reference and dispelled feelings of being different or unusual. Second, it proved useful for children to hear from their parents the fright they had experienced. Older children, the teenage segment, worked out their own solutions to their anxieties through these sessions. For example, rearranging the furniture in their bedrooms changed the association of the room with the earthquake and made it more comfortable to sleep there.

The group sessions also provided a setting in which the parents exchanged ideas directly and found solutions to their own problems in the experiences of others. Under the guidance of the clinic's staff, whose goal was to reestablish normal family patterns, some of these solutions included involving the children as much as possible in family decision-making concerning both earthquake-related and other problems; being firm but not harsh in returning children to their usual routines; suggesting that parents, especially fathers, spend more time with the family; presenting factual information about earthquakes to dispel some of the myths associated with this phenomenon of the environment; and advising parents to share their concern about the earthquake with their children in a reassuring and nonfrightening manner. Some children, for instance, felt that the earthquake was a form of punishment for their being bad.

Based upon staff observations, principal clinical findings included the following:[1342]

1. More girls than boys were affected by the earthquake. Rather than being more afraid, it is possible that girls can express their fears more easily.
2. The age of the population emotionally affected changed as time progressed. Between the ages of 3 and 8, an almost equal number of boys and girls were seen during the first weeks, but this shifted to a predominately girl population from ages 10 to 14 about 2 months after the earthquake. There is some indication that girls in the prepubertal and pubertal stages of life have additional anxieties and that they were unable to deal with these anxieties when the impact of the earthquake was added.
3. Families that were counseled immediately after the earthquake recovered more quickly than those who came in later. For instance, children returned to school sooner and working parents were able to return to their jobs sooner, needing a relatively shorter period to stay with their children.
4. Most parents were surprised that children, especially boys, whom they usually considered competent and unafraid, were affected emotionally.

5. The willingness to share feelings, including fear and anxiety, was shown to be of great therapeutic value, and people were brought closer together as alienation from other families and children was counteracted by the group work at the clinic.

6. Parents were able to resume their roles and perhaps avoid more serious disturbances on the basis of the information given to them by the clinic on how to reassure and work with their children.

7. The large number of parents that brought children to the clinic for help shows a desire to use mental health services when they are known to be available in times of crisis. This compares favorably with parents who have more chronic conditions. It suggests that in time of crisis people should be offered assistance.

8. The group discussions included mental health professionals and people who were knowledgeable about earthquake problems. This arrangement was instrumental in stopping rumors of impending disaster from spreading through a large segment of the community and prevented the development of other irrational fears.

9. In approximately 10 percent of the cases interviewed, anxiety associated with the earthquake triggered the expression of dormant anxieties that were unnoticed before. For example, one child asked the clinic's assistance to help reunite his divorced parents. The earthquake provided the vehicle for discussing his real anxiety.

10. In about 50 percent of the cases, the reaction of parents to their own anxieties directly affected the extent of the children's reactions. For the other half, however, there seemed to be no visible anxiety expressed by the parents.

The San Fernando Valley Child Guidance Clinic was funded 1 year after the earthquake by a Mental Health Small Research Grant to develop a program to deal with mental problems in the event that there was another serious earthquake.[1423] The objectives of the study were fourfold: (1) to gather vital statistics concerning the population of the area, (2) to test the premise that a child with fear due to an earthquake will be helped by some type of professional counseling service, (3) to ascertain through field study how the local population found out about the services of the clinic, and (4) to make mental health facility plans for any future earthquake disaster.

Upon completion of the study, there were several major conclusions.[1423]

1. The people who made use of the professional services of the clinic following the 1971 San Fernando earthquake were of a "higher class" than those who normally used the clinic's services. This could possibly be attributed to a higher recognition level of stress in children by parents of this class, or that these people responded to specialized care, as announced by the mass media following the earthquake, instead of seeking some type of private assistance.

2. The majority of the 1971 earthquake patients were younger than regular patients, and the ratio of females to males was reversed. This resulted in a large number of female patients. This was attributed to the fact that in our society the male is supposed to suppress his fears.

3. The families using the clinic reported more signs of anxiety in their children than did the control group who did not seek professional help. Some fears persisted for over a year.

4. Because of professional help, a decrease in fears among children was reported for those who visited the clinic as opposed to the children in the control group.

5. The best method for reaching the people to inform them that clinical assistance was available was by the mass media, especially radio.

6. There is a need for mental health services following a destructive earthquake, and there must be planning, coordination, and training to deliver mental health services to those people who cannot cope with the disaster. This should come from a master plan, preferably at the state level. The plan must take into account the population to be reached and the methods for reaching that population. More specifically, the program must be capable of going to the people, rather than expecting the people to go to a clinic or center. Disaster-relief personnel (e.g.,

firemen, police, teachers) should be trained in conducting anxiety-reducing group discussions. This would help to diminish anxieties resulting not only from the disaster itself, but also from the inactivity of people who are in the state of waiting for assistance.

The ENKI Research Institute[1341] observed and evaluated the behavior of the patients and attendant staff at the psychiatric unit of Olive View Community Hospital in resonse to the earthquake.* Perhaps the most important finding was that, even though the first story of the building had completely collapsed and all emergency lights were inoperative, virtually all of the patients had "complete contact with reality," and their primary goal was to vacate the second story to safety. To accomplish this, male patients carried female patients, while others formed human chains for leading themselves out of the building. Once out-of-doors, some patients did become hysterical, but once other patients gave them tasks to perform their contact with reality was usually restored.

Approximately 70 of the patients were transferred to Camarillo State Hospital.** This group began to develop severe anxieties because they were not given important information about earthquake-related damage in the Los Angeles region. They had to depend upon rumors and their observations at Olive View Hospital, a facility suffering an extraordinary amount of damage. Consequently, many patients concluded that the entire metropolitan area had suffered similar damage. As might be expected, this caused them to become extremely anxious about their families. A related problem was that it took some families up to 4 days to make contact with the patients because no rosters had been prepared when the patients were moved from Olive View Hospital.

At Camarillo State Hospital, the Olive View patients were separated and placed in wards where space was available, even though the Olive View staff was willing to remain and care for their patients as a group. The Camarillo staff discouraged discussions about the earthquake. By the end of the third or fourth day, the Olive View patients were experiencing great anxieties and frustrations because of this ban. This placed the professional staff "in the position of reinforcing anxieties rather than assisting in their relief."[1342] However, the patients realized, generally for the first time, that the staff from Olive View Hospital "was interested in their problems and this rapport carried over into their further treatment."[1342]

Aftershocks and swarms can disrupt the entire life pattern of a community. Because a main shock usually strikes without warning, there is no preceding apprehension or fear, but this is not true with aftershocks. Lantis[1414] reports that many Alaskan citizens' anxiety symptoms increased as the aftershocks continued after the 1964 earthquake, and that for 2 years aftershocks "still set them trembling or startled them awake at night, whereas before Good Friday 1964 they scarcely had noticed minor earth tremors."

More than 1 month after the 1971 San Fernando earthquake, a strong aftershock ($M_L = 4.0$) caused property damage in Granada Hills. Some of residents thought that this event was larger than the original $M_L = 6.4$ earthquake. In addition, the several hundred aftershocks associated with this event caused some individuals to develop a "*preparatory vulnerability*" wherein a person would freeze with fear instead of seeking safety. This effect seemed to be more common in children and new residents. These aftershocks caused some residents to leave the state permanently, whereas they were able to tolerate the main shock without making this decision.[1425]

---

* Many of the patients had been admitted voluntarily and were not severely disturbed.
** Many of these patients were chronic cases.

Recently (1972), an earthquake swarm totally disrupted life in the city of Ancona, Italy.* The following account aptly describes the level of fear.[1426]

> The life of Ancona virtually collapsed in the face of the long sequence of damaging, frightening earthquakes. Although daytime activities continued to resemble normal times, the city became a ghost town after dark, when earthquake-wary citizens moved into the open, slept in railroad coaches and city parks, or left the city. Fishermen put out to sea at sundown, to ride out the night on the Adriatic, and church services moved out-of-doors.

VIII. SAFETY AND SURVIVAL IN AN EARTHQUAKE

The U.S. Geological Survey and the Office of Emergency Preparedness have recently released a pamphlet, in English and Spanish versions, dealing with earthquake safety and survival.[1427] The above agencies stress that an individual can minimize dangers to himself/herself, family, and community by learning and adhering to a few simple rules. The following material is summarized from *Safety and Survival in an Earthquake*.[1427]

Before an Earthquake

1. The home should be checked for hazards. Because of the possibility of fire, all gas appliances should be secured, and where possible, flexible connections should be used. All shelves should be fastened securely to the walls, and all large or heavy objects should be placed on lower shelves or the floor.

2. Parents should learn all they can about an earthquake and relate this to their children. Discuss earthquakes calmly, and never tell a frightening story to a child.

3. Any person old enough should be instructed on how to shut off gas, electricity, and water systems.

4. Whenever possible, instructions in first aid should be given to family members, as medical facilities may be crowded or inoperative when treatment is needed; all immunizations should be up to date.

5. In the event of a complete power failure, flashlights, transistor radios, and any portable equipment should be available if needed.

During an Earthquake

1. If indoors, watch for falling objects and stay away from windows or other glass fixtures.

2. To avoid being hit by any of the above, try to get under a bed or table, or stand under a doorway with arms crossed over your head.

3. If in a high-rise building, get under a desk or table immediately. Do not run for an exit as the stairway may be blocked or damaged, or the power may fail in elevators.

4. If you happen to be in a crowded store, do not run for an exit. Many people may panic, resulting in possible injury due to irrational behavior. If you feel you must leave a building, remain calm and select an exit that appears to be the least crowded.

5. If out-of-doors, try to get to an open area away from structures, power lines, chimneys, etc. Do not run wildly through streets or crowds. If driving a vehicle stop it, preferably in an open area.

---

*   None of the 2500 shocks in the swarm exceeded $M_L = 4.9$. The population of the city was approximately 100,000 in 1972.[1426]

After an Earthquake
1. Check for injuries to your family and neighbors. Treat all injuries that you can, but do not attempt to move anyone who is seriously injured unless he/she may be in possible danger of further injury.
2. Immediately following injury treatment, check for fire or any fire hazards (electrical and gas systems). If there is any damage, shut the systems off at the mains, and if possible, report the damage to the appropriate utility company.
3. Do not come into contact with any downed electrical lines.
4. Do not light matches or turn on lights if a gas leak is suspcted.
5. Because power may be off for some time, meals should be carefully planned, i.e., use foods that may spoil before others.
6. The telephone should be used only for emergencies.
7. A radio should be turned on for broadcasts concerning the earthquake.
8. Care should be exercised when inspecting home damage. Open cupboard and closet doors carefully.
9. Do not go into hazardous areas just for the sake of sightseeing.
10. If near a waterfront area, keep posted on possible tsunami activity.
11. Aftershocks are common. Be prepared for these as some may be fairly intense, causing additional damage.

Based largely upon its experiences in the 1971 San Fernando earthquake, the San Fernando Valley Child Guidance Clinic has published a booklet in English and Spanish to aid parents, teachers, and other adults in dealing with a child's reactions following an earthquake.[1424] The following material is summarized from *Coping with Children's Reaction to Earthquakes and Other Disasters.*[1424]

1. The family should remain together. Parents should not leave their children alone while they go elsewhere.
2. Verbal reassurance is a necessity. The parents should demonstrate strength to help reassure children, but parents should not refrain from talking about their own fears. This should stimulate communication between the family members, bringing out the child's apprehensions. An adult should continually explain what has happened so that a child knows exactly what is taking place.
3. Children should be made to feel at ease while talking about their fears, regardless of location (home, school).
4. If old enough, children should be included in cleanup activities. It is important to avoid inactivity.
5. The level of discipline can be lessened, but parents must be firm in all decisions made for the child's benefit.
6. Most problems occur at bedtime (refusal to sleep alone, difficulty in falling to sleep, nightmares). Short-term solutions include permitting a child to stay up longer or to sleep in the parents' bedroom or with another child, but these situations should not persist for more than 3 to 4 days. During this period, parents and children want "nearness," but a definite time period should be established.
7. Parents must realize that children will have anxieties, and these should be treated with understanding rather than with punishment. For example, a child may resort to "childish" behavior such as thumb-sucking and bed-wetting following an earthquake. These are usually short-lived and parents should not overreact.
8. Parents are usually able to solve a child's problems. However, if it appears that attempts are failing, professional help should be sought. It is important to seek aid quickly to avoid a longer or more difficult recovery.

Chapter 3

FUTURE PROSPECTS

I. THE EARTHQUAKE THREAT

On the grand scale, the earthquake represents one of the earth's most awesome natural hazards, and its potential impact continues to grow in importance as urban centers spread over more and more of the planet's active seismic zones (Figure 3 in Volume I, Chapter 1). Very recent history vividly demonstrates how vulnerable we actually are to earthquakes; in 1976, there were approximately 690,000 earthquake-attributable deaths, making it the second worst year for earthquake casualties in recorded history (Table 1 in Volume I, Chapter 1).[1428]

It is undeniable that the adoption and implementation of measures to mitigate the earthquake hazard have lagged at all organizational levels in *every* country at risk. The problem continues to be most acute in *developing countries*. There, the average citizen is likely to be no more safe from the earthquake hazard today than were citizens in the historic past. For example, in the People's Republic of China, the structural design of rural housing has changed little since the 16th century, and more than 600 million people currently live in mud and masonry homes with heavy tile or mud roofs.[680,868] In many developing countries, the average citizen often has no alternative, because of economic conditions, but to rebuild a home damaged or destroyed by a seismic event with the same brittle, adobe brick that may have injured or killed family members (Figure 1). In addition, if a country is ill-prepared to deal with an earthquake disaster, survivors may be forced to live in deplorable conditions for an extended period of time (Figure 2).

In the U.S., we have been lulled into a state of complacency largely because of our relatively small earthquake losses. Fewer than 1700 Americans have lost their lives to earthquake disasters (Table 2 in Volume I, Chapter 1), and property damage has been approximately $1.9 billion (Table 3 in Volume I, Chapter 1). However, our development is predominantly a phenomenon of the last 100 years, and as a consequence, the potential exists for catastrophic losses from future earthquakes. For example, a great earthquake in the San Francisco Bay region could result in more than 10,000 deaths* and $20 billion in property damage. Earthquakes in southeastern Missouri, such as those that occurred in 1811 and 1812, could now cause damage as far as 600 km away.[1356,1428] A great earthquake centered anywhere in the U.S. would likely cause an economic disruption in all 50 states.[1428]

Another problem contributing to complacency is the popular notion that the earthquake hazard is confined largely to California and Alaska. However, since 1663, damaging earthquakes also have occurred within or near the borders of 22 additional states: Hawaii, Washington, Nevada, Montana, Idaho, Wyoming, Utah, New Mexico, Missouri, Arkansas, Illinois, Indiana, Tennessee, Kentucky, South Carolina, Mississippi, Pennsylvania, New York, Massachusetts, Vermont, New Hampshire, and Maine (Figure 88 in Volume I, Chapter 2).[1429] A recent study suggests that 39 states occupy regions of major to moderate earthquake risk; these states had a combined population of 70 million in 1970.[1430] Another 115 million Americans are exposed to a less significant, but not negligible, seismic risk.[5] In addition, of the $150 billion in yearly U.S. construction investments, approximately $50 billion is for construction in regions having a moderate to high seismic risk.[1430] In Appendix B, "felt earthquakes" are given for the

---

* The life loss could be in excess of 100,000 if six or more dams failed.[1356]

FIGURE 1.   Following the February 4, 1976 Guatemala earthquake ($M_s = 7.5$), many people in devastated villages had no choice but to rebuild their homes with nonearthquake-resistant adobe brick. (Courtesy of Arnold Bleicher, Yavapi County Planning Department, Prescott, Ariz.)

FIGURE 2.   One of the many "disaster squatter settlements" established in Guatemala City following the February 4, 1976 earthquake ($M_s = 7.5$) that was centered 156 m northeast of the capital city. This particular "settlement" of plastic, tin, and cardboard occupied a traffic island; others were sited on vacant lots and city parks. (Courtesy of Arnold Bleicher, Yavapi County Planning Department, Prescott, Ariz.)

years 1969 through 1977; note that in 1976 earthquakes were felt in 33 states. Figure 3 is a recently released *seismic risk map* of the contiguous U.S. that illustrates the hazard to be expected from ground shaking.[1431,1432] Although the earthquake hazard in the east is lower because of the relative infrequency of large events, the total area that has experienced destructive ground motion during the last 250 years is greater than in the western states.[1433]

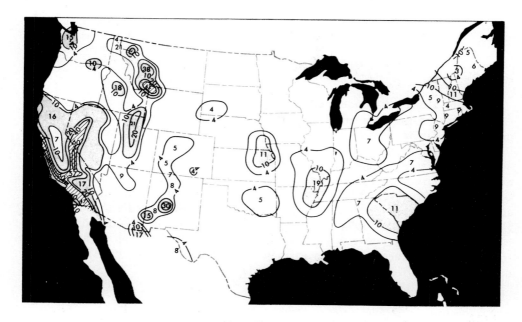

FIGURE 3. Map of the earthquake hazard to be expected from ground shaking in the contiguous U.S. The numbered contours are horizontal accelerations in hard rock expressed as percentages of the earth's gravity. The seismic hazard is depicted at a constant probability level: there is a 90% probability that it will not be exceeded in 50 years. The percentages within each contour are the maximum expected accelerations. (From Algermissen, S. T. and Perkins, D. M., *Earthquake Inf. Bull.*, 9, 20, 1977.)

II. RECENT ACTIONS IN THE U.S. TO MITIGATE EARTHQUAKE EFFECTS

There are five basic strategies that can be initiated by individuals, groups, and various levels of government, as appropriate, to mitigate the effects of an earthquake.[14,30]

Preparation — Preparation for an earthquake includes having plans for warning, response and recovery. These steps reduce the economic and social dislocations by community preparedness prior to the event, providing relief during the emergency period, and assisting in redevelopment and recovery.

Land Use — By considering the regional and local variation of seismic risk in local and State land use plans, the vulnerability of new development can be reduced. Each of the principal sources of earthquake damage (e.g., ground shaking, fault movement, ground failure) is affected by the type of soil and geologic properties of the site, and the position of the site with regard to the location of the earthquake. Effective local and State land use controls can either prevent occupation of a hazardous site, or characterize the hazards at the site so that facilities can be appropriately designed and built.

Building Codes, Standards and Design Practices — The principle that the public has a right to control private and public property for the minimum safety of occupants lies behind public regulation of building. Such regulations are generally adopted as laws of local communities, e.g., building codes. Building codes and standards may be applied to new construction or elimination of hazardous old buildings. In addition, conditions have been adopted for the receipt of financing, e.g., the Minimum Property Standards of FHA, or for the receipt of relief, e.g., flood plain zoning. Such regulations and conditions enable the community to express values and establish priorities.

Insurance and Relief — Economic impacts can be moderated by insurance, loan programs, and public and private relief efforts. Historically, the public and government have responded to the suffering of disaster victims both through the provision of immediate economic aid and long-term economic assistance. Insurance provides one means to spread the economic risk of the disaster.

Information and Education — Through information and education, individuals acquire the background for making decisions at Federal, State, local, corporate and family levels. The acceptance and effectiveness of any mitigation measures — many of which require an economic commitment — will depend critically on the public's perception of the necessity and utility of the measures, as well as on the reliability of the technological information upon which they are based.

All of the above mitigation measures are now being utilized to some extent in the U.S. Clearly, at the state and local level, the most aggressive programs are found in California; there, several far-reaching mitigation measures were adopted following the disastrous 1933 Long Beach and 1971 San Fernando earthquakes.* Most of the mitigation measures enacted by the legislature shortly after the San Fernando earthquake were introduced by members of the California Legislature's *Joint Committee on Seismic Safety* (1969 to 1974), chaired by Senator Alfred E. Alquist, and executive actions resulted from recommendations made by the *Governor's Earthquake Council* (1971 to 1974).[1434,1435] In 1974, the *Seismic Safety Commission* was created by the legislature; both of the above groups had recommended that such an organization be formed "to continue working on earthquake hazard programs".[1435] The following quotation summarizes the commission's mandate and structure:[1435]

. . . Responsibilities and Powers
The Commission has a broad range of responsibilities for the improvement of seismic safety. These include setting long range goals and priorities; requesting State agencies to prepare standards and criteria; recommending changes in existing programs to improve the reduction of hazards; reviewing reconstruction efforts after damaging earthquakes; helping to coordinate the seismic safety activities of all levels of government; and providing information, encouraging research, and sponsoring training.
The Commission is to advise the Governor and Legislature on budgetary aspects of State earthquake programs and needed legislation. It may hold hearings on matters important to seismic safety. In addition, the Strong Motion Instrumentation Board, which oversees the placement of instruments to collect critical earthquake engineering data, and the Building Safety Board, which is responsible for earthquake standards for hospital construction in the State, report annually to the Commission on their programs.

Organization and Structure
The Commission has 17 members. Fifteen, including the Chairman, are appointed by the Governor. One is appointed by the Senate Rules Commitee and one is appointed by the Speaker of the Assembly. The Commissioners represent a wide range of earthquake-related professions and disciplines, providing a multidisciplinary approach to solving the complex problems of seismic safety. Members of the Commission include geologists; structural, civil, and soils engineers; architects; planners; representatives of local government; State legislators; and others . . .

At the federal level, six major elements of research have been defined for establishing a basis for the five mitigation strategies defined at the beginning of this section: (1) *fundamental earthquake studies,* (2) *prediction,* (3) *induced seismicity,* (4) *hazards assessment,* (5) *engineering,* and (6) *research for utilization.* The U.S. Geological Survey (USGS) and the National Science Foundation (NSF) are the two agencies most responsible for conducting or sponsoring research for these elements in the federal government's recently established *Earthquake Hazards Reduction Program* (described later). The USGS has the primary responsibility for prediction, induced seismicity, and hazards assessment, while NSF is responsible for engineering and research for utilization. Both agencies share responsibility for fundamental studies.[1428] The nature of the research elements are summarized in Appendix M.

In addition to the programs of the USGS and NSF, there are several federal programs that contribute to the mitigation of earthquake hazards. Several are described by Hamilton.[1428]

---

* Specific hazard mitigation programs used in California are discussed in Volume II, Chapters 3 and 4 and Volume III, Chapters 1 and 2.

. . . The Reactor Hazards Research Program of the USGS and the Site Safety Research Program of the Nuclear Regulatory Commission (NRC) are of particular importance owing to their contributions to earthquake hazards assessment. The programs of the National Aeronautics and Space Administration (NASA) for long baseline geodetic positioning and of the National Oceanic and Atmospheric Administration/National Geodetic Survey (NOAA/NGS) for level and triangulation surveying contribute to studies of plate tectonics and prediction. Tsunami studies are conducted by NOAA and NSF. The various Federal construction agencies, such as the Army Corps of Engineers, Bureau of Reclamation, and Veterans Administration, conduct engineering design studies.

Coordination among these various programs is essential and has been accomplished effectively. A few specific examples will illustrate this point. All the agencies mentioned above have a liaison representative on the USGS Earthquake Studies Advisory Panel. NOAA and the USGS have a bilateral coordinating committee. USGS, NSF, and NRC grants and contracts managers participate in each others' evaluation meetings. Agencies transfer funds to other agencies to obtain needed expertise on mission programs. Overall, the level of cooperation is high . . .

In retrospect, it was not until the mid-1960s that a number of programs began to be proposed by blue-ribbon panels, comprised of experts from academia, the private sector, and all levels of government, to expand research on measures to mitigate the effects of earthquakes.[5,633,976,1407,1428,1430,1436-1451] An expanded research program, based on the recommendations of the *U.S. Task Force on Earthquake Hazard Reduction,*[1400] was implemented shortly after the 1971 San Fernando earthquake.[1428] However, as noted by Hamilton,[1428] a number of supporters of earthquake research in Congress "did not believe that the increased effort was adequate," and legislation to fund an increased amount of earthquake research "was introduced in several sessions of Congress through the mid-1970s, but none was passed." A sense of urgency to expand federal research occurred in 1976 when it was discovered that a large land area northeast of Los Angeles had risen between 1959 and 1974 (Figure 11 in Volume II, Chapter 1). This uplift, called the Palmdale bulge, was viewed with concern because a similar anomaly had preceded the 1964 Niigata, Japan and 1971 San Fernando earthquakes. The USGS and NSF received a special $2 million appropriation in 1976 to conduct and sponsor several types of investigations in the uplifted region. In the same year, the President's Science Advisor convened an advisory panel to formulate a plan of research for the USGS and NSF. The panel's report, *Earthquake Prediction and Hazard Mitigation-Options for USGS and NSF Programs*, also known as the *Newmark Report,*[1430] stressed the following points:[1428]

1. A balance had to be established among the six elements of mitigation research.
2. The research should range from the most basic to the most applied.
3. Milestones for the research and anticipated public benefits were identified.
4. Three options for increased funding were proposed (Table 1).

Drawing upon the Newmark Report, Option B, or the intermediate level of funding, was sought by the President for his Fiscal Year (FY) 1978 budget requests to Congress for the USGS and NSF earthquake research program. Congress modified the USGS budget for two of the program elements — prediction and hazards assessment; additional appropriations of $1.5 million and $800,000 were allocated to increase the prediction research effort in foreign countries and to accelerate the defining of criteria for the safe siting of dams, respectively. Allocations for the NSF fundamental earthquake and engineering elements were consistent with Option B, while funding for the research for utilization element was $2.5 million below the Option B level specified in the Newmark Report (Tables 1 and 2).[1428] Note that the funding for the USGS and NSF in FY 1978 was more than double that allocated in FY 1977 (Table 1).

Paralleling the Administration's plan to have an expanded earthquake mitigation program, the Congress passed the *Earthquake Hazards Reduction Act of 1977 (Public*

TABLE 1

Three Options of Funding Proposed in the Newmark Report[1430] and the Amounts Authorized in the Earthquake Hazards Reduction Act for Increased Funding of Earthquake Research

	USGS	NSF	Coordinating Office	Total
FY 77	11.2	10.2	—	21.4
FY 78				
Option A	18.5	19.2	—	37.7
Option B	27.9	25.8	—	53.7
Option C	44.8	40.0	—	84.8
Public Law 95-124	27.5	27.5	1.0	56.0
FY 79				
Option A	27.8	22.2	—	50.0
Option B	37.8	32.4	—	70.2
Option C	53.2	42.9	—	96.1
Public Law 95-124	35.0	35.0	2.0	72.0
FY 80				
Option A	39.8	35.1	—	74.9
Option B	46.1	38.9	—	85.0
Option C	59.5	45.7	—	105.2
Public Law 95-124	40.0	40.0	2.0	82.0

Note: Figures are in millions of dollars.

From Hamilton, R. M., Earthquake Hazards Reduction Program — Fiscal Year 1978 Studies Supported by the U.S. Geological Survey, U.S. Geological Survey Circular 780, 1978, 5.

TABLE 2

Earthquake Hazards Reduction Program FY 1978 Funding as Enacted

Program element	USGS[a]	NSF[a]	Percentage of program
Fundamental studies	2,650	5,300	14.9
Prediction	15,764[b]	—	29.4
Induced seismicity	1,200	—	2.2
Hazards assessment	10,607	—	19.8
Engineering	—	15,500	29.0
Research for utilization	—	2,500	4.7
Totals	30,221	23,000	

[a] Figures are in thousands of dollars.
[b] Includes $1,500,000 specifically for prediction studies in foreign countries.

From Hamilton, R. M., Earthquake Hazards Reduction Program — Fiscal Year 1978 Studies Supported by the U.S. Geological Survey, U.S. Geological Survey Circular 780, 1978, 6.

Law 95-124); the President signed it into law on October 7, 1977. The intent of the law is ''to reduce the risks of life and property from future earthquakes in the United States through the establishment and maintenance of an effective earthquake hazards reduction program.''[1452,1453] Highlights of the act include the following:

1. The act directed the President to establish and maintain the *National Earthquake Hazards Reduction Program* with the involvement of the U.S. Geological Survey, National Science Foundation, Department of Defense, Department of Housing and Urban Development, National Aeronautics and Space Administration, National Oceanic and Atmospheric Administration, National Bureau of Standards, Energy Research and Development Administration, Nuclear Regulatory Commission, and the National Fire Prevention and Control Administration. The research aspect of the program is to be comprised of the following elements: (1) basic research into the causes and mechanisms of earthquakes, (2) earthquake prediction, (3) induced seismicity, (4) development of seismic zoning guidelines and preparation of seismic risk analyses for emergency planning and community preparedness purposes, (5) development of methods for designing, constructing, and rehabilitating manmade works so as to resist earthquakes, (6) exploring possible and economic adjustments that could reduce earthquake vulnerability, and (7) investigate foreign experiences with all aspects of earthquakes.

2. The act required the President to develop an *Implementation Plan* that specifies year-to-year targets through 1980 plus specific roles for federal agencies and recommended roles for state and local units of government, private organizations, and individuals needed to carry out the plan. The President assigned the responsibility of preparing the plan to the Office of Science and Technology Policy, and on June 22, 1978, the President transmitted the Implementation Plan to Congress in accordance with the act. The plan names the new *Federal Emergency Management Agency* as the lead organization for managing the program and lists milestones and federal agency responsibilities for the next 3 years.[1454] The highest priorities identified by the President for immediate action are[1454]

   . . . . establishment of the National Earthquake Prediction Council by the U.S. Geological Survey;

   . . . . completion of federal, state, and local contingency plans for responding to earthquake disasters in densely populated areas of highest seismic risk;

   . . . . development of seismic resistant design and construction standards for application in federal construction, and encouragement for the adoption of improved seismic provisions in state and local building codes; and

   . . . . estimation of the hazard exposed to life by possible damage to existing federal facilities from future earthquakes . . . .

3. The President is required to submit to Congress an annual report describing the progress made in reducing the risk of earthquake hazards.

4. The act specified authorizations for increased funding for the USGS and NSF. The funding amounts are summarized in Table 1.

The complete text of the Earthquake Hazards Reduction Act is presented in Appendix N.

References

REFERENCES

1. **Berlin, G. L., Pitrone, D. J., Salisbury, H. G., and Samara, M. F.,** Urban seismology, *CRC Crit. Rev. Environ. Control,* 5 (3), 275, 1975.
2. **Hansen, W. R. and Eckel, E. B.,** A summary description of the Alaska earthquake — its setting and effects, in The Alaska Earthquake March 27, 1964: Field Investigations and Reconstruction Effort, Geological Survey Professional Paper 541, U.S. Government Printing Office, Washington, D.C., 1966, 1.
3. **Saarien, T. F.,** *Environmental Perception,* National Council for the Social Studies, Washington, D.C., 1970.
4. Committee on Earthquake Engineering Research, *Earthquake Engineering Research,* National Academy of Sciences, Washington, D.C., 1969.
5. Panel on the Public Policy Implications of Earthquake Prediction, *Earthquake Prediction and Public Policy,* National Academy of Sciences, Washington, D.C., 1975.
6. **Press, F.,** Natural-hazards reduction, in Earth Science in the Public Service, Geological Survey Professional Paper 921, U.S. Government Printing Office, Washington, D.C., 1974, 71.
7. **Kuo, T. C.,** On the Shensi earthquake of January 23, 1556, *Acta Geophys. Sinica,* 6, 49, 1957.
8. **Kendrick, T. D.,** *The Lisbon Earthquake,* J. B. Lippincott, New York, 1955.
9. **Clark, B.,** America's greatest earthquake: Mississippi River town of New Madrid, Mo., *Read. Dig.,* 94, 110, 1969.
10. **Bronson, W.,** *The Earth Shook, The Sky Burned,* Doubleday, Garden City, N.Y., 1959.
11. **Sutherland, M.,** *The Damndest Finest Ruins,* Ballantine, New York, 1971.
12. **Thomas, G. and Witts, M. M.,** *The San Francisco Earthquake,* Stein & Day, New York, 1971.
13. **Imamura, A.,** A diary on the great earthquake, *Bull. Seismol. Soc. Am.,* 14, 1, 1924.
14. **Christopherson, E.,** *The Night the Mountain Fell, The Story of the Montana-Yellowstone Earthquake,* Earthquake Book, Missoula, Mont., 1960.
15. Anon., Agony of Agadir, a city shaken to death, *Life,* 48, 36, 1960.
16. **Graves, W. P. E.,** Alaska earthquake, *Natl. Geogr. Mag.,* 126, 112, 1964.
17. **Thomas, Mrs. L., Jr.,** An Alaskan family's night of terror, *Natl. Geogr. Mag.,* 126, 142, 1964.
18. **Blank, J. P.,** Earthquake: the horror that hit Peru, *Read. Dig.,* 97, 77, 1970.
19. **McDowell, B.,** Earthquake in Guatemala, *Natl. Geogr. Mag.,* 149, 810, 1976.
20. California Division of Mines and Geology, Dear division, *Min. Inf. Ser.,* 22, 75, 1969.
21. National Oceanic and Atmospheric Administration, Earthquakes, U.S. Government Printing Office, Washington, D.C., 1971.
22. **Halacy, D. S., Jr.,** *Earthquakes: A Natural History,* Bobbs-Merrill, Indianapolis, 1974.
23. **Flamstead, J.,** A letter concerning the natural causes of earthquakes, *Earthquake Inf. Bull.,* 7, 14, 1975.
24. **Kirkland, W. G.,** Introduction, in The Agadir, Morocco Earthquake, February 20, 1960, American Iron and Steel Institute, New York, 1962, 8.
25. National Earthquake Information Center, 4 earthquake, major offshore earthquakes recall the Aztec myth, *Earthquake Inf. Bull.,* 2, 4, 1970.
26. National Earthquake Information Center, Ruaumoko god of volcanoes and earthquakes, *Earthquake Inf. Bull.,* 5, 31, 1973; originally published in *Proc. 3rd. World Conf. on Earthquake Engineering,* International Association for Earthquake Engineering, Auckland, New Zealand, 1965, iv.
27. **Richter, C. F.,** *Elementary Seismology,* W. H. Freeman, San Francisco, 1958.
28. **Reid, H. F.,** The mechanics of the earthquake, in *The California Earthquake of April 18, 1906, Report of the State Earthquake Investigation Commission,* Vol. 2, Carnegie Institution of Washington, Washington, D. C., 1910.
29. **Tsuboi, C.,** Investigation of the deformation of the earth's crust found by precise geodetic means, *Jpn. J. Astron. Geophys.,* 10, 93, 1933.
30. **Mescherikov, J. A.,** Recent crustal movements in seismic regions: Geodetic and geomorphic data, *Tectonophysics,* 6, 29, 1968.
31. **Lensen, G.,** Elastic and non-elastic surface deformation in New Zealand, *Bull. N. Z. Soc. Earthquake Eng.,* 3, 131, 1970.
32. **Scholz, C. H.,** Crustal movements in tectonic areas, *Tectonophysics,* 14, 201, 1972.
33. **Nason, R. D.,** Continuing fault movement after earthquakes, *EOS Trans. Am. Geophys. Union,* 50, 252, 1969.
34. **Greensfelder, R. W. and Crice, D.,** Geodimeter fault movement investigations in California, *Calif. Geol.,* 24, 105, 1971.
35. **Savage, J. C.,** The California geodimeter network: measuring movement along the San Andreas fault, *Earthquake Inf. Bull.,* 6, 3, 1974.

36. **Bacon, C. F., Bennett, J. H., Chase, G. W., Rodgers, D. A., Toppozada, T. R., and Wells, W. M.,** Crustal movement studies near Parkfield, California, *Calif. Geol.,* 28, 202, 1975.
37. **Benioff, H.,** A linear strain seismograph, *Bull. Seismol. Soc. Am.,* 25, 283, 1935.
38. **Swolfs, H. S., Brechtel, C. E., and Brace, W. F.,** Direct measurement of premonitory changes in stress associated with a microearthquake, *EOS Trans. Am. Geophys. Union,* 56, 911, 1975.
39. **Evison, F. F.,** Earthquakes and faults, *Bull. Seismol. Soc. Am.,* 53, 873, 1963.
40. **Eaton, J. P., Lee, W. H. K., and Pakiser, L. C.,** Use of microearthquakes in the study of the mechanics of earthquake generation along the San Andreas fault in central California, *Tectonophysics,* 9, 259, 1970.
41. **Bonilla, M. G.,** Surface faulting and related effects, in *Earthquake Engineering,* Wiegel, R. L., Ed., Prentice-Hall, Englewood Cliffs, N.J., 1970, 47.
42. **U.S. Geological Survey,** Active Faults of California, U.S. Government Printing Office, Washington, D.C., 1972.
43. **Tarr, R. S. and Martin, L.,** The Earthquakes at Yakutat Bay, Alaska in September 1899, Geological Survey Professional Paper 69, U.S. Government Printing Office, Washington, D.C., 1912.
44. **von Hake, C. A.,** Earthquake history of Nevada, *Earthquake Inf. Bull.,* 6, 26, 1974.
45. **Wesson, R. L., Helley, E. J., Lajoie, K. R., and Wentworth, C. M.,** Faults and future earthquakes, in Studies for Seismic Zonation of the San Francisco Bay Region, Borcherdt, R. D., Ed., Geological Survey Professional Paper 941-A, U.S. Government Printing Office, Washington, D.C., 1975, A5.
46. **Wallace, R. E.,** Notes on stream channels offset by the San Andreas fault, southern Coastal Ranges, California, in Proc. Conf. on Geologic Problems of San Andreas Fault System, *Stanford Univ. Pub. Geol. Sci.,* 11, 6, 1968.
47. **Oakeshott, G. B.,** San Andreas fault: geologic and earthquake history, *Calif. Geol.,* 19, 159, 1966.
48. **California Division of Mines and Geology,** Fort Tejon earthquake model, *Calif. Geol.,* 29, 65, 1976.
49. **Tocher, D.,** The Alaska earthquake of July 10, 1958: movement in the Fairweather fault and field investigation of southern epicentral region, *Bull. Seismol. Soc. Am.,* 50, 267, 1960.
50. **Tocher, D. and Miller, D. J.,** Field observations on effects of Alaska earthquakes of July 10, 1958, *Science,* 129, 394, 1959.
51. **Clark, W. B. and Hauge, C. J.,** The earth quakes you can reduce the danger, *Calif. Geol.,* 24, 203, 1971.
52. **Morrison, N. L.,** Vertical crustal movements determined from surveys before and after San Fernando earthquake, in San Fernando, California, Earthquake of February 9, 1971, Vol. 3, U.S. Government Printing Office, Washington, D.C., 1973, 295.
53. **Meade, B. K. and Miller, R. W.,** Horizontal crustal movements determined from surveys after San Fernando earthquake, in San Fernando, California, Earthquake of February 9, 1971, Vol. 3, U.S. Government Printing Office, Washington, D.C., 1973, 243.
54. **Plafker, G.,** Surface Faults on Montague Island Associated with the 1964 Alaska Earthquake, Geological Survey Professional Paper 543-G, U.S. Government Printing Office, Washington, D.C., 1967.
55. **Atomic Energy Commission,** Nuclear power plants, seismic and geologic siting criteria, *Fed. Regist.,* 36, 22601, 1971.
56. **Town and Country Planning Branch,** Town planning and earthquake faults, *Town Country Plan. Bull.,* 7, 1, 1965.
57. **Bolt, B. A., Horn, W. L., Macdonald, G. A., and Scott, R. F.,** *Geological Hazards,* Springer-Verlag, New York, 1975.
58. **Bishop, C. C. and Knox, R. D.,** Tri-cities geologic hazards investigation, in *The Seismic Safety Study for the General Plan,* California Council on Intergovernmental Relations, Sacramento, 1973, 81.
59. **Brown, R. D.,** Active Faults, Probable Active Faults, and Associated Fracture Zones, San Mateo County, California, U.S. Geological Survey Miscellaneous Field Studies Map MF-355, 1972.
60. **California Division of Mines and Geology,** Dear division , *Min. Inf. Ser.,* 23, 33, 1970.
61. **Dunbar, W. S., Boore, D. M., and Thatcher, W. R.,** Strain accumulation and release mechanisms of the White Wolf fault, *EOS Trans. Am. Geophys. Union,* 56, 1067, 1975.
62. **Scholz, C. H. and Fitch, T. J.,** Strain and creep in central California, *J. Geophys. Res.,* 74, 6649, 1970.
63. **Koch, T. W.,** Analysis and effect of current movement on an active fault in Buena Vista Hills oil field, Kern County, California, *Bull. Am. Assoc. Pet. Geol.,* 17, 694, 1933.
64. **Louderback, G. D.,** Faults and earthquakes, *Bull. Seismol. Soc. Am.,* 32, 305, 1942.
65. **Steinbrugge, K. V. and Zacher, E. G.,** Fault creep and property damage, in creep on the San Andreas fault, *Bull. Seismol. Soc. Am.,* 50, 389, 1960.
66. **Tocher, D.,** Creep on the San Andreas fault — creep rate and related measurements at Vineyard, California, in creep on the San Andreas fault, *Bull. Seismol. Soc. Am.,* 50, 396, 1960.
67. **Brown, R. D., Jr. and Wallace, R. E.,** Current and historic fault movement along the San Andreas fault between Paicines and Camp Dix, California, in Proc. Conf. on Geologic Problems of the San Andreas Fault System, *Stanford Univ. Publ. Geol. Sci.,* 11, 22, 1968.

68. **Nason, R. D.**, Measurements and theory of fault creep slippage in central California, *R. Soc. N. Z. Trans.*, 9, 181, 1971.

69. **Savage, J. C. and Burford, R. O.**, Geodetic determination of relative plate motion in central California, *J. Geophys. Res.*, 78, 832, 1973.

70. **Rogers, T. H.**, An active fault in the city of Hollister, *Min. Inf. Ser.*, 22, 159, 1969.

71. **Rogers, T. H. and Nason, R. D.**, Active displacement in the Calaveras fault zone at Hollister, California, *Bull. Seismol. Soc. Am.*, 61, 399, 1971.

72. **Blanchard, F. B. and Laverty, G. L.**, Displacements in the Claremont Water Tunnel at the intersection with the Hayward fault, *Bull. Seismol. Soc. Am.*, 56, 291, 1966.

73. **Bolt, B. A. and Marion, W. C.**, Instrumental measurement of slippage on the Hayward fault, *Bull. Seismol. Soc. Am.* 56, 305, 1966.

74. **Bonilla, M. G.**, Deformation of railroad tracks by slippage on the Hayward fault in the Niles district of Fremont, California, *Bull. Seismol. Soc. Am.*, 56, 281, 1966.

75. **Cluff, L. S. and Steinbrugge, K. V.**, Hayward fault slippage in the Irvington-Niles district of Fremont, California, *Bull. Seismol. Soc. Am.* 56, 257, 1966.

76. **Radbruch, D. H. and Lennert, B. J.**, Damage to culvert under Memorial Stadium, University of California, Berkeley, caused by slippage in the Hayward fault zone, *Bull. Seismol. Soc. Am.*, 56, 295, 1966.

77. **Radbruch, D. H.**, New evidence of historic fault activity in Alameda, Contra Costa, and Santa Clara Counties, California, in Proc. Conf. on Geologic Problems of San Andreas Fault System, *Stanford Univ. Publ. Geol. Sci.*, 11, 46, 1968.

78. **Sharp, R. V.**, Map Showing Recent Tectonic Movement on the Concord Fault, Contra Costa and Solano Counties, California, U.S. Geological Survey Miscellaneous Field Studies Map MF-505, 1973.

79. **Burke, D. B. and Helley, E. J.**, Map Showing Evidence for Recent Fault Activity in the Vicinity of Antioch, Contra Costa County, California, U.S. Geological Survey Miscellaneous Field Studies Map MF-533, 1973.

80. **Clark, M. M.**, Most Recently Active Breaks Along the Garlock and Related Faults, U.S. Geological Survey Open-File Map, in preparation.

81. **Bassett, A. M.**, Remarks, in Proc. Conf. on Geologic Problems of San Andreas Fault System, *Stanford Univ. Publ. Geol. Sci.*, 11, 282, 1968.

82. National Earthquake Information Center, Creep measured on Turkish fault, *Earthquake Inf. Bull.*, 4, 12, 1972.

83. **David, J. L.**, Monitoring movement of California's San Andreas fault, *Earthquake Inf. Bull.*, 3, 10, 1971.

84. National Earthquake Information Center, Every little movement has a meaning all its own, *Earthquake Inf. Bull.*, 4, 22, 1972.

85. **Oakeshott, G. B.**, Parkfield earthquakes Monterey and San Luis Obispo Counties, California, June 27—29, 1966, *Min. Inf. Ser.*, 19, 155, 1966.

86. **Smith, S. W. and Wyss, M.**, Displacement on the San Andreas fault initiated by the 1966 Parkfield earthquake, *Bull. Seismol. Soc. Am.*, 58, 1955, 1968.

87. **Scholz, C. H., Wyss, M., and Smith, S. W.**, Seismic and aseismic slip on the San Andreas fault, *J. Geophys. Res.*, 74, 2049 1969.

88. **Clark, M. M.**, Surface rupture along the Coyote Creek fault, in The Borrego Mountain Earthquake of April 9, 1968, Geological Survey Professional Paper 787, U.S. Government Printing Office, Washington, D.C., 1972, 55.

89. **Båth, M.**, *Introduction to Seismology,* John Wiley & Sons, New York, 1973; first published as *Introduktion till Seismologin,* Natur och Kultur, Stockholm, 1970.

90. **Bolt, B. A.**, *Nuclear Explosions and Earthquakes, The Parted Veil,* W. H. Freeman, San Francisco, 1976.

91. **Watkins, J. S., Bottino, M. L., and Morisawa, M.**, *Our Geological Environment,* W. B. Saunders, Philadelphia, 1975.

92. U.S. Geological Survey, Imperial Valley earthquake recorded on ground and in the air, *Earthquake Inf. Bull.*, 7, 11, 1975.

93. California Division of Mines and Geology, Earthquake sounds captured, *Calif. Geol.*, 29, 91, 1976.

94. **Hill, D. P., Fischer, F. G., Lahr, K. M., and Coakley, J. M.**, Earthquake sounds generated by body waves from local earthquakes, *EOS Trans. Am. Geophys. Union,* 56, 1023, 1975.

95. **Nuttli, O. W.**, The Mississippi Valley earthquakes of 1811 and 1812: intensities, ground motion and magnitudes, *Bull. Seismol. Soc. Am.*, 63, 227, 1973.

96. U.S. Geological Survey, Earthquakes, U.S. Government Printing Office, Washington, D.C., 1969.

97. **Conrad, V.,**Laufzeitkurven des Tauernbebens vom 28, November 1923, *Mitt. Erdbeben Komm. Akag. Wiss. Wien,* 29, 1, 1925.

98. **Mohorovicic, A.**, Das Beben vom 8. X. 1909, *Jahrb. Meteorol. Observ. Zagreb,* 9, 4, 1910.

99. **Meissner, R.**, The "Moho" as a transition zone, *Geophys. Surv.*, 1, 195, 1973.

100. **Gutenberg, B.** The layer of relatively low wave velocity at a depth of 80 kilometers, *Bull. Seismol. Soc. Am.,* 38, 121, 1948.

101. **Gutenberg, B.,** Low-velocity layers in the earth, ocean, and atmosphere, *Science,* 131, 959, 1969.

102. **Anderson, D. L.,** The plastic layer of the earth's mantle, *Sci. Am.,* 207, 52, 1962.

103. **Putnam, W. C. and Bassett, A. B.,** *Geology,* 2nd ed., Oxford University Press, New York, 1971.

104. **Stacey, F. D.,** Physical properties of the earth's core, *Geophys. Surv.,* 1, 99, 1972.

105. **Oldham, R. D.,** The constitution of the earth, as viewed by earthquakes, *Q. J. Geol. Soc. London,* 62, 456, 1906.

106. **Lehmann, I., P.,** *Publ. Bur. Central Seismol. Int. Ser. A,* 14, 37, 1936.

107. U.S. Geological Survey, The Interior of the Earth, U.S. Government Printing Office, Washington, D.C., 1975.

108. **Wilson, J. T.,** Mao's almanac, 3,000 years of killer earthquakes, *Sat. Rev. Sci.,* p. 60, February 19, 1972.

108a. National Earthquake Information Center, Han Dynasty seismometer, *Earthquake Inf. Bull.,* 6, 38, 1974.

109. **Arvidson, R. M.,** On some mental effects of earthquakes, *Am. Psychol.,* 24, 605, 1969.

110. **Galitzin, B.,** *Vorlesungen über Seismometrie,* (German translation from the Russian version). Teubner, Leipzig, 1914.

111. **Benioff, H.,** A new vertical seismograph, *Bull. Seismol. Soc. Am.,* 22, 155, 1932.

112. **Anderson, J. A. and Wood, H. O.,** Description and theory of the torsion seismometer, *Bull. Seismol. Soc. Am.* 15, 1, 1925.

113. **Press, F., Ewing, M., and Lehner, F.,** A long-period seismograph system, *EOS Trans. Am. Geophys Union,* 39, 106, 1958.

114. **Coffman, J. L.,** Earthquake Investigation in the United States, revised ed., U.S. Government Printing Office, Washington, D.C., 1969.

115. **Bolt, B. A.,** Causes of earthquakes, in *Earthquake Engineering,* Wiegel, R. L., Ed., Prentice-Hall, Inc., Englewood Cliffs, N.J., 1970, 21.

116. National Earthquake Information Center, High-gain seismographs, *Earthquake Inf. Bull.,* 4, 18, 1972.

117. National Earthquake Information Center, California earthquake: February 9, 1971, *Earthquake Inf. Bull.,* 3, 3, 1971.

118. California Division of Mines and Geology, First earthquake devices installed, *Calif. Geol.,* 26, 12, 1973.

119. National Earthquake Information Center, Simple instrument—sophisticated view, *Earthquake Inf. Bull.,* 3, 16, 1971.

120. **Morrill, B. J.,** Evidence of record vertical accelerations at Kagel Canyon during the earthquake, in The San Fernando, California, Earthquake of February 9, 1971, Geological Survey Professional Paper 733, U.S. Government Printing Office, Washington, D.C., 1971, 177.

121. **Whitcomb, H. S., Jr.,** Standard seismic point of view, *Earthquake Inf. Bull.,* 3, 16, 1971.

122. **Peterson, J. and Orsini, N. A.,** Seismic research observatories: upgrading the worldwide data network, *EOS Trans. Am. Geophys. Union,* 57, 284, 1976.

123. **Alexander, S. S.,** The seismological potential of the new Seismic Research Observatory (SRO) network, *EOS Trans. Am. Geophys. Union,* 57, 284, 1976.

124. **Berger, J., Agnew, D., and Farrell, W. E.,** A worldwide network of very long period seismometers, *EOS Trans. Am. Geophys. Union,* 56, 1024, 1975.

125. **Berger, J., Buland, R., Agnew, D., Gilbert, F., and Farrell, W.,** Project IDA: a global wideband seismic array, *EOS Trans. Am. Geophys. Union,* 57, 284, 1976.

126. **Hamilton, R. M.,** Earthquake studies in China — a massive earthquake prediction effort is underway, *Earthquake Inf. Bull.,* 7, 3, 1975.

127. **Hill, D. P.,** New cooperative seismograph networks established in southern California, *Earthquake Inf. Bull.,* 6, 8, 1974.

128. **Richter, C. F.,** An instrumental earthquake magnitude scale, *Bull. Seismol. Soc. Am.,* 25, 1, 1935.

129. **Hayes, R. C.,** Measurement of earthquake intensity, *N. Z. J. Sci. Technol. Sect. B,* 22, 202, 1941.

130. **Gutenberg, B. and Richter, C. F.,** On seismic waves (third paper), *Gerlands Beitr. Geophys.,* 47, 73, 1936.

131. **Gutenberg, B. and Richter, C. F.,** Earthquake magnitude, intensity, energy, and acceleration, *Bull. Seismol. Soc. Am.,* 32, 163, 1942.

132. **Gutenberg, B.,** Amplitudes of surface waves and magnitudes of shallow earthquakes, *Bull. Seismol. Soc. Am.,* 35, 3, 1945.

133. **Båth, M.,** Earthquake magnitude determination from the vertical component of surface waves, *EOS Trans. Am. Geophys. Union,* 33, 81, 1952.

134. **Gutenberg, B.,** Amplitudes of P, PP, and S and magnitude of shallow earthquakes, *Bull Seismol. Soc. Am.,* 35, 57, 1945.

135. **Gutenberg, B.,** Magnitude determination for deep-focus earthquakes, *Bull. Seismol. Soc. Am.,* 35, 117, 1945.
136. **Gutenberg, B.,** The energy of earthquakes, *Q. J. Geol. Soc. London,* 112, 1, 1956.
137. National Earthquake Information Center, An interview with Charles F. Richter, *Earthquake Inf. Bull.,* 3, 10, 1971.
138. U.S. Geological Survey, Magnitude and intensity: measures of earthquake size and severity, *Earthquake Inf. Bull.,* 6, 21, 1974.
139. **Stover, C. W., Simon, R. B., and Person, W. J.,** Earthquakes in the United States, January—March 1974, *U.S. Geol. Circular,* No. 723-A, 1976.
140. **Båth, M.,** Earthquake energy and magnitude, in *Physics and Chemistry of the Earth,* Vol. 7, Ahrens, L. H. et al., Eds., Pergamon Press, London, 1966, 115.
141. **Gutenberg, B. and Richter, C. F.,** Magnitude and energy of earthquakes, *Ann. Geofis.,* 9, 1, 1956.
142. **Evernden, J. F.,** Magnitude determinations at regional and near-regional distances in the United States, *Bull. Seismol. Soc. Am.,* 57, 591, 1967.
143. **Nuttli, O. W.,** Ground motion and magnitude relations for central U.S. earthquakes, *Earthquake Notes,* 36, 26, 1965.
144. **Basham, P. W.,** A new magnitude formula for short period continental Rayleigh waves, *Geophys. J. R. Astron. Soc.,* 23, 255, 1971.
145. **Evernden, J. F., Best, W. J., Pomeroy, P. W., McEvilly, T. V., Savino, J. M., and Sykes, L. R.,** Discrimination between small-magnitude earthquakes and explosions, *J. Geophys. Res.,* 76, 8042, 1971.
146. **Nuttli, O. W.,** Seismic wave attentuation and magnitude relations for eastern North America, *J. Geophys. Res.,* 78, 876, 1973.
147. **Press, F. and Ewing, M.,** Two slow surface waves across North America, *Bull. Seismol. Soc. Am.,* 42, 219, 1952.
148. **Eaton, J. P., O'Neill, M. E., and Murdock, J. N.,** Aftershocks of the 1966 Parkfield-Cholame, California, earthquake: a detailed study, *Bull. Seismol. Soc. Am.,* 60, 1151, 1970.
149. **Bisztricsany, E.,** A new method for the determination of the magnitude of earthquakes, *Geofiz. Kozl.,* 7, 69, 1958.
150. **Sole'vev, S. L.,** Seismicity of Sakalin, *Bull. Earthquake Res. Inst. Tokyo Univ.,* 43, 95, 1965.
151. **Tsumura, K.,** Determination of earthquake magnitude from total duration of oscillation, *Bull. Earthquake Res. Inst. Tokyo Univ.,* 45, 7, 1967.
152. **Lee, W. H. K., Bennett, R. E., and Meagher, K. L.,** A Method of Estimating Magnitude of Local Earthquakes from Signal Duration, U.S. Geological Survey Open File Report, Washington, D. C., 1972.
153. **Crosson, R. S.,** Small earthquakes, structure, and tectonics of the Puget Sound region, *Bull. Seismol. Soc. Am.,* 62, 1133, 1972.
154. **Teng, T. L., Real, C. R., and Henyey, T. L.,** Microearthquakes and water-flooding in Los Angeles, *Bull. Seismol. Soc. Am.,* 63, 859, 1973.
155. **Real, C. R. and Teng, T. L.,** Local Richter magnitude and total signal duration in southern California, *Bull. Seismol. Soc. Am.,* 63, 1809, 1973.
156. **Herrmann, R. B.,** The use of duration as a measure of seismic moment and magnitude, *Bull. Seismol. Soc. Am.,* 65, 899, 1975.
157. **Draper, N. R. and Smith, H.,** *Applied Regression Analysis,* John Wiley & Sons, New York, 1966.
158. Committee of Structural Steel Producers, Agadir and the earthquake, in *The Agadir, Morocco Earthquake February 29, 1960,* American Iron and Steel Institute, New York, 1962, 14.
159. **Person, W. J.,** Earthquakes November—December 1964, *Earthquake Inf. Bull.,* 7, 24, 1975.
160. **Allen, C. A., Engen, G. R., Hanks, T. C., Nordquist, J. M., and Thatcher, W. R.,** Main shock and larger aftershocks of the San Fernando earthquake, February 9 through March 1, 1971, in The San Fernando, California, Earthquake of February 9, 1971, Geological Survey Professional Paper 733, U.S. Government Printing Office, Washington, D.C., 1971, 17.
161. **Townley, S. D. and Allen, M. W.,** Descriptive catalog of earthquakes of the Pacific Coast of the United States, 1769 to 1928, *Bull. Seismol. Soc. Am.,* 29, 1, 1939.
162. **Gutenberg, B. and Richter, C. F.,** *Seismicity of the Earth,* 1st ed., Princeton University Press, Princeton, N.J., 1949.
163. **Gutenberg, B. and Richter, C. F.,** *Seismicity of the Earth,* 2nd ed., Princeton University Press, Princeton, N.J., 1954; also reprinted by Hafner, New York, 1965.
164. **Gutenberg, B. and Richter, C. F.,** Earthquake magnitude, intensity, energy and acceleration, *Bull. Seismol. Soc. Am.,* 46, 105, 1956.
165. **Båth, M.,** The energies of seismic body waves and surface waves, in *Contributions in Geophysics: In Honor of Beno Gutenberg,* Benioff, H., Ewing, M., Howell, B. F., Jr., and Press, F., Eds., Pergamon Press, New York, 1958, 1.
166. **Tocher, D.,** Earthquake energy and ground breakage, *Bull. Seismol. Soc. Am.,* 48, 147, 1958.

167. Iida, N., Earthquake magnitude, earthquake fault and source dimensions, *Nagoya Univ. J. Earth Sci.,* 13, 115, 1965.
168. Albee, A. L. and Smith, J. L., Geologic criteria for nuclear power plant location, *Trans., Soc. Mining Eng.,* 238, 430, 1967.
169. Bolt, B. A., Duration of strong ground motion, in *Proc. World Conf. on Earthquake Engineering,* International Association for Earthquake Engineering, Rome, 1973, 1304.
170. Wyss, M., Brune, J. N., and Allen, C. R., Slippage on the Superstition Hills, Imperial, Banning-Mission Creek and Coyote Creek faults associated with the Borrego Mountain earthquake of 9 April 1968, *EOS Trans. Am. Geophys. Union,* 50, 252, 1969.
171. Allen, C. R., Wyss, M., Brune, J. N., Grantz, A., and Wallace, R. E., Displacements on the Imperial, Superstition Hills, and San Andreas faults triggered by the Borrego Mountain earthquake, in The Borrego Mountain Earthquake of April 9, 1968, Geological Survey Professional Paper 787, U.S. Government Printing Office, Washington, D.C., 1972, 87.
172. California Division of Mines and Geology, Fault moves faults, *Min. Inf. Ser.,* 22, 28, 1969.
173. Mogi, K., Some discussions on aftershocks, foreshocks and earthquake swarms — the fracture of a semi-infinite body caused by an inner stress origin and its relation to earthquake phenomena, *Bull. Earthquake Res. Inst. Tokyo Univ.,* 41, 615, 1963.
174. Mogi, K., Earthquakes and fractures, *Tectonophysics,* 5, 35, 1967.
175. Scholz, C. H., Microearthquakes on the San Andreas fault and aftershocks of the San Fernando earthquake, in The San Fernando, California, Earthquake of February 9, 1971, Geological Survey Professional Paper 733, U.S. Government Printing Office, Washington, D.C., 1971, 33.
176. Allen, C. R., Hanks, T. C., and Whitcomb, J. H., San Fernando earthquake: seismological studies and their tectonic implications, in San Fernando, California, Earthquake of February 9, 1971, Vol. 3, Benfer, N. A., Coffman, J. L., Bernick, J. R., and Dees, L. T., Eds., U.S. Government Printing Office, Washington, D.C., 1973, 13.
177. Press, F. and Jackson, D., Alaskan earthquake, 27 March 1964 — vertical extent of faulting and elastic strain energy release, *Science,* 147, 867, 1965.
178. Arabasz, W. J., Richins, W. D., and Langer, C. J., Detailed characteristics of the March 1975 Idaho-Utah border earthquake sequence, *EOS Trans. Am. Geophys. Union,* 56, 1022, 1975.
179. Morrison, P., Stump, B., and Uhrhammer, R., The Oroville earthquake sequence and its characteristics, *EOS Trans. Am. Geophys. Union,* 56, 1023, 1975.
180. Johnson, C. E. and Hadley, D. M., Tectonic implications of the Brawley earthquake swarm, Imperial Valley, California, January 1975, *EOS Trans. Am. Geophys. Union,* 56, 1022, 1975.
181. Hagiwara, T. and Rikitake, T., Japanese program on earthquake prediction, *Science,* 157, 761, 1967.
182. Rikitake, T., Earthquake prediction, *Earth Sci. Rev.,* 4, 245, 1968.
183. Ichikawa, M., The Japan Meteorological Agency observation of the Matsushiro earthquake swarm, *EOS Trans. Am. Geophys. Union,* 50, 396, 1969.
184. Tufty, B., *1001 Questions Answered About Natural Land Disasters,* Dodd Mead, New York, 1969.
185. Kanamori, H. and Anderson, D. L., Theoretical basis of some empirical relations in seismology, *Bull. Seismol. Soc. Am.,* 65, 1073, 1975.
186. Utsu, T., Aftershocks and earthquake statistics. III, *J. Fac. Sci. Hokkaido Univ. Ser. 4,* 3, 379, 1971.
187. Utsu, T., A three-parameter formula for magnitude distribution of earthquakes, *J. Phys. Earth,* 22, 71, 1974.
188. Riznichenko, Y.V., The investigation of seismic activity by the method of earthquake summation, *Izv. Acad. Sci. USSR Geophys. Ser.,* 969, 1964.
189. Okada, M., Magnitude-frequency relationships of earthquakes and estimation of supremum, *J. Meteorol. Res.,* 22, 8, 1970.
190. Otsuka, M., A simulation of earthquake occurrence. Part 2. Magnitude-frequency relation of earthquakes, *Zisin Ser. II,* 24, 215, 1971.
191. Sacuiu, I. and Zorilescu, D., Statistical analysis of seismic data on earthquakes in the area of Vrancea focus, *Bull. Seismol. Soc. Am.,* 60, 1089, 1970.
192. Purcaru, I. and Zorilescu, D., A magnitude frequency relation for the lognormal distribution of earthquake magnitude, *Pure Appl. Geophys.,* 87, 43, 1971.
193. Gutenberg, B. and Richter, C. F., Seismicity of the Earth, GSA Special Paper No. 34, Geological Society of America, New York, 1941.
194. National Earthquake Information Center, Magnitude-determining the size of an earthquake, *Earthquake Inf. Bull.,* 2, 26, 1970.
195. Gorshkov, G. P. and Shenkarev, G. A., On correlating seismic scales, in *Problems in Engineering Seismology,* Academie Nauk SSSR Instituta, Fiziki Zemli Trudy, 1958, 44.
196. de Rossi, M. S., Programma dell' observatorio ed archivo centrale geodinamico, *Boll. Vulcanismo Ital.,* 10, 3, 1883.

197. **Forel, F. A.,** Les tremblements de terre etudies par la commission seismologique suisse pendant l'annee 1881; 2me rapport, Arch. Sci. Phys. Nat., 11, 147, 1884.
198. **Wood, F. J.,** Seismicity of Alaska, in The Prince William Sound, Alaska, Earthquake of 1964 and Aftershocks, Wood, F. J., Ed., Vol. 1, U.S. Government Printing Office, Washington, D.C., 1968, 11.
199. **Mercalli, G.,** Sulle modificazioni proposte alla scala sismica De Rossi-Forel, *Boll. Soc. Sismol. Ital.,* 8, 184, 1902.
200. **Cancani, A.,** Sur l'emploi d'une double échelle seismique des intensities, empirique et absolute, *Gerlands Beitr. Geophys.,* 2, 281, 1904.
201. **Sieberg, A.,** *Erdbebenkunde,* Fischer, Jena, 1923.
202. **Wood, H. O. and Neumann, F.,** Modified Mercalli Intensity Scale of 1931, *Bull. Seismol. Soc. Am.,* 21, 277, 1931.
203. **Simon, R. B.,** Role of the California seismic observer, *Calif. Geol.,* 29, 56, 1976.
204. **Barosh, P. J.,** Use of Seismic Intensity Data to Predict the Effects of Earthquakes and Underground Nuclear Explosions in Various Geologic Settings, *U.S. Geol. Surv. Bull.,* No. 1279, 1969.
205. **Lomnitz, C.,** *Global Tectonics and Earthquake Risk,* Elsevier, New York, 1974.
206. **Scott, N. H.,** Felt area and intensity of San Fernando earthquake, in San Fernando, California, Earthquake of February 9, 1971, Vol. 3, U.S. Government Printing Office, Washington, D.C., 1973, 23.
207. **Wilson, J. T.,** A comparison of isoseismal maps with accelerations calculated from specific earthquake sources, *EOS Trans. Am. Geophys. Union,* 53, 447, 1972.
208. **Mitchell, B. J.,** Radiation and attenuation of Rayleigh waves from the southeastern Missouri earthquake of October 21, 1965, *J. Geophys. Res.,* 78, 886, 1973.
209. **Nuttli, O. W. and Zollweg, J. E.,** The relation between felt area and magnitude for central United States earthquakes, *Bull. Seismol. Soc. Am.,* 64, 73, 1974.
210. **Ergin, K.,** Observed intensity-epicentral distance relations in earthquakes, *Bull. Seismol. Soc Am.,* 59, 1227, 1969.
211. **Medvedev, S. V.,** Determination of the intensity scale of earthquakes, in *Earthquakes in the U.S.S.R.,* Academie Nauk SSSR Council on Seismology, 1961, 103; translated by U.S. Atomic Energy Commission, AEC-tr-5424, 1962, 124.
212. **Esteva, L. and Rosenblueth, E.,** Espectros de temblores a distancias moderades y grandes, *Bol. Soc. Mex. Ing. Sismica,* 2, 1, 1964.
213. **Toppozada, T. R.,** Earthquake magnitude as a function of intensity data in California and western Nevada, *Bull. Seismol. Soc. Am.,* 65, 1223, 1975.
214. **Roberts, E. B.,** Magnitude and intensity scales, *Bull. Seismol. Soc. Am.,* 47, 13, 1957.
215. **Derr, J. S.,** Locating the world's earthquakes, *Earthquake Inf. Bull.,* 6, 11, 1974.
216. **Derr, J. S.,** personal communication, 1976.
217. National Earthquake Information Center, Old globe retires at the NEIC, *Earthquake Inf. Bull.,* 4, 12, 1972.
218. **Tarr, A. C.,** Seismology enters the computer age, *Earthquake Inf. Bull.,* 2, 14, 1970.
219. **Stewart, S. W.,** Computer systems for automatic earthquake detection, *Earthquake Inf. Bull.,* 6, 17, 1974.
220. **Coffman, J. L., von Hake, C. A., Spence, W., Carver, D. L., Covington, P. A., Dunphy, G. J., Irby, W. L., Person, W. J., and Stover, C. W.,** United States Earthquakes, 1973, U.S. Government Printing Office, Washington, D.C., 1975.
221. U.S. Geological Survey, Earthquake information services available from NOAA, *Earthquake Inf. Bull.,* 6, 31, 1974.
222. U.S. Geological Survey, Seismic Engineering Program Report, October—December 1975, *U.S. Geol. Surv. Circ.,* No. 717-D, 1975.
223. U.S. Geological Survey, Automated earthquake data file, *Earthquake Inf. Bull.,* 7, 31, 1975.
224. U.S. Geological Survey, Where to obtain information about historic earthquakes, *Earthquake Inf. Bull.,* 6, 35, 1974.
225. **McKenzie, D. P. and Parker, R. L.,** The North Pacific: an example of tectonics on a sphere, *Nature (London),* 216, 1276, 1967.
226. **Isacks, B., Oliver, J., and Sykes, L. R.,** Seismology and the new global tectonics, *J. Geophys. Res.,* 73, 5855, 1968.
227. **Morgan, W. J.,** Rises, trenches, great faults, and crustal blocks, *J. Geophys. Res.,* 73, 1959, 1968.
228. **Le Pichon, X.,** Sea-floor spreading and continental drift, *J. Geophys. Res.,* 73, 3661, 1968.
229. **Wilson, J. T.,** *Continents Adrift, Readings from Scientific American,* W. H. Freeman, San Francisco, 1970.
230. **Tarling, D. and Tarling, M.,** *Continental Drift,* Doubleday, Garden City, N.Y., 1971.
231. **Bird, J. M. and Isacks, B.,** Eds., *Plate Tectonics,* American Geophysical Union, Washington, D.C., 1972.

232. **Tarling, D. H. and Runcorn, S. K.,** Eds., *Implications of Continental Drift to the Earth Sciences,* Academic Press, New York, Vol. 1, 1972; Vol. 2, 1973.
233. **Hallam, A.,** *A Revolution in the Earth Sciences,* Oxford University Press, London, 1973.
234. **Cox, A.,** Ed., *Plate Tectonics and Geomagnetic Reversals,* W. H. Freeman, San Francisco, 1973.
235. **Sullivan, W.,** *Continents in Motion—The New Earth Debate,* McGraw-Hill, New York, 1974.
236. **Kahle, C. F.,** Ed., *Plate Tectonics—Assessments and Reassessments,* Memoir 23, American Association of Petroleum Geologists, Tulsa, Okla., 1974.
237. **Glen, W.,** *Continental Drift and Plate Tectonics,* Charles E. Merrill, Columbus, Ohio, 1975.
238. **Taylor, F. B.,** Bearing of the Tertiary mountain belt on the origin of the earth's plan, *Geol. Soc. Am. Bull.,* 21, 179, 1910.
239. **Wegener, A.,** *The Origin of Continents and Oceans,* translated by Biram, J., 4th ed., Dover, New York, 1966.
240. **Holmes, A.,** *Principles of Physical Geology,* Thomas Nelson, London, 1944.
241. **Cox, A.,** The beginning: marine geology, introduction and reading list, in *Plate Tectonics and Geomagnetic Reversals,* Cox, A., Ed., W. H. Freeman, San Francisco, 1973, 10.
242. **Hopkins, W.,** Researchers in physical geology, *Philos. Trans. R. Soc. London,* 129, 381, 1839.
243. **Nopcsa, F.,** The influence of geological and chronological factors on the distribution of non-marine reptiles and Stegocephalia, *Q. J. Geol. Soc. London,* 90, 76, 1934.
244. **Du Toit, A. L.,** *Our Wandering Continents,* Hafner, New York, 1937.
245. **Fisher, R. L. and Revelle, R.,** The trenches of the Pacific, *Sci. Am.,* 193, 36, 1955.
246. **Wiseman, J. D. H. and Seymour Sewall, R. B.,** The floor of the Arabian Sea, *Geol. Mag.,* 74, 219, 1937.
246a. **Vening Meinesz, F. A.,** Maritime gravity surveys in the Netherlands East Indies, tentative interpretation of the results, *Proc. Acad. Sci. Amsterdam,* 33, 56, 1930.
246b. **Vening Meinesz, F. A., Umbgrove, J. H. F., and Kuenen Ph. H.,** *Gravity Expeditions at Sea, 1923—1932,* Vol. 2, Netherlands Geodetic Commission, Delft, 1934.
246c. **Vening Meinesz, F. A.,** *Gravity Expeditions at Sea, 1923—1938,* Vol. 4, Netherlands Geodetic Commission, Delft, 1948.
247. **Ewing, M. and Heezen, B. C.,** *Antarctica in the International Geophysical Year,* Geophys. Monogr. 1, American Geophysical Union, Washington, D.C., 1956.
248. **Menard, H. W.,** Development of median elevations in the ocean basins, *Geol. Soc. Am. Bull.,* 69, 1179, 1958.
249. **Menard, H. W.,** Geology of the Pacific sea floor, *Experienta,* 15, 205, 1959.
250. **Rothe, J. P.,** La zone séismique médiane Indo-Atlantique, *Proc. R. Soc. London Ser. A.,* 222, 387, 1954.
251. **Heezen, B. C.,** The rift in the ocean floor, *Sci. Am.,* 203, 98, 1960.
252. **Bullard, E. C., Maxwell, A. E., and Revelle, R.,** Heat flow through the deep sea floor, *Adv. Geophys.,* 3, 153, 1956.
253. **von Herzen, R.,** Heat flow values from the southern Pacific, *Nature,* 183, 882, 1959.
254. **Bruun, A. F.,** The Philippine Trench and its bottom fauna, *Nature,* 168, 692, 1951.
255. **Gaskell, T. F., Swallow, J. C., and Ritchie, G. S.,** Further notes on the greatest oceanic sounding and the topography of the Marianas Trench, *Deep Sea Res.,* 1, 60, 1953.
256. **Ewing, M. and Worzel, J. L.,** Gravity anomalies and the structure of the West Indies. I, *Geol. Soc. Am. Bull.,* 65, 165, 1954.
257. **Raitt, R. W., Fisher, R. L., and Mason, R. G.,** Tonga Trench, in *Crust of the Earth,* GSA Special Paper 62, Geological Society of America, New York, 1955, 237.
258. **Ewing, M. and Heezen, B. C.,** Puerto Rico Trench topographic and geophysical data, in *Crust of the Earth,* GSA Special Paper 62, Geological Society of America, New York, 1955, 255.
259. **Benioff, H.,** Seismic evidence for the fault origin of oceanic deeps, *Geol. Soc. Am. Bull.,* 60, 1837, 1949.
260. **Benioff, H.,** Orogenesis and deep crustal structure; additional evidence from seismology, *Geol. Soc. Am. Bull.,* 65, 385, 1954.
261. **Benioff, H.,** Seismic evidence for crustal structure and tectonic activity, in *Crust of the Earth,* GSA Special Paper 62, Geological Society of America, New York, 1955, 61.
262. **Cox, A., Dalrymple, G. B., and Doell, R. R.,** Reversals of the earth's magnetic field, *Sci. Am.,* 216, 44, 1967.
263. **Greer, K. M.,** A review of palaeomagnetism, *Earth Sci. Rev.,* 6, 369, 1970.
264. **Irving, E.,** Palaeomagnetic and palaeoclimatological aspects of polar wandering, *Pure Appl. Geophys.,* 33, 23, 1956.
265. **Runcorn, S. K.,** Palaeomagnetic comparisons between Europe and North America, *Proc. Geol. Assoc. Can.,* 8, 77, 1956.
266. **Du Bois, P. M.,** Comparison of paleomagnetic results for selected rocks of Great Britain and North America, *Adv. Phys.,* 6, 177, 1957.

267. **Creer, K. M., Irving, E., Nairn, A. E. M., and Runcorn, S. K.,** Palaeomagnetic results from different continents and their relation to the problems of continental drift, *Ann. Geophys.,* 14, 492, 1958.

268. **Creer, K. M.,** Preliminary palaeomagnetic measurements from South America, *Ann. Geophys.,* 14, 373, 1958.

269. **Clegg, J. A., Deutsch, E. R., and Griffiths, D. H.,** Rock magnetism in India, *Philos. Mag.,* 1, 419, 1956.

270. **Deutsch, E. R., Radakrishnanurty, C., and Sahasrabudhe, P. W.,** Palaeomagnetism of the Deccan Traps, *Ann. Geophys.,* 15, 39, 1959.

271. **Irving, E. and Green, R.,** Polar movement relative to Australia, *R. Astron. Soc. Geophys. J.,* 1, 64, 1958.

272. **Irving, E.,** Palaeomagnetic pole positions: a survey and analysis, *R. Astron. Soc. Geophys. J.,* 2, 51, 1959.

273. **Kawai, N., Ito, H., and Kume, S.,** Deformation of the Japanese Islands as inferred from rock magnetism, *Geophys. J.,* 6, 124, 1961.

274. **Irving, E.,** *Paleomagnetism and Its Application to Geological and Geophysical Problems,* John Wiley & Sons, New York, 1964.

275. **Nairn, A. E. M., Frost, D. V., and Light, B. G.,** Palaeomagnetism of certain rocks from Newfoundland, *Nature,* 183, 596, 1959.

276. **Dietz, R. S.,** Continent and ocean basin evolution by spreading of the sea floor, *Nature,* 190, 854, 1961.

277. **Hess, H. H.,** History of the ocean basins, in *Petrologic Studies: A Volume in Honor of A. F. Buddington,* Engel, A. E. J., James, H. L., and Leonard, B. F., Eds., Geological Society of America, New York, 1962, 599.

278. **Egyed, L.,** The change of the earth's dimensions determined from paleogeographical data, *Pure Appl. Geophys.,* 33, 42, 1956.

279. **Brunhes, B.,** Recherches sur la direction d'aimentation des roches volcaniques. I, *J. Physique,* 5, 705, 1906.

280. **Mercanton, P. L.,** Inversion de l'inclinaison magnétique terrestre aux âges géologiques, *J. Geophys. Res.,* 31, 187, 1926.

281. **Matuyama, M.,** On the direction of magnetisation of basalt in Japan, Tyôsen, and Manchuria, *Proc. Jpn. Acad.,* 5, 203, 1929.

282. **Cox, A.,** Introduction and reading list, in *Plate Tectonics and Geomagnetic Reversals,* W. H. Freeman, San Francisco, 1973, 138.

283. **Bullard, E. C.,** The stability of a homopolar dynamo, *Proc. Cambridge Philos. Soc.,* 51, 744, 1955.

284. **Elsasser, W. M.,** Hydromagnetism. I. A review, *Am. J. Phys.,* 23, 590, 1955.

285. **Cox, A., Doell, R. R., and Dalrymple, G. B.,** Geomagnetic polarity epochs: Sierra Nevada. II, *Science,* 142, 382, 1963.

286. **Cox, A., Doell, R. R., and Dalrymple, G. B.,** Reversals of the earth's magnetic field, *Science,* 144, 1534, 1964.

287. **Doell, R. R. and Dalrymple, G. B.,** Geomagnetic polarity epochs: a new polarity event and the age of the Brunhes-Matuyamu boundary, *Science,* 152, 1060, 1966.

288. **Cox, A. and Dalrymple, G. B.,** Statistical analysis of geomagnetic reversal data and the precision of potassium-argon dating, *J. Geophys. Res.,* 72, 2603, 1967.

289. **Cox, A., Doell, R. R., and Dalrymple, G. B.,** Radiometric time scale for geomagnetic reversals, *Q. J. Geol. Soc. London,* 124, 53, 1968.

290. **Cox, A.,** Geomagnetic reversals, *Science,* 163, 237, 1969.

291. **McDougall, I. and Tarling, D. H.,** Dating geomagnetic polarity zones, *Nature (London),* 202, 171, 1964.

292. **Chamalaun, F. H. and McDougall, I.,** Dating geomagnetic polarity epochs in Teunion, *Nature (London),* 210, 1212, 1966.

293. **McDougall, I. and Chamalaun, F. H.,** Geomagnetic polarity scale of time, *Nature (London),* 212, 1415, 1966.

294. **McDougall, I. and Chamalaun, F. H.,** Isotopic dating and geomagnetic polarity studies on volcanic rocks from Mauritius, Indian Ocean, *Geol. Soc. Am. Bull.,* 80, 1419, 1969.

295. **Harrison, C. G. A. and Funnell, B. M.,** Relationship of paleomagnetic reversals and micropalaeontology in two late Cenozoic cores from the Pacific Ocean, *Nature (London),* 204, 566, 1964.

296. **Opdyke, N. D., Glass, B., Hays, J. D., and Foster, J.,** Paleomagnetic study of Antarctic deep-sea cores, *Science,* 154, 349, 1966.

297. **Harrison, C. G. A.,** The paleomagnetism of deep-sea sediments, *J. Geophys. Res.,* 71, 3033, 1966.

298. **Opdyke, N. D., Burckle, L. H., Hays, J. D., and Saito, T.,** Extension of the magnetic stratigraphy to the middle Miocene in deep-sea sediments, *Abstr. Program Geol. Soc. Am.,* 2, 642, 1970.

299. **Foster, J. H. and Opdyke, N. D.,** Upper Miocene to Recent magnetic stratigraphy in deep-sea sediments, *J. Geophys. Res.,* 75, 4465, 1970.

300. **Gromme, C. S. and Hay, R. L.,** Geomagnetic polarity epochs: age and duration of the Olduvai normal polarity event, *Earth Planet. Sci. Lett.,* 10, 179, 1971.
301. **Opdyke, N. D.,** Paleomagnetism of deep-sea cores, *Rev. Geophys. Space Phys.,* 10, 213, 1972.
302. **Heezen, B. C., Ewing, M., and Miller, E. T.,** Trans-Atlantic profile of total magnetic intensity and topography, *Deep Sea Res.,* 1, 25, 1953.
303. **Mason, R. G.,** A magnetic survey off the west coast of the United States between latitudes 32° and 36° N, longitudes 121° and 128°W, *R. Astron. Soc. Geophys. J.,* 1, 320, 1958.
304. **Mason, R. G. and Raff, A. D.,** A magnetic survey off the west coast of North America 32° N to 42° N, *Geol. Soc. Am. Bull.,* 72, 1259, 1961.
305. **Raff, A. D. and Mason, R. G.,** Magnetic survey off the west coast of North America 40° N latitude to 50° N latitude, *Geol. Soc. Am. Bull.,* 72, 1267, 1961.
306. **Raff, A. D.,** The magnetism of the ocean floor, *Sci. Am.,* 205, 146, 1961.
307. **Heirtzler, J. R.,** Vema Cruise No. 16 Magnetic Measurements, Tech. Rep. No. 2, Lamont Geological Observatory, Columbia University, Palisades, N.Y., 1961.
308. **Adams, R. D. and Christoffel, D. A.,** Total magnetic field surveys between New Zealand and the Ross Sea, *J. Geophys. Res.,* 67, 805, 1962.
309. **Vacquier, V.,** Magnetic evidence for horizontal displacements in the floor of the Pacific Ocean, in *Continental Drift,* Runcorn, S. K., Ed., Academic Press, New York, 1962, 135.
310. **Vine, F. J. and Matthews, D. H.,** Magnetic anomalies over oceanic ridges, *Nature (London),* 199, 947, 1963.
311. **Vine, F. J. and Wilson, J. T.,** Magnetic anomalies over a young oceanic ridge off Vancouver Island, *Science,* 150, 485, 1965.
312. **Vine, F. J.,** Spreading of the ocean floor: new evidence, *Science,* 154, 1405, 1966.
313. **Pittman, W. C., III and Heirtzler, J. R.,** Magnetic anomalies over the Pacific-Antarctic Ridge, *Science,* 154, 1164, 1966.
314. **Pittman, W. C., III, Herron, E. M., and Heirtzler, J. R.,** Magnetic anomalies in the Pacific and sea floor spreading, *J. Geophys. Res.,* 73, 2069, 1968.
315. **Dickson, G. O., Pittman, W. C., III, and Heirtzler, J. R.,** Magnetic anomalies in the South Atlantic and ocean floor spreading, *J. Geophys. Res.,* 73, 2087, 1968.
316. **Le Pichon, X. and Heirtzler, J. R.,** Magnetic anomalies in the Indian Ocean and sea floor spreading, *J. Geophys. Res.,* 73, 2101, 1968.
317. **Hayes, D. E.,** Nature and implications of asymmetric sea-floor spreading, different rates for different plates, *Geol. Soc. Am. Bull.,* 87, 994, 1976.
318. **Heirtzler, J. R., Dickson, G. O., Herron, E. M., Pitman, W. C., and Le Pichon, X.,** Marine magnetic anomalies, geomagnetic field reversals, and motions of the ocean floor and continents, *J. Geophys. Res.,* 73, 2119, 1968.
319. **Wilson, J. T.,** Cabot fault, an Appalachian equivalent of the San Andreas and Great Glen faults and some implications for continental displacement, *Nature (London),* 195, 135, 1962.
320. **Wilson, J. T.,** Continental drift, *Sci. Am.,* 208, 86, 1963.
321. **Wilson, J. T.,** The Cabot and Great Glen faults, *Nature (London),* 197, 680, 1963.
322. **Wilson, J. T.,** Evidence from islands on the spreading of the ocean floor, *Nature (London),* 197, 536, 1963.
323. **Wilson, J. T.,** A possible origin of the Hawaiian Islands, *Can. J. Phys.,* 41, 863, 1963.
324. **Unger, J. D.,** Scientists probe earth's secrets at the Hawaiian Volcano Observatory, *Earthquake Inf. Bull.,* 6, 3, 1974.
325. **Wilson, J. T.,** A new class of faults and their bearing on continental drift, *Nature (London),* 207, 343, 1965.
326. **Sykes, L. R.,** Mechanism of earthquakes and nature of faulting on the mid-ocean ridges, *J. Geophys. Res.,* 72, 2131, 1967.
327. **Talwani, M., Le Pichon, X., and Heirtzler, J. R.,** East Pacific Rise: the magnetic pattern and the fracture zones, *Science,* 150, 1109, 1965.
328. **Sykes, L. R.,** Seismological evidence for transform faults, sea-floor spreading, and continental drift, in *History of the Earth's Crust,* Phinney, R. A., Ed., Princeton University Press, Princeton, N.J., 1968, 20.
329. **Tobin, D. G. and Sykes, L. R.,** Seismicity and tectonics of the northeast Pacific Ocean, *J. Geophys. Res.,* 73, 3821, 1968.
330. **Banghar, A. R. and Sykes, L. R.,** Focal mechanisms of earthquakes in the Indian Ocean and adjacent regions, *J. Geophys. Res.,* 74, 632, 1969.
331. **Molnar, P. and Sykes, L. R.,** Tectonics of the Caribbean and Middle American regions from focal mechanisms and seismicity, *Geol. Soc. Am. Bull.,* 80, 1639, 1969.
332. **Peterson, M. W. A., Edgar, N. T., Cita, M., Gartner, S., Jr., Goell, R., Nigrini, C., and Von der Bosch, C.,** Shipboard scientific procedures, in Initial Reports of the Deep Sea Drilling Project, Vol. 2, U.S. Government Printing Office, Washington, D.C., 1970, 452.

333. **Peterson, M. W. A., Edgar, N. T., Von der Borch, C. C., and Rex, R. W.,** Cruise log summary and discussion, in Initial Reports of the Deep Sea Drilling Project, Vol. 2, U.S. Government Printing Office, Washington, D.C., 1970, 413.

334. **Maxwell, H. E., Von Herzen, R. P., Andrews, J. E., Boyce, R. E., Milow, E. D., Hsu, K. J., Percival, S. F., and Saito, T.,** Summary and conclusions, in Initial Reports of the Deep Sea Drilling Project, Vol. 2, U.S. Government Printing Office, Washington, D.C., 1970, 441.

335. **Maxwell, A. E., Von Herzen, R. P., Hsu, K. J., Andrews, J. E., Saito, T., Percival, S. F., Milow, E. D., and Boyce, R. E.,** Deep-sea drilling in the South Atlantic, *Science,* 168, 1047, 1970.

336. **Davies, T. A. and Edgar, N. T.,** Some geophysical aspects of deep-sea drilling, *Geophys. Surv.* 1, 391, 1974.

337. **Heirtzler, J. R. and Le Pichon, X.,** FAMOUS: a plate tectonic study of the lithosphere, *Geology,* 2, 273, 1974.

338. **Heirtzler, J. R. and Bryon, W. B.,** The floor of the Mid-Atlantic rift, *Sci. Am.,* 233, 178, 1975.

339. **Anderson, S. H.,** Man's first look at the mid-ocean ridge, *NOAA,* 5, 12, 1975.

340. **Hurley, P. M.,** The confirmation of continental drift, *Sci. Am.,* 218, 52, 1968.

341. **Bullard, E. C., Everett, J. E., and Smith, A. G.,** Fit of continents around Atlantic, in Symposium on Continental Drift, *Philos. Trans. R. Soc. London,* 258, 41, 1965.

342. **Smith, A. G. and Hallam, A.,** The fit of the southern continents, *Nature (London),* 225, 139, 1970.

343. **Dietz, R. S. and Holden, J. C.,** Reconstruction of Pangaea: breakup and dispersion of continents, Permian to Present, *J. Geophys. Res.,* 75, 4939, 1970.

344. **Kurten, B.,** Continental drift and evolution, *Sci. Am.,* 220, 54, 1969.

345. **Bartholomai, A. and Howie, A.,** Vertebrate fauna from the Lower Trias of Australia, *Nature (London),* 225, 1063, 1970.

346. **Elliot, D. H., Colbert, E. H., Breed, W. J., Jensen, J. A., and Powell, J. S.,** Triassic tetrapods from Antarctica: evidence for continental drift, *Science,* 169, 1197, 1970.

347. **Colbert, E. H.,** Antarctic fossils and the reconstruction of Gondwanaland, *Nat. Hist.,* 81, 66, 1972.

348. **Kitching, J. W., Collinson, J. W., Elliot, D. H., and Colbert, E. H.,** Lystrosauras zone (Triassic) fauna from Antarctica, *Science,* 175, 524, 1972.

349. **Colbert, E. H.,** *Wandering Lands and Animals,* E. P. Dutton, New York, 1973.

350. **Milner, A. R. and Panchen, A. L.,** Geographical variation in the tetrapod faunas of the Upper Carboniferous and Lower Permian, in *Implications of Continental Drift to the Earth Sciences,* Vol. 1, Tarling, D. H. and Runcorn, S. K., Eds., Academic Press, London, 1973, 353.

351. **Cox, C. B.,** The distribution of Triassic terrestrial tetrapod families, in *Implications of Continental Drift to the Earth Sciences,* Vol. 1, Tarling, D. H. and Runcorn, S. K., Eds., Academic Press, London, 1973, 369.

352. **Cracraft, J.,** Vertebrate evolution and biogeography in the old world tropics: implications of continental drift and palaeoclimatology, in *Implications of Continental Drift to the Earth Sciences,* Vol. 1, Tarling, D. H. and Runcorn, S. K., Eds., Academic Press, London, 1973, 373.

353. **Colbert, E. H.,** Continental drift and the distribution of fossil reptiles, in *Implications of Continental Drift to the Earth Sciences,* Vol. 1, Tarling, D. H. and Runcorn, S. K., Eds., Academic Press, London, 1973, 395.

354. **Maheshwari, H. K.,** Permian wood from Antarctica and revision of some Lower Gondwana wood taxa, *Palaeontographica,* 138, 1, 1972.

355. **Plumstead, E. P.,** The enigmatic glossopteris flora and uniformitarianism, in *Implications of Continental Drift to the Earth Sciences,* Vol. 1, Tarling, D. H. and Runcorn, S. K., Eds., Academic Press, London, 1973, 413.

356. **Chaloner, W. G. and Creber, G. T.,** Growth rings in fossil woods as evidence of past climates, in *Implications of Continental Drift to the Earth Sciences,* Vol. 1, Tarling, D. H. and Runcorn, S. K., Eds., Academic Press, London, 1973, 425.

357. **Melville, R.,** Continental drift and plant distribution, in *Implications of Continental Drift to the Earth Sciences,* Vol. 1, Tarling, D. H. and Runcorn, S. K., Eds., Academic Press, London, 1973, 439.

358. **Schopf, J. M.,** Coal, climate and global tectonics, in *Implications of Continental Drift to the Earth Sciences,* Vol. 1, Tarling, D. H. and Runcorn, S. K., Eds., Academic Press, London, 1973, 609.

359. **Keast, A.,** Continental drift and the evolution of the biota on southern continents, *Q. Rev. Biol.,* 46, 335, 1971.

360. **Keast, A.,** Contemporary biotas and the separation sequence of the southern continents, in *Implications of Continental Drift to the Earth Sciences,* Vol. 1, Tarling, D. H. and Runcorn, S. K., Eds., Academic Press, London, 1973, 309.

361. **Flakes, L. A. and Crowell, J. C.,** Late Paleozoic glaciation. I. South America, *Geol. Soc. Am. Bull.,* 80, 1007, 1969.

362. **Crowell, J. C. and Flakes, L. A.,** Phanerozoic glaciation and the causes of ice ages, *Am. J. Sci.,* 268, 193, 1970.

363. Flakes, L. A. and Crowell, J. C., Late Paleozoic glaciation. II. Africa exclusive of the Karroo Basin, *Geol. Soc. Am. Bull.*, 81, 2261, 1970.

364. Crowell, J. C. and Flakes, L. A., Late Paleozoic glaciation. IV. Australia, *Geol. Soc. Am. Bull.*, 82, 2515, 1971.

365. Crowell, J. C. and Flakes, L. A., Late Paleozoic glaciation of Australia, *J. Geol. Soc. Aust.*, 17, 115, 1971.

366. Flakes, L. A., Kemp, E. M., and Crowell, J. C., Late Paleozoic glaciation. VI. Asia, *Geol. Soc. Am. Bull.*, 86, 454, 1975.

367. Stoneley, R., The Niger Delta region in the light of the theory of continental drift, *Geol. Mag.*, 103, 385, 1966.

368. Hurley, P. M., de Almeida, F. F. M., Melcher, G. C., Cordani, U. G., Rand, J. R., Kawashita, K., Vandoros, P., Pinson, W. H., Jr., and Fairbairn, H. W., Test of continental drift by comparison of radiometric ages, *Science*, 157, 495, 1967.

369. Siedner, G. and Miller, J. A., K-Ar age determinations on basaltic rocks from South-West Africa and their bearing on continental drift, *Earth Planet. Sci. Lett.*, 4, 451, 1968.

370. Allard, G. O. and Hurst, V. J., Brazil-Gabon geologic link supports continental drift, *Science*, 163, 528, 1969.

371. Hertz, N., Anorthosite belts, continental drift and the anorthosite event, *Science*, 164, 944, 1969.

372. Reyment, R. A., Ammonite biostratigraphy, continental drift and oscillatory transgressions, *Nature (London)*, 224, 137, 1969.

373. Gough, D. I., Opdyke, N. D., and McElhinny, M. W., The significance of palaeomagnetic results from Africa, *J. Geophys. Res.*, 69, 2509, 1964.

374. Briden, J. C., A new palaeomagnetic result from the Lower Cretaceous of east central Africa, *Geophys. J. R. Astron. Soc.*, 12, 375, 1967.

375. Strangway, D. W. and Vogt, P. R., Aeromagnetic tests for continental drift in Africa and South America, *Earth Planet. Sci. Lett.*, 7, 429, 1970.

376. Creer, K. M., Mitchell, J. G., and Abou Deeb, J., Palaeomagnetism and radiometric age of the Jurassic Chon Aike Formation from Santa Cruz Province, Argentina: implications for the opening of the South Atlantic, *Earth Planet. Sci. Lett.*, 14, 131, 1972.

377. Gidskehaug, A., Creer, K. M., and Mitchell, J. G., Palaeomagnetism and K-Ar ages of the South-West African basalts and their bearing on the time of initial rifting of the South Atlantic Ocean, *Geophys. J. R. Astron. Soc.*, 42, 1, 1975.

378. Kelleher, J., Sykes, L., and Oliver, J., Possible criteria for predicting earthquake locations and their application to major plate boundaries of the Pacific and the Caribbean, *J. Geophys. Res.*, 78, 2547, 1973.

379. Dewey, J. F., Plate tectonics, *Sci. Am.*, 226, 56, 1972.

380. McKenzie, D. P. and Morgan, W. J., Evolution of triple junctions, *Nature (London)*, 224, 125, 1969.

381. Garfunkel, Z., Growth, shrinking, and long-term evolution of plates and their implications for the flow pattern in the mantle, *J. Geophys. Res.*, 80, 4425, 1975.

382. Forsyth, D. W., A model of the evolution of the upper mantle beneath mid-ocean ridges, *EOS Trans. Am. Geophys. Union*, 57, 328, 1976.

383. Dickson, W. R., Plate tectonics in geologic history, *Science*, 174, 107, 1971.

384. Herron, E. M., Sea-floor spreading and the Cenozoic history of the east-central Pacific, *Geol. Soc. Am. Bull.*, 83, 1671, 1972.

385. Sykes, L. R., Earthquake swarms and sea-floor spreading, *J. Geophys. Res.*, 75, 6598, 1970.

386. Francis, T. J. G., A new interpretation of the 1968 Fernandina Caldera collapse and its implications for the mid-oceanic ridges, *Geophys. J. R. Astron. Soc.*, 39, 301, 1974.

387. Laughton, A., The birth of an ocean, *New Sci.*, 27, 218, 1966.

388. McKenzie, D., Davies, D., and Molnar, P., Plate tectonics of the Red Sea and East Africa, *Nature (London)*, 226, 1, 1970.

389. Freund, R., Letters: plate tectonics of the Red Sea and East Africa, *Nature (London)*, 228, 453, 1970.

390. Laughton, A. S., A new bathymetric chart of the Red Sea, *Philos. Trans. R. Soc. London Ser. A*, 267, 21, 1970.

391. Fairhead, D. and Girdler, R., The seismicity of the Red Sea, Gulf of Aden and Afar Triangle, *Philos. Trans. R. Soc. London Ser. A*, 267, 49, 1970.

392. Allan, T. D., Magnetic and gravity fields over the Red Sea, *Philos. Trans. R. Soc. London Ser. A*, 267, 153, 1970.

393. Davies, D. and Tramontini, C., The deep structure of the Red Sea, *Philos. Trans. R. Soc. London Ser. A*, 267, 181, 1970.

394. Phillips, J. D., Magnetic anomalies in the Red Sea, *Philos. Trans. R. Soc. London Ser. A*, 267, 205, 1970.

395. **Girdler, R. and Darracott, B.**, African poles of rotation, *Comments Earth Sci. Geophys.*, 2, 131, 1972.
396. **Girdler, R. and Styles, P.**, Two-stage Red Sea floor spreading, *Nature (London)*, 247, 7, 1974.
397. **Richardson, E. S. and Harrison, C. G. A.**, Opening of the Red Sea with two poles of rotation, *Earth Planet. Sci. Lett.*, 30, 135, 1976.
398. **Ross, D. A.**, The Red Sea: an ocean in the making, *Nat. His.*, 85, 74, 1976.
399. **Toksöz, M. N., Sleep, N. H., and Smith, A. T.**, Evolution of the downgoing lithosphere and the mechanics of deep focus earthquakes, *Geophys. J. R. Astron. Soc.*, 35, 285, 1973.
400. **Toksöz, M. N.**, The subduction of the lithosphere, *Sci. Am.*, 233, 80, 1975.
401. **Isacks, B. and Molnar, P.**, Distribution of stresses in the descending lithosphere from a global survey of focal-mechanism solutions of mantle earthquakes, *Rev. Geophys. Space Phys.*, 9, 103, 1971.
402. **Stauder, W.**, Tensional character of earthquake foci beneath the Aleutian Trench with relation to sea floor spreading, *J. Geophys. Res.*, 73, 7693, 1968.
403. **Kanamori, H.**, Seismological evidence for a lithospheric normal faulting — the Sanriku earthquake of 1933, *Phys. Earth Planet. Inter.*, 4, 289, 1971.
404. **Hanks, T. C.**, The Kuril Trench-Hokkaido Rise system: large shallow earthquakes and simple models of deformation, *Geophys. J. R. Astron. Soc.*, 23, 173, 1971.
405. **Plafker, G.**, Alaskan earthquake of 1964 and Chilean earthquake of 1960: implications for arc tectonics, *J. Geophys. Res.*, 77, 901, 1972.
406. **Stauder, W.**, Mechanism of the Rat Island earthquake sequence of 4 February 1965 with relation to island arcs and sea-floor spreading, *J. Geophys. Res.*, 73, 3847, 1968.
407. **Barazangi, M. and Dorman, J.**, World seismicity maps compiled from ESSA, Coast and Geodetic Survey, epicenter data, 1961-1967, *Bull. Seismol. Soc. Am.*, 59, 369, 1969.
408. **Isacks, B. and Molnar, P.**, Mantle earthquake mechanisms and the sinking of the lithosphere, *Nature (London)*, 223, 1121, 1969.
409. **Isacks, B. L. and Barazangi, M.**, High frequency shear waves guided by a continuous lithosphere descending beneath western South America, *Geophys. J. R. Astron. Soc.*, 33, 129, 1973.
410. **Kelleher, J. and McCann, W.**, Buoyant zones, great earthquakes, and unstable boundaries of subduction, *J. Geophys. Res.*, 81, 4885, 1976.
411. **Ericksen, G.**, Earthquakes and sea-floor spreading in the Andean region, South America, *Earthquake Inf. Bull.*, 5, 23, 1973.
412. **Thompson, T. L.**, Plate tectonics in oil and gas exploration of continental margins, *Am. Assoc. Petrol. Geol. Bull.*, 60, 1463, 1976.
413. **Dewey, J. F. and Bird, J. M,.** Mountain belts and the new global tectonics, *J. Geophys. Res.*, 75, 2625, 1970.
414. **Fitch, T. J.**, Earthquake mechanisms of the Himalayan, Burmese, and Andaman regions and continental tectonics in Central Asia, *J. Geophys. Res.*, 75, 2699, 1970.
415. **Kono, M.**, Gravity anomalies in east Nepal and their implications to the crustal structure of the Himalayas, *Geophys. J. R. Astron. Soc.*, 39, 283, 1974.
416. **Bird, P. and Toksöz, M. N.**, Himalayan orogeny modelled with finite elements, *EOS Trans. Am. Geophys. Union*, 57, 334, 1976.
417. **McKenzie, D. P.**, Plate tectonics of the Mediterranean region, *Nature (London)*, 226, 239, 1970.
418. **Atwater, T. and Menard, H. W.**, Magnetic lineations in the northeast Pacific, *Earth Planet. Sci. Lett.*, 7, 445, 1970.
419. **Atwater, T.**, Implications of plate tectonics for the Cenozoic evolution of western North America, *Geol. Soc. Am. Bull.*, 81, 3513, 1970.
420. **Anderson, D. L.**, The San Andreas fault, *Sci. Am.*, 225, 52, 1971.
421. **Huffman, F. O.**, Miocene and post-Miocene offset on the San Andreas fault in central California, *Geol. Soc. Am. Abst. Programs*, 2, 104, 1970.
422. **Dickson, W. R., Cowan, D. S., and Schweickert, R. A.**, Test of new global tectonics: discussion, *Am. Assoc. Petrol. Geol. Bull.*, 56, 375, 1972.
423. **Gilliland, W. N. and Meyer, G. P.**, Two classes of transform faults, *Geol. Soc. Am. Bull.*, 87, 1127, 1976.
424. **Dietz, R. S. and Holden, J. C.**, The breakup of Pangaea, *Sci. Am.*, 223, 30, 1970.
424a. **Griggs, D. T.**, A theory of mountain building, *Am. J. Sci.*, 237, 611, 1939.
425. **Bostrom, R. C.**, Westward displacement of the lithosphere, *Nature (London)*, 234, 536, 1971.
426. **Moore, G. W.**, Westward tidal lag as the driving force of plate tectonics, *Geology*, 1, 99, 1973.
427. **Daněs, Z. F.**, Mainstream mantle convection: a geologic analysis of plate motion: discussion, *Am. Assoc. Petrol. Geol. Bull.*, 57, 410, 1973.
428. **Cullen, D. J.**, Tidal friction as a possible factor in continental drift, *Comments Earth Sci. Geophys.*, 3, 181, 1973.
429. **Runcorn, S. K.**, Towards a theory of continental drift, *Nature (London)*, 193, 313, 1962.

430. **Takeuchi, H. and Sakata, M.,** Convection in the mantle with variable viscosity, *J. Geophys. Res.,* 75, 921, 1970.
431. **Morgan, W. J.,** Convection plumes in the lower mantle, *Nature (London),* 230, 42, 1971.
432. **Torrance, K. E. and Turcotte, D. L.,** Structure of convention cells in the mantle, *J. Geophys. Res.,* 76, 1156, 1971.
433. **Turcotte, D. L. and Oxburgh, E. R.,** Mantle convection and the new global tectonics, *Annu. Rev. Fluid Mech.,* 4, 33, 1972.
434. **Morgan, W. J.,** Deep mantle convection plumes and plate motions, *Am. Assoc. Petrol. Geol. Bull.,* 56, 203, 1972.
435. **Schubert, G. and Turcotte, D. L.,** One-dimensional model of shallow-mantle convection, *J. Geophys. Res.* 77, 945, 1972.
436. **McKenzie, D. P., Roberts, J., and Weiss, N.,** Numerical models of convection in the earth's mantle, *Tectonophysics,* 19, 89, 1973.
437. **Richter, F.,** Convection and large-scale circulation in the mantle, *J. Geophys. Res.,* 78, 8735, 1973.
438. **McKenzie, D. P., Roberts, J. M., and Weiss, N. O.,** Convection in the earth's mantle, *J. Fluid Mech.,* 62, 465, 1974.
439. **Houston, M. H., Jr. and De Bremaecker, J. Ce.,** Numerical models of convection in the upper mantle, *J. Geophys. Res.,* 80, 742, 1975.
440. **Richter, F. M. and Parsons, B.,** On the interaction of two scales of convection in the mantle, *J. Geophys. Res.,* 80, 2529, 1975.
441. **Parmentier, E. M. and Turcotte, D. L.,** Numerical experiments on the structure of mantle plumes, *J. Geophys. Res.,* 80, 4417, 1975.
442. **Kanasewich, E. R.,** Plate tectonics and planetary convection, *Can. J. Earth Sci.,* 13, 331, 1976.
443. **Elsasser, W. M.,** Sea-floor spreading as thermal convection, *J. Geophys. Res.,* 76, 1101, 1971.
444. **Elsasser, W. M.,** Convection and Stress Propagation in the Upper Mantle, Tech. Rep. 5, Princeton University, Princeton, N.J., 1967.
445. **Elsasser, W. M.,** Convection and stress propagation in the upper mantle, in *The Application of Modern Physics to the Earth and Planetary Interiors,* Runcorn, S. K., Ed., Interscience, New York, 1969, 223.
446. **Richter, F.,** Dynamical models for sea-floor spreading, *Rev. Geophys. Space Phys.,* 11, 223, 1973.
447. **Tullis, T. E. and Chapple, W. M.,** What makes the plates go, *EOS Trans. Am. Geophys. Union,* 54, 468, 1973.
448. **Sykes, L. R. and Sbar, M. L.,** Intraplate earthquakes, lithospheric stresses and the driving mechanism of plate tectonics, *Nature,* 245, 298, 1973.
449. **Orowan, E.,** Continental drift and the origin of mountains, *Science,* 146, 1003, 1964.
450. **Hales, A. L.,** Gravitational sliding and continental drift, *Earth Planet. Sci. Lett.,* 6, 31, 1969.
451. **Lliboutry, L.,** Sea-floor spreading, continental drift, and lithosphere sinking with an asthenosphere at melting point, *J. Geophys. Res.,* 74, 6525, 1969.
452. **Jacoby, W. B.,** Instability in the upper mantle and global movements, *J. Geophys. Res.,* 75, 5671, 1970.
453. **Artemjev, M. E. and Artyushkov, E. V.,** Structure and isostacy of the Baikal rift and the mechanism of rifting, *J. Geophys. Res.,* 76, 1197, 1971.
454. **Wilson, J. T. and Burke, K.,** Plate tectonics and plume mechanics, *EOS Trans. Am. Geophys. Union,* 54, 238, 1973.
455. **Forsyth, D. and Uyeda, S.,** On the relative importance of the driving forces of plate motion, *Geophys. J. R. Astron. Soc.,* 43, 163, 1975.
456. **Tarr, A. C.,** Global seismicity studies — projections of the hyperspace, *Earthquake Inf. Bull.,* 4, 4, 1972.
457. **Sykes, L. R.,** The seismicity and deep structure of island arcs, *J. Geophys. Res.,* 71, 2981, 1966.
458. **Oliver, J. and Isacks, B.,** Deep earthquake zones, anomalous structures in the upper mantle, and the lithosphere, *J. Geophys. Res.,* 72, 4259, 1967.
459. **Sykes, L. R.,** Deep-focus earthquakes in the New Hebrides region, *J. Geophys. Res.,* 69, 5353, 1964.
460. **Chapman, M. E. and Solomon, S. C.,** North American-Eurasian plate boundary in northeast Asia, *J. Geophys. Res.,* 81, 921, 1976.
461. **Scholz, C. H., Kozynski, T. A., and Hutchins, D. G.,** Evidence for incipient rifting in southern Africa, *Geophys. J. R. Astron. Soc.,* 44, 135, 1976.
462. **Nowroozi, A. A.,** Focal mechanisms of earthquakes in Persia, Turkey, West Pakistan and Afghanistan and plate tectonics of the Middle East, *Bull. Seismol. Soc. Am.,* 62, 823, 1972.
463. **Rowan, C.,** Rigid plates, buffer plates and sub-plates — comment on active tectonics of the Mediterranean region by D. P. McKenzie, *Geophys. J. R. Astron. Soc.,* 33, 369, 1973.
464. American Geological Institute, Chinese plate movements, *Geotimes,* 21, 26, 1976.
465. National Earthquake Information Center, Earthquake history of Alaska, *Earthquake Inf. Bull.,* 2, 30, 1970.

466. **Carey, S. W.,** The tectonic approach to continental drift, in *Continental Drift: A Symposium,* Carey, S. W., Ed., University of Tasmania, Hobart, 1958, 177.

467. **Wise, D. U.,** An outrageous hypothesis for the tectonic pattern of the North American Cordillera, *Geol. Soc. Am. Bull.,* 74, 357, 1963.

468. **Hamilton, W. and Myers, W. B.,** Cenozoic tectonics of western United States, *Rev. Geophys.,* 4, 509, 1966.

469. **Woodford, A. O. and McIntyre, D. B.,** Curvature of the San Andreas fault, California, *Geology,* 4, 573, 1976.

470. **Smith, R. B. and Sbar, M. L.,** Contemporary tectonics and seismicity of the western United States with emphasis on the intermountain seismic belt, *Geol. Soc. Am. Bull.,* 85, 1205, 1974.

471. California Division of Mines and Geology, Eastern U.S. earthquakes, *Calif. Geol.,* 26, 221, 1973.

472. **Gutenberg, B.,** Tilting due to glacial melting, *J. Geol.,* 41, 449, 1933.

473. **Daly, R. A.,** *The Changing World of the Ice Age,* Yale University Press, New Haven, Conn., 1934.

474. **Leet, L. D.,** Mechanics of earthquakes where there is no faulting, *Bull. Seismol. Soc. Am.,* 32, 93, 1942.

475. **Page, R. A., Molnar, P. H., and Oliver, J.,** Seismicity in the vicinity of the Ramapo fault, New Jersey-New York, *Bull. Seismol. Soc. Am.,* 58, 681, 1968.

476. **Voight, B.,** Evolution of North American Ocean: relevance of rock-pressure measurements, in *North Atlantic — Geology and Continental Drift Memoir 12,* Kay, M., Ed., American Association of Petroleum Geologists, Tulsa, Okla., 1969, 955.

477. **Woollard, G. P.,** Tectonic activity in North America, in *The Earth's Crust and Upper Mantle,* Geophysical Monograph 13, Hart, P. J., Ed., American Geophysical Union, Washington, D.C., 1969, 125.

478. **McGinnis, L. D.,** *Earthquakes and Crustal Movement Related to the Water Load in the Mississippi Valley Region,* Illinois State Geological Survey, Urbana, 1963.

479. **Oliver, J. and Isacks, B.,** Seismicity and tectonics of the eastern United States, *Earthquake Notes,* 43, 30, 1972.

480. **Bollinger, G. A.,** Seismicity and crustal uplift in southeastern United States, *Am. J. Sci.,* 273A, 396, 1973.

481. **Brown, L. D. and Oliver, J. E.,** Vertical crustal movements from leveling data and their relation to geologic structure in the eastern United States, *Rev. Geophys. Space Phys.,* 14, 13, 1976.

482. **Mendiguren, J. A.,** Focal mechanism of a shock in the middle of the Nazca plate, *J. Geophys. Res.,* 76, 3861, 1971.

483. **Sbar, M. L. and Sykes, L. R.,** Contemporary compressive stress and seismicity in eastern North America: an example of intraplate tectonics, *Geol. Soc. Am. Bull.,* 84, 1861, 1973.

484. **Artyushkov, E. V.,** Stresses in the lithosphere caused by crustal thickness inhomogeneities, *J. Geophys. Res.* 78, 7675, 1973.

485. **McKenzie, D. P.,** Plate tectonics, in *The Nature of the Solid Earth,* Robertson, E. C., Ed., McGraw-Hill, New York, 1972, 323.

486. **Turcotte, D. L. and Oxburgh, E. R.,** Mid-plate tectonics, *Nature,* 244, 337, 1973.

487. **Oxburgh, E. R. and Turcotte, D. L.,** Membrane tectonics and the east African rift, *Earth Planet. Sci. Lett.,* 22, 133, 1974.

488. **Turcotte, D. L.,** Membrane tectonics, *Geophys. J. R. Astron. Soc.,* 36, 33, 1974.

488a. **Kane, M. F.,** Correlation of major eastern U.S. earthquake centers with mafic/ultramafic masses, *EOS Trans. Am. Geophys. Union,* 57, 963, 1976.

488b. **Kane, M. F.,** personal communication, 1976.

489. **Alfors, J. T., Burnett, J. L., and Gay, T. E., Jr.,** Urban Geology Master Plan for California, Bull. 198, California Division of Mines and Geology, Sacramento, 1973.

490. **Algermissen, S. T., Rinehart, W. A., and Dewey, J.,** *A Study of Earthquake Losses in the San Francisco Bay Area,* National Oceanic and Atmospheric Administration, Washington, D.C., 1972.

491. **Plafker, G., Bonilla, M. G., and Bonis, S. B.** The Guatemalan Earthquake of February 4, 1976, A Preliminary Report, Geological Survey Professional Paper 1002, U.S. Government Printing Office, Washington, D.C., 1976.

492. **Brown, R. D., Jr., Ward, P. L., and Plafker, G.,** Geologic and Seismologic Aspects of the Managua, Nicaragua, Earthquakes of December 23, 1972, Geological Survey Professional Paper, 838, U.S. Government Printing Office, Washington, D.C., 1973.

493. **Brown, R. D., Jr., Ward, P. L., and Plafker, G.,** Map showing faults and fractures related to the Managua earthquakes of December 23, 1972, in Geologic and Seismologic Aspects of the Managua, Nicaragua, Earthquakes of December 23, 1972, Geological Survey Professional Paper 838, U.S. Government Printing Office, Washington, D.C., 1973, plate I.

494. **Bonilla, M. G., Buchanan, J. M., Castle, R. O., Clark, M. M., Frizzell, V. A., Gulliver, R. M., Miller, F. K., Pinkerton, J. P., Ross, D. C., Sharp, R. V., Yerkes, R. F., and Ziony, J. I.,** Surface faulting, in The San Fernando, California, Earthquake of February 9, 1971, Geological Survey Professional Paper 733, U.S. Government Printing Office, Washington, D.C., 1971, 55.

495. **Youd, T. L. and Olsen, H. W.,** Damage to constructed works, associated with soil movements and foundation failures, in The San Fernando, California, Earthquake of February 9, 1971, Geological Survey Professional Paper 733, U.S. Government Printing Office, Washington, D.C., 1971, 126.

496. **Imamura, A.,** *Theoretical and Applied Seismology,* Maruzen, Tokyo, 1937.

497. **Plafker, G.,** Tectonic deformation associated with the 1964 Alaska earthquake, *Science,* 148, 1675, 1965.

498. **Grantz, A., Plafker, G., and Kachadoorian, R.,** Alaska's Good Friday Earthquake, March 27, 1964, A Preliminary Geologic Evaluation, Geological Survey Circular 491, U.S. Geological Survey, Washington, D.C., 1964.

499. **Plafker, G., Kachadoorian, R., Eckel, E. B., and Mayo, L. R.,** Effects of the earthquake of March 27, 1964 on various communities, in The Alaska Earthquake March 27, 1964: Effects on Communities, Geological Survey Professional Paper 542, U.S. Government Printing Office, Washington, D.C., 1969, G1.

500. **Ovenshine, A. T., Larson, D. E., and Bartsch-Winkler, S. R.,** Subsidence, inundation, and sedimentation: environmental consequences of the 1964 Alaska earthquake in the Portage, Alaska, area, *Earthquake Inf. Bull.,* 6, 3, 1974.

501. **Housner, G. W.,** Strong ground motion, in *Earthquake Engineering,* Wiegel, R. L., Ed., Prentice-Hall, Englewood Cliffs, N.J., 1970, 75.

502. **Benioff, H.,** Mechanism and strain characteristics of the White Wolf fault as indicated by the aftershock sequence, in Earthquakes in Kern County, California, during 1952, Oakeshott, G. B., Ed., Bull. 171, California Division of Mines and Geology, Sacramento, 1955, 199.

503. **Espinosa, A. F., Husid, R., and Quesada, A.,** Intensity distribution and source parameters from field observations, in The Guatemalan Earthquake of February 4, 1976, A Preliminary Report, Geological Survey Professional Paper 1002, U.S. Government Printing Office, Washington, D.C., 1976, 52.

504. **Savage, J. C.** Radiation from a realistic model of faulting, *Bull. Seismol. Soc. Am.,* 56, 577, 1966.

505. **Schnabel, P. B. and Seed, H. B.,** Accelerations in rock for earthquakes in the western United States, *Bull. Seismol. Soc. Am.,* 63, 501, 1973.

506. **Nason, R.,** Increased seismic shaking above a thrust fault, in *San Fernando, California, Earthquake of February 9, 1971,* Vol. 3, Benfer, N. A., Coffman, J. L., Bernick, J. R., and Dees, L. T., Eds., U.S. Department of Commerce, Washington, D.C., 1973, 123.

507. **Savage, J. C. and Hastie, L. M.,** Surface deformation associated with dip-slip faulting, *J. Geophys. Res.,* 71, 4897, 1966.

508. **Bolt, B. A.,** Elastic waves in the vicinity of the earthquake source, in *Earthquake Engineering,* Wiegel, R. L., Ed., Prentice-Hall, Englewood Cliffs, N.J., 1970, 1.

509. Office of Emergency Preparedness, Disaster protection — earthquakes, in *Disaster Preparedness,* Vol. 1, U.S. Government Printing Office, Washington, D.C., 1972, 73.

510. **Steinbrugge, K. V. and Cloud, W. K.,** Epicentral intensities and damage in the Hebgen Lake, Montana, earthquake of August 17, 1959, *Bull. Seismol. Soc. Am.,* 52, 181, 1962.

511. **Ambraseys, N. N.,** The Buyin-Zara (Iran) earthquake of September 1962 — a field report, *Bull. Seismol. Soc. Am.,* 53, 705, 1963.

512. **Miller, E. A.,** Geologic and soil conditions, in *The Santa Rosa, California, Earthquakes of October 1, 1969,* U.S. Government Printing Office, Washington, D.C., 1970, 47.

513. **Page, R. A., Boore, D. M., and Dieterich, J. H.,** Estimation of bedrock motion at the ground surface, in Studies for Seismic Zonation of the San Francisco Bay Region, Borcherdt, R. D., Ed., Geological Survey Professional Paper 941-A, U.S. Government Printing Office, Washington, D.C., 1975, 31.

514. **Nicols, D. R.,** General information and glossary — seismic hazards, in *The Seismic Safety Study for the General Plan — Tri-Cities Seismic Safety and Environmental Resources Study,* California Council on Intergovernmental Relations, Sacramento, 1973, 66.

515. **Page, R. A., Boore, D. M., Joyner, W. B., and Coulter, H. W.,** Ground Motion Values for Use in the Seismic Design of the Trans-Alaska Pipeline System, Geological Survey Circular 672, U.S. Geological Survey, Washington, D.C., 1972.

516. **Ambraseys, N. N. and Zátopek, A.,** The Varto Ûstûkran (Anatolia) earthquake of 19 August 1966 summary of a field report, *Bull. Seismol. Soc. Am.,* 58, 47, 1968.

517. **Lawson, A. C., Chairman,** *The California Earthquake of April 18, 1906, Report of the State Earthquake Commission,* Carnegie Institution, Washington, D.C., 1908.

518. **Wood, H. O.**, Distribution of apparent intensity in San Francisco, in *The California Earthquake of April 18, 1906, Report of the State Earthquake Investigation Commission,* Carnegie Institution, Washington, D.C., 1908, 220.

519. **Omote, S.**, The relation between the earthquake damages and the structure of ground in Yokohama, *Bull. Earthquake Res. Inst. Tokyo Univ.,* 27, 13, 1949.

520. **Miyamura, S.**, Notes on the geography of earthquake damage distribution, in *Proc. 7th Pacific Sci. Congr.,* 2, 653, 1953.

521. **Ooba, S.**, Study of the relation between the subsoil conditions and the distribution of the damage percentage of wooden dwelling houses in the Province of Tōtomi in the case of the Tōnankai earthquake of December 7, 1944. *Bull. Earthquake Res. Inst. Tokyo Univ.,* 35, 201, 1957.

522. **Neumann, F.**, *Earthquake Intensity and Related Ground Motion,* Washington University Press, Seattle, 1954.

523. **Duke, C. M. and Leeds, D. J.**, Soil conditions and damage in the Mexico earthquake of July 28, 1957, *Bull. Seismol. Soc. Am.,* 49, 179, 1959.

524. **Bender, R. W.**, The Acapulco, Mexico, earthquakes of May 11 and 19, 1962, in *Proc. 3rd World Conf. on Earthquake Engineering,* Vol. 3, R. E. Owens, Government Printer, Wellington, New Zealand, 1966, 45.

525. **Steinbrugge, K. V. and Cluff, L. S.**, The Caracas, Venezuela, earthquakes of July 29, 1967, *Mineral Inf. Ser.,* 21, 3, 1968.

526. **Hanson, R. D. and Degenkolb, H. J.**, *The Venezuela Earthquake, July 29, 1967,* American Iron and Steel Institute, New York, 1969.

527. **Seed, H. B., Idriss, I. M., and Dezfulian, H.**, Relationships Between Soil Conditions and Building Damage in the Caracas Earthquake of July 29, 1967, Report EERC70-2, College of Engineering, University of California, Berkeley, 1970.

528. **Espinosa, A. F. and Algermissen, S. T.**, Ground amplification studies in areas damaged by the Caracas earthquake of July 29, 1967, *EOS Trans. Am. Geophys. Union,* 52, 284, 1971.

529. **Seed, H. B., Whitman, R. V., Dezfulian, H., Dorby, R., and Idriss, I. M.**, Soil conditions and building damage in 1967 Caracas earthquake, *Am. Soc. Civil Eng. Proc. J. Soil Mechanics and Found. Div.,* 98, 787, 1972.

530. **Espinosa, A. F. and Algermissen, S. T.**, Soil amplification studies in areas damaged by the Caracas earthquake of July 29, 1967, in *Contributions to Seismic Zoning,* U.S. Government Printing Office, Washington, D.C., 1973, 65.

531. **Gordan, D. W., Bennett, T. J., Herrmann, R. B., and Rogers, A. M.**, The south-central Illinois earthquake of November 9, 1968: macroseismic studies, *Bull. Seismol. Soc. Am.,* 60, 953, 1970.

532. **Tezcan, S. S. and Ipek, M.**, Long distance effects of the 28 March 1970 Gedis, Turkey earthquake, *Int. J. Earthquake Eng. Struct. Dynamics,* 1, 203, 1973.

533. **Slossen, J. E.**, Damage to residential housing by the earthquake of February 9, 1971, *Geol. Soc. Am. Abstr. with Programs,* 4, 238, 1972.

534. **Humphrey, R. L.**, The effects of the earthquake and fire on various structures and structural materials, in The San Francisco Earthquake and Fire of April 18, 1906, Geological Survey Bulletin 324, U.S. Government Printing Office, Washington, D.C., 1907, 14.

535. **Borcherdt, R. D., Joyner, W. B., Warrick, R. E., and Gibbs, J. F.**, Response of local geologic units to ground shaking, in Studies for Seismic Zonation of the San Francisco Bay Region, Borcherdt, R. D., Ed., Geological Survey Professional Paper 941-A, U.S. Government Printing Office, Washington, D.C., 1975, 52.

536. **Dezfulian, H. and Seed, H. B.**, Seismic response of soil deposits underlain by sloping rock boundaries, *J. Soil Mech. Found. Div. Am. Soc. Civ. Eng.,* Proceedings Paper 7659, 96, 1893, 1970.

537. **Murphy, J. R. and Hewlett, R. A.**, Analysis of seismic response in the city of Las Vegas, Nevada: a preliminary microzonation, *Bull. Seismol. Soc. Am.,* 65, 1575, 1975.

538. **Oldham, R. D.**, Report on the great earthquake of 12th June 1897, *Mem. Geol. Survey India,* 29, 1, 1899.

539. **Granilla, V. P. and Callaghan, E.**, The earthquake of December 20, 1932, at Cedar Mountain, Nevada, and its bearing on the genesis of Basin Range structure, *J. Geol.,* 42, 1, 1934.

540. **Hadley, J. B.**, Landslides and related phenomena accompanying the Hebgen Lake earthquake of August 17, 1959, in The Hebgen Lake, Montana, Earthquake of August 17, 1959, Geological Survey Professional Paper 435, U.S. Government Printing Office, Washington, D.C., 1964, 107.

541. **Nason, R. D.**, Shattered earth at Wallaby Street, Sylmar, in The San Fernando, California, Earthquake of February 9, 1971, Geological Survey Professional Paper 733, U.S. Government Printing Office, Washington, D.C., 1971, 97.

542. **Barrows, A. G., Kahle, J. E., Weber, F. H., Jr., and Saul, R. B.**, Map of Surface Breaks Resulting from the San Fernando, California, Earthquake of February 9, 1971, Preliminary Report II, Plate I, California Division of Mines and Geology, Sacramento, 1971.

543. **Barrows, A. G., Kahle, J. E., Weber, F. H., Jr., and Saul, R. B.**, Map of surface breaks resulting from the San Fernando, California, earthquake of February 9, 1971, in *San Fernando, California, Earthquake of February 9, 1971,* Vol. 3, U.S. Department of Commerce, Washington, D.C., 1973, 127.

544. **Trifunac, M. D. and Hudson, D. E.**, Analysis of Pacoima Dam accelerogram, in *San Fernando, California, Earthquake of February 9, 1971,* Vol. 3, Benfer, N. A., Coffman, J. L., Bernick, J. R., and Dees, L. T., Eds., U.S. Department of Commerce, Washington, D.C., 1973, 375.

545. **Maley, R. P. and Cloud, W. K.**, Preliminary strong-motion results from the San Fernando earthquake of February 9, 1971, in The San Fernando, California, Earthquake of February 9, 1971, Geological Survey Professional Paper 733, U.S. Government Printing Office, Washington, D.C., 1971, 63.

546. **Bouchon, M.**, Effect of topography on surface motion, *Bull. Seismol. Soc. Am.,* 63, 615, 1973.

547. **Davis, L. L. and West, L. R.**, Observed effects of topography on ground motion, *Bull. Seismol. Soc. Am.,* 63, 283, 1973.

548. **Eckel, E. B.**, The Alaska Earthquake, March 27, 1964: Lessons and Conclusions, Geological Survey Professional Paper 546, U.S. Government Printing Office, Washington, D.C., 1970.

549. **Heck, N. H.**, *Earthquakes,* Hafner, New York, 1965.

550. **Hansen, W. R.**, Effects of the earthquake of March 27, 1964, at Anchorage, Alaska, in The Alaska Earthquake, March 27, 1964 — Effects on Communities, Geological Survey Professional Paper 542, U.S. Government Printing Office, Washington, D.C., 1965, A1.

551. **Terzaghi, K.**, Mechanism of landslides, in *Application of Geology to Engineering Practice,* Geological Society of America, New York, 1950, 83.

552. **Nilsen, T. H. and Brabb, E. E.**, Landslides, in Studies for the Seismic Zonation of the San Francisco Bay Region, Borcherdt, R. D., Ed., Geological Survey Professional Paper 941-A, U.S. Government Printing Office, Washington, D.C., 1975, A75.

553. Office of Emergency Preparedness, Landslides, in *Disaster Preparedness,* Vol. 1, U.S. Government Printing Office, Washington, D.C., 1972, 87.

554. **Youd, T. L.**, Liquefaction, Flow, and Associated Ground Failure, Geological Survey Circular 688, U.S. Geological Survey, Washington, D. C., 1973.

555. **Youd, T. L.**, Liquefaction, *Earthquake Inf. Bull.,* 5, 10, 1973.

556. **Rogers, T. H.**, Bear Valley-Melendy Ranch earthquakes of 24—27 February 1972, *Calif. Geol.,* 25, 138, 1972.

557. **Nolan, T. B.**, Foreword, in The Hebgen Lake, Montana, Earthquake of August 17, 1959, Geological Survey Professional Paper 435, U.S. Government Printing Office, Washington, D.C., 1964, v.

558. **Lemke, R. W.**, Effects of the earthquake of March 27, 1964, at Seward, Alaska, in The Alaska Earthquake, March 27, 1964: Effects on Communities, Geological Survey Professional Paper 542, U.S. Government Printing Office, Washington, D.C., 1969, E1.

559. **Coulter, H. W. and Migliaccio, R. R.**, Effects of the earthquake of March 27, 1964, at Valdez, Alaska, in The Alaska Earthquake, March 27, 1964: Effects on Communities, Geological Survey Professional Paper 542, U.S. Government Printing Office, Washington, D.C., 1969, C1.

560. **Plafker, G., Ericksen, G. E., and Concha, J. F.**, Geological aspects of the May 31, 1970, Peru earthquake, *Bull. Seismol. Soc. Am.,* 61, 543, 1971.

561. **Cluff, L. S.**, Peru earthquake of May 31, 1970; engineering geology observations, *Bull. Seismol. Soc. Am.,* 61, 511, 1971.

562. **Witkind, I. J.**, Events on the night of August 17, 1959 — the human story, in The Hebgen Lake, Montana, Earthquake of August 17, 1959, Geological Survey Professional Paper 435, U.S. Government Printing Office, Washington, D.C., 1964, 1.

563. **Kachadoorian, R.**, Effects of the earthquake of March 27, 1964 at Whittier, Alaska, in The Alaska Earthquake, March 27, 1964: Effects on Communities, Geological Survey Professional Paper 542, U.S. Government Printing Office, Washington, D.C., 1969, B1.

564. **Miller, D. J.**, Giant waves in Lituya Bay, Alaska, in Shorter Contributions to General Geology 1959, Geological Survey Professional Paper 354, U.S. Government Printing Office, Washington, D.C., 1961, 51.

565. **Stanley, K. W.**, Beach changes on Homer Spit, in The Alaska Earthquake, March 27, 1964: Effects on Communities, Geological Survey Professional Paper 542, U.S. Government Printing Office, Washington, D.C., 1969, D20.

566. **Waller, R. M.**, Effects of the earthquake of March 27, 1964, in the Homer area, Alaska, in The Alaska Earthquake, March 27, 1964: Effects on Communities, Geological Survey Professional Paper 542, U.S. Government Printing Office, Washington, D.C., 1969, D1.

567. **Ferrians, O. J., Jr.**, Effects of the Earthquake of March 27, 1964 in the Copper River Basin Area, Alaska, Geological Survey Professional Paper 543, U.S. Government Printing Office, Washington, D.C., 1966.

568. **Tuthill, S. J. and Laird, W. M.,** Geomorphic Effects of the Earthquake of March 27, 1964 in the Martin-Bering Rivers Area, Alaska, Geological Survey Professional Paper 543-B, U.S. Government Printing Office, Washington, D.C., 1966.

569. **Seed, H. B.,** Soil problems and soil behavior, in *Earthquake Engineering,* Wiegel, R. L., Ed., Prentice-Hall, Englewood Cliffs, N.J., 1970, 227.

570. **Numata, R.,** Pillars of a bridge projected above ground level by the shock of the Kwanto earthquake, *Reikishi-chiri,* 43, 197, 1925.

571. **Kirov, K.,** Contribution to the study of the earthquakes of 14 and 18 April 1928 in south Bulgaria, *Proc. Bulg. Acad. Sci.,* 29, 1935.

572. **Jankof, K.,** Changes in ground level produced by the earthquakes of April 14 and 18, 1928, in southern Bulgaria, in *Tremblements de Terre en Bulgarie,* Institut Meteorologique Central de Bulgarie, Sofia, 1945, 131.

573. **Gee, E.,** The Dhubri earthquake of 3rd July 1930, *Mem. Geol. Surv. India,* 65, 12, 1934.

574. **Lister-Jackson, A.,** Report, Eastern Bengal Railway, *Mem. Geol. Surv. India,* 65, 99, 1934.

575. **Poddar, M.,** Preliminary report of the Assam earthquake 15 August 1950, *Bull. Geol. Surv. India Ser. B.,* 2, 13, 1952.

576. **Steinbrugge, K. V. and Flores, R.,** A strucutral engineering viewpoint, *Bull. Seismol. Soc. Am.,* 53, 225, 1963.

577. **Duke, E. M. and Leeds, D. J.,** Response of soils, foundations, and earth structures, *Bull. Seismol. Soc. Am.,* 53, 309, 1963.

578. **Kawakami, F. and Asada, A.,** Damage to the ground and earth structures by the Niigata earthquake, *Soil & Foundation,* 6, 18, 1966.

579. **Yamada, G.,** Damage to earth structures and foundations by the Niigata earthquake, *Soil & Foundation,* 6, 4, 1966.

580. **Seed, H. B. and Idriss, I. M.,** Analysis of soil liquefaction: Niigata earthquake, *Am. Soc. Civil Eng. Proc. J. Soil Mech. Found. Div.,* 94, 83, 1967.

581. **Kawasumi, H.,** Ed., *General Report on the Niigata Earthquake of 1964,* Tokyo Electrical Engineering College Press, Tokyo, 1968.

582. **Ambraseys, N. N. and Sarma, S.,** Liquefaction of soils induced by earthquakes, *Bull. Seismol. Soc. Am.,* 59, 651, 1969.

583. **Housner, G. W.,** The mechanism of sandblows, *Bull. Seismol. Soc. Am.,* 48, 155, 1958.

584. **Spaeth, M. G.,** *Communication Plan for Tsunami Warning System,* 8th ed., U.S. Department of Commerce, National Oceanic and Atmospheric Administration, Silver Spring, Md., 1975.

585. **Wiegel, R. L.,** Tsunamis, in *Earthquake Engineering,* Wiegel, R. L., Ed., Prentice-Hall, Englewood Cliffs, N.J., 1970, 253.

586. Office of Emergency Preparedness, Tsunamis, in *Disaster Preparedness,* Vol. 3, U.S. Government Printing Office, Washington, D.C., 1972, 99.

587. National Oceanic and Atmospheric Administration, *Tsunami: the Great Waves,* U.S. Government Printing Office, Washington, D.C., 1975.

588. **Lamb, H.,** *Hydrodynamics,* 6th ed., Cambridge University Press, Cambridge, 1932.

589. Office of Emergency Preparedness, Tsunamis, in *Disaster Preparedness,* Vol. 1, U.S. Government Printing Office, Washington, D.C., 1972, 91.

590. **Fraser, G. D., Eaton, J. P., and Wentworth, C. K.,** The tsunami of March 9, 1957, on the island of Hawaii, *Bull. Seismol. Soc. Am.,* 49, 79, 1959.

591. **Cox, D. C. and Mink, J. F.,** The tsunami of 23 May 1960 in the Hawaiian Islands, *Bull. Seismol. Soc. Am.,* 53, 1191, 1963.

592. **Hwang, L.-S., Butler, H. L., and Divoky, D. J.,** Tsunami model: generation and open-sea characteristics, *Bull. Seismol. Soc. Am.,* 62, 1579, 1972.

593. **Heck, N. H.,** Japanese earthquakes, *Bull. Seismol. Soc. Am.,* 34, 117, 1944.

594. **Macdonald, G. A., Shepard, F. P., and Cox, D. C.,** The tsunami of April 1, 1946, in the Hawaiian Islands, *Pac. Sci.,* 1, 21, 1947.

595. **Cox, D. C.,** Variation of intensity of the 1964 tsunami on Hawaiian shores, *Proc. Hawaii. Acad. Sci.,* 22, 8, 1947.

596. **Shepard, F. P., Macdonald, G. A., and Cox, D. C.,** Tsunami of April 1, 1946, *Bull. Scripps Inst. Oceanogr.,* 5, 391, 1950.

597. **Macdonald, G. A. and Wentworth, C. K.,** The tsunami of November 1, 1952, on the island of Hawaii, *Bull. Seismol. Soc. Am.,* 44, 463, 1954.

598. **Wiegel, R. L.,** *Oceanographical Engineering,* Prentice-Hall, Englewood Cliffs, N.J., 1964.

599. **Heck, N. H.,** List of seismic sea waves, *Bull. Seismol. Soc. Am.,* 37, 269, 1947.

600. **Briggs, P.,** *Will California Fall into the Sea?,* David McKay, New York, 1972.

601. **Nakano, T., Kadamura, H., Mizutani, T., Okuda, M., and Sekiguchi, T.,** Natural hazards: report from Japan, in *Natural Hazards,* White, G. F., Ed., Oxford University Press, New York, 1974, 231.

602. **Spaeth, M. G. and Berkman, S. C.,** The Tsunami of March 28, 1964, as Recorded at Tide Stations, Environmental Science Services Administration, Tech. Report C&GS 33, U.S. Government Printing Office, Washington, D.C., 1967.

603. **Tilling, R. I., Koyanagi, R. Y., Lipman, P. W., Lockwood, J. P., Moore, J. G., and Swanson, D. W.,** Earthquake and Related Catastrophic Events Island of Hawaii, November 29, 1975: A Preliminary Report, Geological Survey Circular 740, U.S. Geological Survey, Reston, Va., 1976.

604. **Powers, H. A.,** The Aleutian tsunami at Hilo, Hawaii, April 1, 1946, *Bull. Seismol. Soc. Am.,* 36, 355, 1946.

605. **Kachadoorian, R. and Plafker, G.,** Effect of the earthquake of March 27, 1964 on the communities of Kodiak and nearby islands, in The Alaska Earthquake, March 27, 1964: Effects on Communities, Geological Survey Professional Paper 542, U.S. Government Printing Office, Washington, D.C., 1969, F1.

606. National Weather Service, *Tsunami Watch and Warning,* U.S. Department of Commerce, Washington, D.C., 1975.

607. National Earthquake Information Center, At Palmer Seismological Observatory the countdown ends at earthquake + 15, *Earthquake Inf. Bull.,* 3, 20, 1971.

608. **Wadati, K., Hirono, T., and Hisamoto, S.** On the Tsunami Warning Service in Japan, in *Proc. 10th Pacific Science Congr.,* L'Institut Géographique National, Paris, 1963, 138.

609. **McCulloch, D. S.,** Slide-Induced Waves, Seiching, and Ground Fracturing Caused by the Earthquake of March 27, 1964, at Kenai Lake, Alaska, Geological Survey Professional Paper 543-A, U.S. Government Printing Office, Washington, D.C., 1966.

610. **Forel, F. A.,** *Le Leman-Monographie Limnologique,* F. Rouge, Lausanne, Switzerland, 1895.

611. **McGarr, A. and Vorhis, R. C.,** Seismic seiches from the March 1964 Alaska earthquake, in The Alaska Earthquake, March 27, 1964: Effects on the Hydrologic Regimen, Geological Survey Professional Paper 544, U.S. Government Printing Office, Washington, D.C., 1968, E1.

612. **Kvale, A.,** Seismic seiches in Norway and England during the Assam earthquake of August 15, 1950, *Bull. Seismol. Soc. Am.,* 45, 93, 1955.

613. **McGarr, A.,** Excitation of seiches in channels by seismic waves, *J. Geophys. Res.,* 70, 847, 1965.

614. National Earthquake Information Service, Seismic seiches, *Earthquake Inf. Bull.,* 8, 13, 1976.

615. **Plafker, G.,** *Tectonics of the March 27, 1964 Alaska Earthquake,* Geological Survey Professional Paper 543-I, U.S. Government Printing Office, Washington, D.C., 1969.

616. **Davison, C.,** *Great Earthquakes,* Murby, London, 1936.

617. **Stermitz, F.,** Effects of the Hebgen Lake earthquake on surface water, in Effects of the Hebgen Lake, Montana, Earthquake of August 17, 1959, Geological Survey Professional Paper 435, U.S. Government Printing Office, Washington, D.C., 1964, 139.

618. **Donn, W. L.,** Alaskan earthquake of 27 March 1964 — remote seiche stimulation, *Science,* 145, 261, 1964.

619. **Miller, W. D. and Reddell, D. L.,** Alaskan earthquake damages Texas High Plains water wells, *EOS Trans. Am. Geophys. Union,* 45, 659, 1964.

620. **Waller, R. M., Thomas, H. E., and Vorhis, R. C.,** Effects of the Good Friday earthquake on water supplies, *Am. Water Works Assoc. J.,* 57, 123, 1965.

621. **Vorhis, R. C.,** Hydrologic effects of the earthquake of March 27, 1964 outside Alaska, in The Alaska Earthquake, March 27, 1964: Effects on the Hydrologic Regimen, Geological Survey Professional Paper 544, U.S. Government Printing Office, Washington, D.C., 1968, C1.

622. **Sewell, J. S.,** The effects of the earthquake and fire on buildings, engineering structures, and structural materials, in The San Francisco Earthquake and Fire of April 18, 1906, Geological Survey Circular 324, U.S. Government Printing Office, Washington, D.C., 1907, 62.

623. **Cheminant, L. B.,** High-pressure fire protection in San Francisco, *Bull. Seismol. Soc. Am.,* 19, 80, 1929.

624. National Earthquake Information Center, The great San Francisco earthquake, *Earthquake Inf. Bull.,* 3, 15, 1971.

625. **Hanson, R. D.,** Behavior of liquid-storage tanks, in *The Great Alaska Earthquake of 1964 — Engineering,* National Academy of Sciences, Washington, D.C., 1973, 331.

626. **Steinbrugge, K. V., Schader, E. E., Bigglestone, H. C., and Weers, C. A.,** *San Fernando Earthquake, February 9, 1971,* Pacific Fire Rating Bureau, San Francisco, 1971.

627. **Rojahn, C.,** Managua, Nicaragua earthquakes of December 23, 1972, *Earthquake Inf. Bull.,* 5, 14, 1973.

628. **Meehan, J. F., Degenkolb, H. J., Morgan, D. F., and Steinbrugge, K. V.,** Engineering aspects, in *Managua, Nicaragua Earthquake of December 23, 1972,* Earthquake Engineering Research Institute, Oakland, Cal., 1973, 27.

629. Advisory Group on Disaster Preparedness, Report of the Advisory Group on Disaster Preparedness, in *Meeting the Earthquake Challenge,* California Legislature/Joint Committee on Seismic Safety, Sacramento, 1974, 57.

630. Borcherdt, R. D., Brabb, E. E., Joyner, W. B., Helley, E. J., Lajoie, K. R., Page, R. A., Wesson, R. L., and Youd, T. L., Predicted geologic effects of a postulated earthquake, in Studies for Seismic Zonation of the San Francisco Bay Region, Borcherdt, R. D., Ed., Geological Survey Professional Paper 941-A, U.S. Government Printing Office, Washington, D.C., 1975, 88.

631. Hamilton, R. M., Perspectives the future of earthquake prediction, in United States Geological Survey Annual Report, Fiscal Year 1975, U.S. Government Printing Office, Washington, D.C., 1976, 7.

632. Press, F., Haiching and Los Angeles: a tale of two cities, EOS Trans. Am. Geophys. Union, 57, 435, 1976.

633. Panel on Earthquake Prediction (Allen, C. R.-Chairman, Edwards, W., Hall, W. J., Knopoff, L., Raleigh, C. B., Savit, C. H., Sykes, L. R., Toksöz, M. N., and Turner, R. H.), Predicting Earthquakes — A Scientific and Technical Evaluation With Implications for Society, National Academy of Sciences, Washington, D.C., 1976.

634. Page, R., The Sitka, Alaska, earthquake of 1972, an expected visitor, Earthquake Inf. Bull., 5, 4, 1973.

635. Sykes, L. R., Seismicity as a guide to global·tectonics and earthquake prediction, Tectonophysics, 13, 393, 1972.

636. Fedotov, S. A., Regularities of the distribution of strong earthquakes of Kamchatka, the Kurile Islands, and northeastern Japan, Tr. Inst. Fiz. Zemli Akad. Nauk SSSR, 36, 66, 1965.

637. Kelleher, J., Sykes, L., and Oliver, J., Possible criteria for predicting earthquake locations and their application of major plate boundaries of the Pacific and Caribbean, J. Geophys. Res. 78, 2547, 1973.

638. Allen, C. R., Saint-Amand, P., Richter, C. F., and Nordquist, J. M., Relationship between seismicity and geologic structure in the southern California region, Bull. Seismol. Soc. Am., 55, 753, 1965.

639. Tobin, D. and Sykes, L. R., Seismicity and tectonics of the northeast Pacific Ocean, J. Geophys. Res., 73, 3821, 1968.

640. Mogi, K., Some features of recent seismic activity in and near Japan, I., Bull. Earthquake Res. Inst., Tokyo Univ., 46, 1225, 1968.

641. Mogi, K., Some features of recent seismic activity in and near Japan. II. Activity before and after great earthquakes, Bull. Earthquake Res. Inst., Tokyo Univ., 47, 395, 1969.

642. Utsu, T., Large earthquakes near Hokkaido and the expectancy of the occurrence of a large earthquake off Nemuro, Rep. Coordinating Committee Earthquake Prediction, 7, 7, 1972.

643. Ando, M., Possibility of a major earthquake in the Tokai district, Japan and its pre-estimated seismotectonic effects, Tectonophysics, 25, 69, 1975.

644. Shimazaki, K., Nemuro-Oki earthquake of June 17, 1973: a lithospheric rebound at the upper half of an interface, Phys. Earth Planet. Interiors, 9, 314, 1974.

645. Sykes, L. R., Aftershock zones of great earthquakes, seismicity gaps and earthquake prediction for Alaska and the Aleutians, J. Geophys. Res., 76, 8021, 1971.

646. Kelleher, J. A. and Savino, J. M., Seismic preconditions of the 1972 Sitka earthquake and several other large earthquakes, EOS Trans. Am. Geophys. Union, 54, 370, 1973.

647. Allen, C., The tectonic environments of seismically active and inactive areas along the San Andreas fault system, Stanford Univ. Publ. Geol. Sci., 11, 70, 1968.

648. Scholz, C. H., Molnar, P., and Johnston, T., Detailed studies of frictional sliding of granite and implications for the earthquake mechanism, J. Geophys. Res., 77, 6392, 1972.

649. Kelleher, J. A., Space-time seismicity of the Alaska-Aleutian seismic zone, J. Geophys. Res., 75, 5745, 1970.

650. Kelleher, J. A., Rupture zones of large South American earthquakes and some predictions, J. Geophys. Res., 77, 2087, 1972.

651. Anderson, D. L., Accelerated plate motions, Science, 187, 1077, 1975.

652. Mogi, K., Sequential occurrences of recent great earthquakes, J. Phys. Earth, 16, 30, 1968.

653. Mogi, K., Migration of seismic activity, Bull. Earthquake Res. Inst., Tokyo Univ., 46, 53, 1968.

654. Dewey, J. W., Turkey's North Anatolian fault . . . a comparison with the San Andreas fault, Earthquake Inf. Bull., 6, 12, 1974.

655. Dewey, J. W., Seismicity of Northern Anatolia, Bull. Seismol. Soc. Am., 66, 843, 1976.

656. Savage, J. C., A theory of creep waves propagating along a transform fault, J. Geophys. Res., 76, 1954. 1971.

657. Nikonov, A. A., The migration of strong earthquakes along major fault zones in Soviet Central Asia, Dokl. Acad. Sci. USSR Earth Sci. Sect., 225, 7, 1975.

658. Nikonov, A. A., Migration of large earthquakes along the great fault zones in Middle Asia, Tectonophysics, 31, T55, 1976.

659. Ellsworth, W. L. and Wesson, R. L., Earthquake prediction along the San Andreas fault south of Paicines, California, EOS Trans. Am. Geophys. Union, 54, 371, 1973.

660. Thatcher, W., Hillman, J. A., and Hanks, T. C., Seismic slip distribution along the San Jacinto fault zone, southern California, and its implications, Geol. Soc. Am. Bull., 86, 1140, 1975.

661. **Gelfand, I. M., Guberman, S. I., Izvekova, M. L., Keilis-Borok, V. I., and Ranzman, E. J.,** Criteria of high seismicity, determined by pattern recognition, *Tectonophysics,* 13, 415, 1972.

662. **Press, F. and Briggs, P.,** Pattern recognition applied to earthquake epicenters in California and Nevada, *EOS Trans. Am. Geophys. Union,* 56, 1150, 1974.

663. **Wesson, R. L., Burford, R. O., and Ellsworth, W. L.,** A conceptual model of seismicity and fault creep along the central San Andreas fault, in *U.S.-Japan Seminar on Earthquake Prediction and Control,* American Geophysical Union, Washington, D.C., 1974, microfiche, 39.

664. **Bufe, C. G., Harsh, P. W., and Burford, R. O.,** Steady-state seismic slip — a precise recurrence model, *EOS Trans. Am. Geophys. Union,* 57, 758, 1976.

665. **Bufe, C. G., Harsh, P. W., and Burford, R. O.,** Steady-state seismic slip — a precise recurrence model, *Geophys. Res. Lett.,* 4, 91, 1977.

666. **Bufe, C. G.,** personal communication, 1977.

667. **Nason, R. D. and Tocher, D.,** Anomalous fault slip rates before and after the April 1961 earthquakes near Hollister, California, *EOS Trans. Am. Geophys. Union,* 52, 278, 1971.

668. **Nason, R. D. and Burford, R. O.,** Pre-earthquake fault creep event at Melendy Ranch, California, *Earthquake Notes,* 43, 25, 1972.

669. **Johnson, T. L.,** A comparison of frictional sliding on granite and dunite surfaces, *J. Geophys. Res.,* 80, 2600, 1975.

670. **Suyehiro, S. and Sekiya, H.,** Foreshocks and earthquake prediction, *Tectonophysics,* 14, 219, 1972.

671. **Suyehiro, S.,** Difference between aftershocks and foreshocks in the relationship of magnitude to frequency of occurrence for the great Chilean earthquake of 1960, *Bull. Seismol. Soc. Am.,* 56, 185, 1966.

671a. **Van Wormer, J. D., Gedney, L. D., Davies, J. N., and Condal, N.,** V_P/V_s and b-values: a test of the dilatancy model for earthquake precursors, *Geophys. Res. Lett.,* 2, 514, 1975.

672. **Tocher, D.,** Seismic history of the San Francisco Bay region, in San Francisco Earthquakes of March 1957, Special Report 57, California Division of Mines and Geology, Sacramento, 1959, 39.

673. **Tobin, D. G. and Sykes, L. R.,** Relationship of hypocenters of earthquakes to the geology of Alaska, *J. Geophys. Res.,* 71, 1659, 1966.

674. **Kelleher, J. and Savino, J.,** Distribution of seismicity before large strike-slip and thrust-type earthquakes, *J. Geophys. Res.,* 80, 260, 1975.

675. **Sykes, L. R. and Raleigh, C. B.,** Earthquake research in China. VII. Premonitory effects of earthquakes, *EOS Trans. Am. Geophys. Union,* 56, 858, 1975.

676. **Ohtake, M., Matumoto, T., and Latham, G. V.,** Temporal changes in seismicity preceding some shallow earthquakes in Mexico and South America, *EOS Trans. Am. Geophys. Union,* 57, 959, 1976.

677. **Wesson, R. L. and Ellsworth, W. L.,** Seismicity preceding moderate earthquakes in California, *J. Geophys. Res.,* 78, 8527, 1973.

678. **Sadovsky, M. A., Nersesov, I. L., Nigmatullaev, S. K., Latynina, L. A., Lukk, A. A., Semenov, A. N., Simbireva, I. G., and Ulomov, V. I.,** The processes preceding strong earthquakes in some regions of Middle Asia, *Tectonophysics,* 14, 295, 1972.

679. **Hagiwara, T.,** Summary of the seismographic observation of the Matsushiro swarm earthquakes, *EOS Trans. Am. Geophys. Union.,* 50, 391, 1969.

680. **Sykes, L. R.,** Earthquake-prediction research outside the United States, in *Predicting Earthquakes — A Scientific and Technical Evaluation With Implications for Society,* National Academy of Sciences, Washington, D.C., 1976, 51.

681. **Lensen, G. J.,** Earth deformation and earthquake prediction, *Bull. New Zealand Soc. Earthquake Eng.,* 2, 444, 1969.

682. **California Division of Mines and Geology,** Prediction data from tiltmeters, *Calif. Geol.,* 25, 161, 1972.

683. **Bacon, C. F., Bennett, J. H., Chase, G. W., Rodgers, D. A., Toppozada, T. R., and Wells, W. M.,** Crustal movement studies near Parkfield, California, *Calif. Geol.,* 28, 202, 1975.

684. **Tsubokawa, J., Ogawa, Y., and Hayashi, T.,** Crustal movements before and after the Niigata earthquake, *J. Geod. Soc.,* (Japan), 10, 165, 1965.

685. **Boulanger, Y. D., Fevnev, A. K., Enman, V. P., Atrushkevick, P. A., and Antonenko, E. M.,** Geodetic studies and forerunners of earthquakes, *Tectonophysics,* 14, 183, 1972.

686. **Castle, R. O., Alt, J. N., Savage, J. C., and Balazs, E. I.,** Elevation changes preceding the San Fernando earthquake of February 9, 1971, *Geology,* 2, 61, 1974.

687. **Castle, R. O., Church, J. P., Elliott, M. R., and Morrison, N. L.,** Vertical crustal movements preceding and accompanying the San Fernando earthquake of February 9, 1971: a summary, *Tectonophysics,* 29, 127, 1975.

688. **Hagiwara, T. and Rikitake, T.,** Japanese program on earthquake prediction, *Science,* 157, 761, 1967.

689. **Wood, M. D. and Allen, R. V.,** Anomalous microtilt preceding a local earthquake, *Bull. Seismol. Soc. Am.,* 61, 1801, 1971.

690. **Johnston, M. J. S. and Mortensen, C. E.,** Tilt precursors before earthquakes on the San Andreas fault, California, *Science,* 186, 1031, 1974.

691. U.S. Geological Survey, Systematic changes in surface tilt observed before earthquakes near the San Andreas fault, California, *Earthquake Inf. Bull.,* 6, 16, 1974.

692. **Stuart, W. D. and Johnston, M. J. S.,** Anomalous tilt before three recent earthquakes, *EOS Trans. Am. Geophys. Union,* 56, 400, 1975.

693. **Mortensen, C. E. and Johnston, M. J. S.,** Anomalous tilt preceding the Hollister earthquake of November 28, 1974, *J. Geophys. Res.,* 81, 3561, 1976.

694. **Alexander, G.,** Scientists said quake would occur and it did, *Los Angeles Times,* January 8, 1977.

695. **Johnston, M. J. S.,** personal communication,1977.

696. **Myren, G. D. and Johnston, M. J. S.,** Preliminary results from a tiltmeter array around the Los Angeles Basin and along the San Jacinto fault, *EOS Trans. Am. Geophys. Union,* 56, 1059, 1975.

697. **Buckley, C. P., Maloney, N. J., Wright, N. A., and Kohlenberger, C. W.,** Variations of crustal tilt patterns preceding southern California earthquakes of 4.0 magnitude or larger, *EOS Trans. Am. Geophys. Union,* 56, 1059, 1975.

698. **Maloney, N. J., Buckley, C. P., Wright, N. A., and Kohlenberger, W.,** Association of earth tilt percursors to earthquakes, *EOS Trans. Am. Geophys. Union,* 56, 1059, 1975.

699. **Sylvester, A. G. and Pollard, D. D.,** Observations of crustal-tilt preceding an aftershock at San Fernando, California, *Bull. Seismol. Soc. Am.,* 62, 927, 1972.

700. **Castle, R. O., Church, J. P., and Elliott, M. R.,** Aseismic uplift in southern California, *Science,* 192, 251, 1976.

701. U.S. Geological Survey, Land swelling in southern California discovered, *Earthquake Inf. Bull.,* 8, 14, 1976.

702. **Real, C. R. and Bennett, J. H.,** Palmdale bulge, *Calif. Geol.,* 29, 171, 1976.

703. U.S. Geological Survey, Palmdale 'Bulge' Changing Shape; Still a Geological Puzzle, news release, U.S. Department of the Interior, Washington, D.C., March 3, 1977.

704. **Brace, W. F. and Orange, A. S.,** Electrical resistivity changes in saturated rock under stress, *Science,* 153, 1525, 1966.

705. **Brace, W. F. and Orange, A. S.,** Electrical resistivity changes in saturated rocks during fracture and frictional sliding, *J. Geophys. Res.,* 73, 1433, 1968.

706. **Brace, W. F. and Orange, A. S.,** Further studies of the effects of pressure on electrical resistivity of rocks, *J. Geophys. Res.,* 73, 5407, 1968.

707. **Barsukov, O. M. and Sorokin, O. H.,** Variations in apparent resistivity of rocks in the seismically active Garm region, Izv. Phys. Solid Earth, 10, 685, 1973.

708. **Bolt, B. A. and Wang, C. Y.,** The present status of earthquake prediction, *Crit. Rev. Solid State Sci.,* 5, 125, 1975.

709. **Sumner, J. S.,** *Geophysics, Geologic Structures and Tectonics,* Wm. C. Brown, Dubuque, Iowa, 1969.

710. **Fedotov, S. A., Sobolev, G. A., Boldyrev, S. A., Gusev, A. A., Kondratenko, A. M., Potapova, O. V., Slavina, L. B., Theophylatov, V. D., Khramov, A. A., and Shirokov, V. A.,** Long- and short-term earthquake prediction in Kamchatka, *Tectonophysics,* 37, 305, 1977.

711. **Honkura, Y., Niblett, E. R., and Kurtz, R. D.,** Changes in magnetic and telluric fields in a seismically active region of eastern Canada preliminary results of earthquake prediction studies, *Tectonophysics,* 34, 219, 1976.

712. **Reddy, I. K., Phillips, R. J., Whitcomb, J. H., and Cole, D. M.,** Magnetotellurics in earthquake prediction, *EOS Trans. Am. Geophys. Union,* 56, 1060, 1975.

713. **Barsukov, O. M.,** Variations of electric resistivity of mountain rocks connected with tectonic causes, *Tectonophysics,* 14, 273, 1972.

714. **Mazzella, A. and Morrison, H. F.,** Electrical resistivity variations associated with earthquakes on the San Andreas fault, *Science,* 185, 855, 1974.

715. **Corwin, R. F., Morrison, H. F., and Chang, M. Y.,** Monitoring earthquake-related variations of electrical resistivity and self-potential on the San Andreas fault, *EOS Trans. Am. Geophys. Union,* 57, 288, 1976.

716. **Rikitake, T. and Yamazaki, Y.,** Strain steps observed by a resistivity variometer, *Tectonophysics,* 9, 197, 1970.

717. **Yamazaki, Y.,** Coseismic resistivity steps, *Tectonophysics,* 22, 159, 1974.

718. **Yamazaki, Y. and Rikitake, T.,** Precursory and coseismic resistivity changes, *EOS Trans. Am. Geophys. Union,* 55, 1112, 1974.

719. **Rikitake, T.,** Geomagnetism and earthquake prediction, *Tectonophysics,* 6, 59, 1968.

720. **Moore, G. W.,** Magnetic disturbances preceding the 1964 Alaska earthquake, *Science,* 203, 508, 1964.

721. **Wilson, E.,** On the susceptibility of feebly magnetic bodies as affected by compression, *Proc. R. Soc. London, Ser. A,* 101, 445, 1922.

722. **Kalashinikov, A. C.,** The possible application of magnetometric methods to the question of earthquake indications, *Tr. Geofiz. Inst. Akad. Nauk SSSR,* 25, 162, 1954.

723. **Kapitsa, S. P.,** Magnetic properties of eruptive rocks exposed to mechanical stresses, *Izv. Akad. Nauk SSSR Ser. Geol.,* 6, 489, 1955.

724. **Nagata, T. and Kinoshita, H.,** Studies of piezo-magnetization I. Magnetization of titaniferous magnetite under uniaxial compression, *J. Geomagn. Geoelectric.,* 17, 121, 1965.

725. **Ohnaka, M. and Kinoshita, H.,** Effect of uniaxial compression on remanent magnetization, *J. Geomagn. Geoelectr.,* 29, 93, 1968.

726. **Kern, J. W.,** Effect of stress on the susceptibility and magnetization of a partially magnetized multi-domain system, *J. Geophys. Res.,* 66, 3807, 1961.

727. **Stacey, F. D.,** Theory of the magnetic susceptibility of stressed rocks, *Phil. Mag.,* 7, 551, 1962.

728. **Stacey, F. D.,** Seismo-magnetic effect and the possibility of forecasting earthquakes, *Nature (London),* 200, 1083, 1963.

729. **Stacey, F. D.,** The seismo-magnetic effect, *Pure Appl. Geophys.,* 58, 5, 1964.

730. **Stacey, F. D. and Westcott, P.,** Seismomagnetic effect — limit of observability imposed by local variations in geomagnetic disturbances, *Nature (London),* 206, 1209, 1965.

731. **Yukutake, T. and Tachinaka, H.,** Geomagnetic variation associated with stress change within a semi-infinite elastic earth caused by a cylindrical source, *Bull. Earthquake Res. Inst., Tokyo Univ.,* 45, 785, 1967.

732. **Nagata, T.,** Basic magnetic properties of rocks under the effects of mechanical stresses, *Tectonophysics,* 9, 167, 1970.

733. **Stacey, F. D. and Johnston, M. J. S.,** Theory of piezo-magnetic effects in titanomagnetic bearing rocks, *Pure Appl. Geophys.,* 95, 50, 1972.

734. **Breiner, S. and Kovach, R. L.,** Local geomagnetic events associated with displacements on the San Andreas fault, *Science,* 158, 116, 1967.

735. **Breiner, S. and Kovach, R. L.,** Local magnetic events associated with displacement along the San Andreas fault (California), *Tectonophysics,* 6, 69, 1968.

736. **Johnston, M. J. S., Smith, B. E., Johnston, J. R., and Williams, F. J.,** A search for tectonomagnetic effects in California and western Nevada, Proceedings of Conference on Tectonic Problems of the San Andreas Fault System, *Stanford Univ. Publ. Geol. Sci.,* 13, 225, 1973.

737. **Johnston, M. J. S.,** Preliminary results from a search for regional tectonomagnetic effects in California and western Nevada, *Tectonophysics,* 23, 267, 1974.

738. **Johnston, M. J. S., Myren, G. D., O'Hara, N. W., and Rodgers, J. H.,** A possible seismomagnetic observation on the Garlock fault, California, *Bull. Seismol. Soc. Am.,* 65, 1129, 1975.

739. **Johnston, M. J. S., Mortensen, C. E., Smith, B. E., and Stuart, W. D.,** Summary and implications of simultaneous observation of tilt and local magnetic field changes prior to a magnitude 5.2 earthquake near Hollister, California, *EOS Trans. Am. Geophys. Union,* 56, 400, 1975.

740. **Smith, B. E. and Johnston, M. J. S.,** A tectonomagnetic effect observed before a magnitude 5.2 earthquake near Hollister, California, *J. Geophys. Res.,* 81, 3556, 1976.

741. **Smith, B. E., Johnston, M. J. S., and Myren, G. D.,** Results from a differential magnetometer array along the San Andreas fault in central California, *EOS Trans. Am. Geophys. Union,* 55, 1113, 1974.

742. **Savarensky, E. F.,** Introductory remarks to the Symposium on Earthquake Prediction, *Tectonophysics,* 23, 221, 1974.

743. **Abdullabekov, K. N., Bezuglaya, L. S., Golovkov, V. P., and Skovorodkin, Y. P.,** On the possibility of using magnetic methods to study tectonic processes, *Tectonophysics,* 14, 257, 1972.

744. **Rikitake, T.,** Earthquake prediction studies in Japan, *Geophys. Surv.,* 1, 4, 1972.

745. **Rikitake, T.,** Japanese national program on earthquake prediction, *Tectonophysics,* 23, 225, 1974.

746. **Yamazaki, Y. and Rikitake, T.,** Local anomalous changes in the geomagnetic field at Matsushiro, *Bull. Earthquake Res. Inst., Tokyo Univ.,* 48, 637, 1970.

747. **Ulomov, V. I. and Mavashev, B. Z.,** Forerunner of a strong tectonic earthquake, *Dokl. Akad. Nauk Moscow SSSR,* 176, 319, 1967.

748. **Hammond, A. L.,** Earthquake predictions: breakthrough in theoretical insight? *Science,* 180, 851, 1973.

749. **King, C.-Y.,** Radon emanation along an active fault, *EOS Trans. Am. Geophys. Union,* 56, 1019, 1975.

750. **King, C.-Y.,** Anomalous radon emanation on San Andreas fault, *EOS Trans. Am. Geophys. Union,* 57, 957, 1976.

751. **King, C.-Y.,** Temporal variations in radon emanation along active faults, *EOS Trans. Am. Geophys. Union,* 58, 434, 1977.

752. **Birchard, C. F. and Libby, W. F.,** An inexpensive radon earthquake prediction concept, *EOS Trans. Am. Geophys. Union,* 57, 957, 1976.

753. **Smith, A. R., Wollenberg, H. A., and Mosier, D. F.,** Radon-222 in subsurface waters as an earthquake predictor: central California studies, *EOS Trans. Am. Geophys. Union,* 57, 957, 1976.

754. **Teng, T. L., Ku, T. L., and McElrath, R. P.,** Groundwater radon measurements along San Andreas fault from Cajon to Gorman, *EOS Trans. Am. Geophys. Union,* 56, 1019, 1975.

755. **Cathey, L.,** Continuous water borne radon monitoring using ionization chambers, *EOS Trans. Am. Geophys. Union,* 58, 434, 1977.

756. **Moore, W. S., Chiang, J. H., Talwani, P., and Stevenson, D. A.,** Earthquake prediction studies at Lake Jocassee: relation of seismicity to radon anomalies in groundwaters, *EOS Trans. Am. Geophys. Union,* 58, 434, 1977.

757. **Kuo, T.-K., Chin, P.-Y., and Feng, H. T.,** Discussion of the change of ground water level preceding a large earthquake from an earthquake source model, *Acta Geophys. Sinica,* 17, 99, 1974.

758. **Bolt, B. A.,** Earthquake studies in the People's Republic of China, *EOS Trans. Am. Geophys. Union,* 55, 108, 1974.

759. **Johnson, A. G., Kovach, R. L., Nur, A., and Booker, J.,** Pore pressure changes during creep events on the San Andreas fault, *J. Geophys. Res.,* 78, 851, 1973.

760. **Johnson, A. G., Kovach, R. L., and Nur, A.,** Fluid-pressure variations and fault creep in central California, *Tectonophysics,* 23, 258, 1974.

761. **Kovach, R. L., Nur, A., Wesson, R. L., and Robinson, R.,** Water-level fluctuations and earthquakes on the San Andreas fault zone, *Geology,* 3, 437, 1975.

762. **Kovach, R. L., Nur, A., and Wesson, R. L.,** Can a well tell?, *EOS Trans. Am. Geophys. Union,* 56, 449, 1975.

763. **Chinnery, M.,** The deformation of the ground around surface faults, *Bull. Seismol. Soc. Am.,* 51, 355, 1961.

764. **Savage, J.,** A theory of creep waves propagating along a transform fault, *J. Geophys. Res.* 76, 1954, 1971.

765. **Sundaram, P. N., Goodman, R. E., and Wang, C.-Y.,** Precursory and coseismic water-pressure variations in stick-slip experiments, *Geology,* 4, 108, 1976.

766. **Kuo, T. C.,** The change of water table preceding earthquakes, *Acta Geophys. Sinica,* 13, 223, 1964.

767. **Arieh, E. and Merzer, A. M.,** Fluctuations in oil flow before and after earthquakes, *Nature (London),* 247, 534, 1974.

768. **Wu, F. T.,** Gas well pressure fluctuations and earthquakes, *Nature (London),* 257, 661, 1975.

769. **Wadati, K.,** On the travel time of earthquake waves. II. *Geophys. Mag.,* 7, 101, 1933.

770. **Kondratenko, A. M. and Nersesov, I. L.,** Some results of the study on change in the velocity of longitudinal wave and relation between the velocities of longitudinal and transverse waves in a focal zone, *Tr. Inst. Fiz. Zemli,* 25, 130, 1962.

771. **Nersevov, I. L., Semenov, A. N., and Simbireva, I. G.,** Space-time distribution of the travel time ratios of transverse and longitudinal waves in the Garm area, in *The Physical Basis of Foreshocks,* Akad. Nauk SSSR, Moscow, 1969.

772. **Semenov, A. N.,** Variations in the traveltime of transverse and longitudinal waves before violent earthquakes, *Bull. Acad. Sci. USSR Phys. Solid Earth,* 3, 245, 1969.

773. **Aggarwal, Y. P., Sykes, L. R., Armbruster, J., and Sbar, M. L.,** Prediction of earthquakes: premonitory changes in seismic velocities prior to earthquakes in the Blue Mountain Lake (BML) area, N. Y., *EOS Trans. Am. Geophys. Union,* 53, 1041, 1972.

774. **Aggarwal, Y. P., Sykes, L. R., Armbruster, J., and Sbar, M. L.,** Premonitory changes in seismic velocities and prediction of earthquakes, *Nature (London),* 241, 101, 1973.

775. **Aggarwal, Y. P. and Sykes, L. R.,** Premonitory velocity changes, time variant stress and space-time distribution of the Blue Mountain Lake (BML) earthquakes, *EOS Trans. Am. Geophys. Union,* 54, 371, 1973.

776. **Richards, P. G. and Aggarwal, Y. P.,** Temporal fluctuations of P and S velocities in an earthquake source region, *EOS Trans. Am. Geophys. Union,* 54, 1134, 1973.

777. **Whitcomb, J. H., Garmany, J. D., and Anderson, D. L.,** Earthquake prediction: variation of seismic velocities before the San Fernando earthquake, *Science,* 180, 632, 1973.

778. **Liaw, H. B., Wu, F. T., Tsai, Y. B., and Yu, S. B.,** Monitoring a t_s/t_p anomaly in Taiwan, *EOS Trans. Am. Geophys. Union,* 56, 1018, 1975.

779. **Wyss, M. and Johnston, A. C.,** A search for teleseismic P residual changes before large earthquakes in New Zealand, *J. Geophys. Res.,* 79, 3283, 1974.

780. **Rynn, J. M. W. and Scholz, C. H.,** Regional variations of t_s/t_p in the South Island, New Zealand, *EOS Trans. Am. Geophys. Union,* 57, 289, 1976.

781. **Ohtake, M.,** Change in the V_p/V_s ratio related with occurrence of some shallow earthquakes in Japan, *J. Phys. Earth,* 21, 173, 1973.

782. **Wyss, M. and Holcomb, D. J.,** Earthquake prediction based on station residuals, *Nature (London),* 245, 139, 1973.

783. **Brown, R. and Zandt, G.,** Time variations of V_p/V_s before strike-slip events, *EOS Trans. Am. Geophys. Union,* 55, 364, 1974.

784. **Suzuki, Z.,** Results of geodetic and seismological observations before and after the southeast Akita earthquake, in *U.S.-Japan Seminar on Earthquake Prediction and Control,* American Geophysical Union, Washington, D.C., 1974, microfiche, 7.

785. **Hasegawa, A., Hasegawa, T., and Hori, S.,** Premonitory variation in seismic velocity related to the southeatern Akita earthquake of 1970, *J. Phys. Earth,* 23, 189, 1975.

786. **Nersesov, I. L., Lukk, A. A., Ponomarev, V. S., Rautian, T. G., Rulev, B. G., Semenov, A. N., and Simbireva, I. G.,** Possibilities of earthquake prediction, exemplified by the Garm area of the Tadzhik, S. S. R., *Akad. Nauk USSR Inst. Fiz. Zembli,* 72, 1973.

787. **Wyss, M.,** Precursors to the Garm earthquake of March 1969, *J. Geophys. Res.,* 80, 2926, 1975.

788. **Malone, S. D.,** Some characteristics of the micro earthquake swarms in the Columbia River Basin, *EOS Trans. Am. Geophys. Union,* 55, 347, 1974.

789. **Holcomb, D. J. and Wyss, M. C.,** Earthquake precursors from P-wave residuals, *Abstr. Progr. Geol. Soc. Am.,* 6, 296, 1974.

790. **Aggarwal, Y. P., Simpson, D. W., and Sykes, L. R.,** Temporal and spatial analysis of premonitory velocity anomalies for the Aug. 3, 1973 Blue Mt. Lake (BML) earthquake, *EOS Trans. Am. Geophys. Union,* 55, 365, 1974.

791. **Aggarwal, Y.P., Sykes, L.R., Simpson, D.W., and Richards, P.G.,** Spatial and temporal variations in t,/t, and in P wave residuals at Blue Mountain Lake, New York: application to earthquake prediction, *J. Geophys. Res.,* 80, 718, 1975.

792. **Talwani, P. and Scheffler, P.,** Anomalous velocity changes following the August 2, 1974 South Carolina earthquake, *EOS Trans. Am. Geophys. Union,* 55, 1150, 1974.

793. **Talwani, P.,** Reservoir induced earthquakes in South Carolina and t,/t, anomalies, *EOS Trans. Am. Geophys. Union,* 57, 289, 1976.

794. **Stevenson, D.A., Talwani, P., and Amick, D.C.,** Recent seismic activity near Lake Jocassee, Oconee County, South Carolina, preliminary results, and a successful earthquake prediction, *EOS Trans. Am. Geophys. Union,* 57, 290, 1976.

795. **Talwani, P., Stevenson, D.A., and Amick, D.C.,** Earthquake prediction studies at Lake Jocassee: a successful test of a prediction model, *EOS Trans. Am. Geophys. Union,* 58, 433, 1977.

796. **Stewart, G.S.,** Prediction of the Pt. Mugu earthquake by two methods, in *Proc. Conf. on Tectonic Problems of the San Andreas Fault System,* Kovack, R.L. and Nur, A., Eds., Stanford University, Stanford, Calif., 1973, 473.

797. **Robinson, R., Wesson, R.L., and Ellsworth, W.L.,** Variation of P-wave velocity before the Bear Valley, California, earthquake of 24 February 1972, *Science,* 184, 1281, 1974.

798. **Robinson, R., Wesson, R.L., Ellsworth, W.L., and Steppe, J.A.,** Variation in P-wave velocity before earthquakes along the central San Andreas fault, *EOS Trans. Am. Geophys. Union,* 55, 695, 1974.

799. **Cramer, C.,** Time variations in teleseismic residuals prior to the magnitude 5.1 Bear Valley earthquake of February 24, 1972, *EOS Trans. Am. Geophys. Union,* 55, 1150, 1974.

800. **Whitcomb, J.H.,** Earthquake prediction: seismic velocity variations in southern California, *EOS Trans. Am. Geophys. Union,* 56, 399, 1975.

801. **Lee, W.H.K. and Healy, J.H.,** A search for seismic precursors of the Hollister earthquake of November 28, 1974, *Abstr. Progr. Geol. Soc. Am.,* 7, 413, 1975.

802. **Gupta, I. N.,** Premonitory variations in S-wave velocity anisotropy before earthquakes in Nevada, *Science,* 182, 1129, 1973.

803. **Ellsworth, W.L.,** Preliminary evaluation of teleseismic travel time residuals preceding the Hawaiian earthquake of 29 November 1975, *EOS Trans. Am. Geophys. Union,* 57, 288, 1976.

804. **McEvilly, T.V. and Johnson, L.R.,** Earthquakes of strike-slip type in central California: evidence on the question of dilatancy, *Science,* 182, 581, 1973.

805. **Bakun, W.H., Stewart, R.M., and Tocher, D.,** Variations in V,/V, in Bear Valley in 1972, *Stanford Univ. Publ. Geol. Sci.,* 8, 453, 1973.

806. **Allen, C.R. and Helmberger, D.V.,** Search for temporal changes in seismic velocities using large explosions in southern California, in *Proc. Conf. on Tectonic Problems of the San Andreas Fault System,* Kovack, R.L. and Nur, A., Eds., Stanford University, Stanford, Calif., 1973, 436.

807. **Allen, C.R. and Helmberger, D.V.,** Search for temporal changes in seismic velocities using large explosions in southern California, in *U.S.-Japan Seminar on Earthquake Prediction and Control,* American Geophysical Union, Washington, D.C., 1974, microfiche, 19.

808. **Cramer, C.H. and Kovack, R.L.,** A search for teleseismic travel-time anomalies along the San Andreas fault zone, *Geophys. Res. Lett.,* 1, 90, 1974.

809. **Ergas, R.A. and Jackson, D.D.,** Search for travel time anomalies in Oroville aftershocks, *EOS Trans. Am. Geophys. Union,* 57, 956, 1976.

810. **Cramer, C.H.,** Teleseismic residuals prior to the November 28, 1974 Thanksgiving day earthquake near Hollister, California, *Bull. Seismol. Soc. Am.,* 66, 1233, 1976.

811. **Kanamori, H. and Fuis, G.**, Variation of P-wave velocity before and after the Galway earthquake ($M_L = 5.2$) and the Goat Mountain earthquakes ($M_L = 4.7, 4.7$), 1975, in the Mojave Desert, California, *Bull. Seismol. Soc. Am.*, 66, 2017, 1976.

812. **Bolt, B.A.**, Constancy of P travel times from Nevada explosions to Oroville Dam Station, 1970-1976, *Bull. Seismol. Soc. Am.*, 67, 27, 1977.

813. **Whitcomb, J.H.**, An update of time-dependent V_p/V_s and $\bar{V}_p$ in an area of the Transverse Ranges of southern California, *EOS Trans. Am. Geophys. Union*, 58, 305, 1977.

814. **Nur, A. and Simmons, G.**, Stress-induced velocity anisotropy in rock: an experimental study, *J. Geophys. Res.*, 74, 6667, 1969.

815. **Nur, A.**, Effects of stress on velocity anisotropy in rocks with cracks, *J. Geophys. Res.*, 76, 2022, 1971.

816. **Whitcomb, J.H.**, Time-dependent $\bar{V}_p$ and V_p/V_s in an area of the Transverse Ranges of southern California, *EOS Trans. Am. Geophys. Union*, 57, 288, 1976.

817. **Alexander, G.**, Caltech scientist offers cautious quake prediction, *Los Angeles Times,* April 21, 1976.

818. **Anon.**, Dilemma of a quake prediction, *Los Angeles Times,* April 29, 1976.

819. **Liddick, B.**, Seismograph vigil continues at Caltech, *Los Angeles Times,* April 29, 1976.

820. **Alexander, G.**, Experts won't accept quake prediction, *Los Angeles Times,* May 1, 1976.

821. **Logan, J.M.**, Animal behaviour and earthquake prediction, *Nature*, 265, 404, 1977.

822. **Anderson, C.J.**, Animals, earthquakes, and eruptions, *Field Mus. Nat. Hist. Publ. Geol. Ser.*, 44, 9, 1973.

823. **Simon, R.B.**, Animal behavior and earthquakes, *Earthquake Inf. Bull.*, 7, 9, 1975.

824. **U.S. Geological Survey**, Animals as Quake Predictors?, Department of Interior News Release, March 21, 1977.

825. **Molnar, P., Hanks, T., Nur, A., Raleigh, B. (Chairman), Wu, F., Savage, J., Scholz, C., Craig, H., Turner, R., and Bennett, G.**, Haicheng Earthquake Study Delegation, Prediction of the Haicheng earthquake, *EOS Trans. Am. Geophys. Union*, 58, 236, 1977.

826. **Everden, J.F.**, Ed., Conference I—Abnormal Animal Behavior Prior to Earthquakes, National Technical Information Service, U.S. Department of Commerce, Springfield, Va., 1977.

827. **Simon, R.B.**, Preliminary report on experiments to record anomalous animal behavior as possible earthquake precursor, *EOS Trans. Am. Geophys. Union*, 57, 759, 1976.

828. **United Press International**, Doctor believes roaches predict earthquakes, *Arizona Republic,* March 6, 1977.

829. **Enterprise Science Service**, Erratic behavior in animals is sign of coming earthquake, Chinese experts say, *Arizona Republic,* February 16, 1975.

830. **Scholz, C.H.**, A physical interpretation of the Haicheng earthquake prediction, *Nature (London)*, 267, 121, 1977.

831. **Reynolds, O.**, On the dilatancy of media composed of rigid particles in contact, *Philos. Mag.*, 20, 469, 1885.

832. **Mead, W.J.**, The geologic role of dilatancy, *J. Geol.*, 33, 685, 1925.

833. **Bridgman, P.**, Volume changes in the plastic stages of simple compression, *J. Appl. Phys.*, 20, 1241, 1949.

834. **Robertson, E.C.**, Creep in Solenhofen limestone under moderate hydrostatic pressure, in *Rock Deformation*, Geol. Soc. Am. Memoir 79, Geological Society of America, New York, 1960, 227.

835. **Matsushima, S.**, On the flow and fracture of igneous rocks, *Disaster Prevention Res. Inst., Tokyo Univ. Bull.*, 36, 2, 1960.

836. **Handin, J., Hager, R.V., Jr., Friedman, M., and Feather, J.N.**, Experimental deformation of sedimentary rocks under confining pressure: pore pressure tests, *Bull. Am. Assoc. Petrol. Geol.*, 47, 717, 1963.

837. **Frank, F.C.**, On dilatancy in relation to seismic sources, *Rev. Geophys.*, 3, 485, 1965.

838. **Brace, W.F.**, Some new measurements of linear compressibility of rocks, *J. Geophys. Res.*, 70, 391, 1965.

839. **Brace, W.F., Paulding, B.W., Jr., and Scholz, C.H.**, Dilatancy in the fracture of crystalline rocks, *J. Geophys. Res.*, 71, 3939, 1966.

840. **Nur, A. and Simmons, G.**, The effect of saturation on velocity in low porosity rocks, *Earth Planet. Sci. Lett.*, 7, 183, 1969.

841. **Nur, A.**, Dilatancy, pore fluids, and premonitory variations of t_s/t_p traveltimes, *Bull. Seismol. Soc. Am.*, 62, 1217, 1972.

842. **Scholz, C., Sykes, L.R., and Aggarwal, Y.P.**, Earthquake prediction: a physical basis, *Science*, 181, 803, 1973.

843. **Anderson, D.L. and Whitcomb, J.H.**, Time-dependent seismology, *J. Geophys. Res.*, 80, 1497, 1975.

844. **Myachkin, V.I., Sobolev, G.A., Dolbilkina, N.A., Morozow, V.N., and Preobrazensky, V.B.**, The study of variations in geophysical fields near focal zones of Kamchatka, *Tectonophysics*, 14, 279, 1972.

845. **Stuart, W.D.,** Diffusionless dilatancy model for earthquake precursors, *Geophys. Res. Lett.,* 1, 261, 1974.
846. **Brady, B.T.,** A physical basis for earthquake precursors in dry brittle rock, *Abstr. Progr. Geol. Soc. Am.,* 6, 285, 1974.
847. **Brady, B.T.,** Theory of earthquake. I. A scale independent theory of rock failure, *Pure Appl. Geophys.,* 112, 701, 1974.
848. **Brady, B.T.,** Theory of earthquakes. II. Inclusion theory of crustal earthquakes, *Pure Appl. Geophys.,* 113, 149, 1975.
849. **Press, F.,** Earthquake prediction, *Sci. Am.,* 232, 14, 1975.
850. **Scholz, C.H.,** The frequency-magnitude relation of microfracturing in rock and its relation to earthquakes, *Bull. Seismol. Soc. Am.,* 58, 399, 1968.
851. **Wyss, M.,** Towards a physical understanding of the earthquake frequency distribution, *Geophys. J. R. Astron. Soc.,* 31, 341, 1973.
852. **Dieterich, J.H.,** A model for earthquake precursors based on premonitory fault slip, *EOS Trans. Am. Geophys. Union,* 56, 1059, 1975.
853. **Dieterich, J.H.,** Preseismic fault slip and earthquake prediction, preprint, 1977.
854. **Kanamori, H. and Cipar, J.J.,** Focal process of the great Chilean earthquake May 22, 1960, *Phys. Earth Planet. Inter.,* 9, 128, 1974.
855. **Bott, M.H.P. and Dean, D.S.,** Stress diffusion from plate boundaries, *Nature (London),* 243, 339, 1973.
856. **Anon.,** Tangshan quake: portrait of a catastrophe, *Sci. News,* 111, 388, 1977.
857. **Lapwood, E.R.,** Earthquake prediction, *Nature,* 266, 220, 1977.
858. **Rikitake, T.,** Japanese program on earthquake prediction, in *U.S.-Japan Seminar on Earthquake Prediction and Control,* American Geophysical Union, Washington, D.C., 1974, microfiche, 1.
859. **Anon.,** U.S.-Japan seminar on earthquake precursors, *EOS Trans. Am. Geophys. Union,* 58, 600, 1977.
860. **Mikumo, T.,** A review of studies on microseismicity in southwest Japan and its tectonic implications, in *U.S.-Japan Seminar on Earthquake Prediction and Control,* American Geophysical Union, Washington, D.C., 1974, microfiche, 33.
861. **Rikitake, T.,** *Earthquake Prediction,* Elsevier, New York, 1976.
862. **Savarensky, E.F.,** On the prediction of earthquakes, *Tectonophysics,* 6, 17, 1968.
863. **U.S. Geological Survey,** Seminar in Tashkent, U.S.S.R.: seismic considerations in housing design and construction, *Earthquake Inf. Bull.,* 8, 10, 1976.
864. **Dieterich, J.H. and Brace, W.F.,** Earthquake-related experimental studies in the USSR, *EOS Trans. Am. Geophys. Union,* 56, 221, 1975.
865. **Sadovsky, M.A. and Nersesov, I.L.,** Forecasts of earthquakes on the basis of complex geophysical features, *Tectonophysics,* 23, 247, 1974.
866. **Fedotov, S.A., Dolbilkina, N.A., Morozov, V.N., Myachkin, V.I., Preobrazhensky, V.B., and Sobolev, G.A.,** Investigations on earthquake prediction in Kamchatka, *Tectonophysics,* 9, 249, 1970.
867. **Fedotov, S.A., Gusev, A.A., Boldyrev, S.A.,** Progress of earthquake prediction in Kamchatka, *Tectonophysics,* 14, 279, 1972.
868. **Hamilton, R.M.,** Earthquake research in China. II. National program, *EOS Trans. Am. Geophys. Union,* 56, 841, 1975.
869. **Press, F. and Bullock, M.,** Earthquake research in China. I. Introduction and overall assessment, *EOS Trans. Am. Geophys. Union,* 56, 838, 1975.
870. **Shapley, D.,** Chinese earthquakes: the Maoist approach on seismoloy, *Science,* 193, 656, 1976.
871. **Hofheinz, R., Jr.,** Earthquake research in China. XI. History and politics of Chinese earthquake studies, *EOS Trans. Am. Geophys. Union,* 56, 873, 1975.
872. **Brace, W.F.,** Earthquake research in China. III. Chinese universities and institutes, *EOS Trans. Am. Geophys. Union,* 56, 844, 1975.
873. **Press, F., Allen, C., Bonilla, M.G., Brace, W.F., Bullock, M., Clough, R.W., Hamilton, R.M., Hofheinz, R., Jr., Kisslinger, C., Knopoff, L., Park, M., Raleigh, C.B., and Sykes, L.R.,** American Seismology Delegation, Earthquake research in China, *EOS Trans. Am. Geophys. Union,* 56, 838, 1975.
874. **Oliver, J.,** Recent earthquake prediction in the U.S.A., *Tectonophysics,* 9, 283, 1970.
875. **Press, F., Benioff, H., Frosch, R.A., Griggs, D.T., Handin, J., Hanson, R.E., Hess, H.H., Housner, G.W., Munk, W.H., Orowan, E., Pakiser, L.C., Sutton, G., and Tocher, D.,** Earthquake Prediction: A Proposal for a Ten Year Program of Research, Office of Science and Technology, Washington, D.C., 1965.
876. **U.S. Geological Survey,** Advisory panel on earthquake studies, *Earthquake Inf. Bull.,* 8, 19, 1976.
877. **U.S. Geological Survey,** Earthquake prediction council established, *Earthquake Inf. Bull.,* 9, 38, 1977.

878. **Gay, T. E. Jr.,** Earthquake evaluation guidelines, *Calif. Geol.,* 30, 158, 1977.

879. **Wehr, E.,** Earthquake prediction due federal boost, *Arizona Republic,* July 14, 1977.

880. **Kisslinger, C. and Rikitake, T.,** U.S.-Japan seminar on earthquake prediction and control, *EOS Trans. Am. Geophys. Union,* 55, 9, 1974.

881. **Wallace, R.E.,** The Talas-Fergana fault, Kirgiz and Kazakh, U.S.S.R., *Earthquake Inf. Bull.,* 8, 4, 1976.

882. **United Press International,** Science pact hailed as sign of better U.S.-Soviet ties, *Arizona Republic,* July 9, 1977.

883. **Emiliani, C., Harrison, C.G.A., and Swanson, M.,** Underground nuclear explosions and the control of earthquakes, *Science,* 165, 1255, 1969.

884. **Healy, J.H. and Hamilton, R.M.,** Seismic activity following the "Benham" nuclear explosion, *EOS Trans. Am. Geophys. Union,* 50, 248, 1969.

885. **Ryall, A., Boucher, G., Savage, W.U., and Jones, A.E.,** Earthquakes following underground nuclear explosions, *EOS Trans. Am. Geophys. Union,* 50, 248, 1969.

886. **Engdahl, E.R., Mickey, W.V., Brockman, S.R., and King, K.W.,** The seismic regime following a large scale nuclear explosion, *EOS Trans. Am. Geophys. Union,* 50, 248, 1969.

887. **Boucher, G., Ryall, A., and Jones, A.E.,** Earthquakes associated with underground nuclear explosions, *J. Geophys. Res.,* 74, 3808, 1969.

888. **Ryall, A. and Savage, W.U.,** Comparison of seismological effects for the Nevada underground test Boxcar with natural earthquakes in the Nevada region, *J. Geophys. Res.,* 74, 4281, 1969.

889. **Kisslinger, C. and Cherry, J.T., Jr.,** Excitation of earthquakes by underground explosions, *EOS Trans. Am. Geophys. Union,* 51, 353, 1970.

890. **Hamilton, R.M., Smith, B.E., Fischer, F.G., and Papanek, P.J.,** Earthquakes caused by underground nuclear explosions on Pahute Mesa, Nevada Test Site, *Bull. Seismol. Soc. Am.,* 62, 1319, 1972.

891. **Engdahl, E.R.,** Seismic effects of the Milrow and Cannikin nuclear explosions, *Bull. Seismol. Soc. Am.,* 62, 1411, 1972.

892. **Engdahl, E.R.,** Milrow and Cannikin seismic effects, *EOS Trans. Am. Geophys. Union,* 53, 442, 1972.

893. **Press, F. and Archambeau, C.,** Release of tectonic strain by underground nuclear explosions, *J. Geophys. Res.,* 67, 337, 1962.

894. **Hoy, R.B.,** Induced faulting, *Bull. Seismol. Soc. Am.,* 53, 845, 1963.

895. **Brune, J.N. and Pomeroy, P.W.,** Surface wave radiation patterns for underground nuclear explosions and small-magnitude earthquakes, *J. Geophys. Res.,* 68, 5005, 1963.

896. **Aki, K.,** A note on surface waves from the Hardhat nuclear explosion, *J. Geophys. Res.,* 69, 1131, 1964.

897. **Toksöz, M.N., Harkrider, D.G., and Ben-Menahem, A.,** Determination of source parameters by amplitude equalization of seismic surface waves, *J. Geophys. Res.,* 70, 907, 1965.

898. **Dickey, D.D.,** Fault displacement as a result of underground nuclear explosions, in *Nevada Test Site,* Eckel, E.B., Ed., Memoir 110, Geological Society of America, New York, 1968, 219.

899. **Hamilton, R.M., McKeown, F.A., and Healy, J.H.,** Seismic activity and faulting associated with a large underground nuclear explosion, *Science,* 166, 601, 1969.

900. **Nuttli, O.W.,** Travel times and amplitudes of S waves from nuclear explosions in Nevada, *Bull. Seismol. Soc. Am.,* 59, 385, 1969.

901. **Aki, K., Reasenberg, P., DeFazio, T., and Tsai, Y.-B.,** Near-field and far-field seismic evidences for triggering of an earthquake by the Benham explosion, *Bull. Seismol. Soc. Am.,* 59, 2197, 1969.

902. **Bucknam, R.C.,** Geologic effects of the Benham underground nuclear explosion, Nevada Test Site, *Bull. Seismol. Soc. Am.,* 59, 2209, 1969.

903. **Dickey, D.D.,** Strain associated with the Benham underground nuclear explosion, *Bull. Seismol. Soc. Am.,* 59, 2221, 1969.

904. **McKeown, F.A. and Dickey, D.D.,** Fault displacements and motion related to nuclear explosions, *Bull. Seismol. Soc. Am.,* 59, 2253, 1969.

905. **Hamilton, R.M. and Healy, J.H.,** Aftershocks of the Benham nuclear explosion, *Bull. Seismol. Soc. Am.,* 59, 2271, 1969.

906. **Smith, S.W.,** Transient and residual strains associated with large underground explosions, *EOS Trans. Am. Geophys. Union,* 50, 248, 1969.

907. **Evans, D.M.,** The Denver area earthquakes and the Rocky Mountain Arsenal disposal well, *Mt. Geol.,* 3, 23, 1966.

908. **Ives, R.L.,** Earthquakes in the Denver area, *Geog. Rev.,* 59, 616, 1969.

909. **Hubbert, M.K. and Rubey, W.W.,** Mechanics of fluid-filled porous solids and its application to overthrust faulting, *Geol. Soc. Am. Bull.,* 70, 115, 1959.

910. **Raleigh, B.,** Can we control earthquakes, *Earthquake Inf. Bull.,* 9, 4, 1977.

911. **Healy, J.H. and Pakiser, L.C.**, Man-made earthquakes and earthquake prediction, *EOS Trans. Am. Geophys. Union,* 52, 171, 1971.

912. **Healy, J.H., Rubey, W.W., Griggs, D.T., and Raleigh, C.B.,** The Denver earthquakes, *Science,* 161, 1301, 1968.

913. **U.S. Geological Survey,** Man-made earthquakes at Denver and Rangely, Colorado, *Earthquake Inf. Bull.,* 5, 4, 1973.

914. **Gibbs, J.F., Healy, J.H., Raleigh, C.B., and Coakley, J.M.,** Seismicity in the Rangely, Colorado, area: 1962-1970. *Bull. Seismol. Soc. Am.,* 63, 1557, 1973.

915. **Raleigh, C.B., Healy, J.H., and Bredehoeft, J.D.,** An experiment in earthquake control at Rangely, Colorado, *Science,* 191, 1230, 1976.

916. **Raleigh, C.B., Healy, J.H., and Bredehoeft, J.D.,** Faulting and crustal stress at Rangely, Colorado, in *Flow and Faulting,* Geophys. Monogr. 16, American Geophysical Union, Washington, D.C., 1972, 275.

917. **Dieterich, J.H. and Raleigh, C.B.,** A laboratory earthquake control experiment, *EOS Trans. Am. Geophys. Union,* 55, 353, 1974.

918. **Ohtake, M.,** Seismic activity induced by water injection at Matsushiro, Japan, *J. Phys. Earth,* 22, 163, 1974.

919. **Fletcher, J.B., Sykes, L.R., and Sbar, M.L.,** Seismic activity associated with hydraulic mining in western New York state, *EOS Trans. Am. Geophys. Union,* 54, 371, 1973.

920. **Fletcher, J.B. and Sykes, L.R.,** Earthquakes related to hydraulic mining and natural seismic activity in western New York, *EOS Trans. Am. Geophys. Union,* 57, 289, 1976.

921. **Wallace, R.E.,** Earthquake recurrence intervals on the San Andreas fault, *Geol. Soc. Am. Bull.,* 81, 2875, 1970.

922. **Shoemaker, E.M., Ed.,** *Continental Drilling,* Carnegie Institution of Washington, Washington, D.C., 1975.

923. **Raleigh, C.B.,** The scientific basis for earthquake prediction, in Earthquake Hazards Act, Hearing before the Subcommittee on Oceans and Atmosphere, 92nd Congress, U.S. Government Printing Office, Washington, D.C., 1972, 69.

924. **Healy, J.H., Raleigh, C.B., Lee, W.H.K., and Byerlee, J.D.,** Man-made earthquakes in the Basin and Range province, *Abstr. Progr. Geol. Soc. Am.,* 6, 191, 1974.

925. **Weisbecker, L.W., Stoneman, W.C., Ackerman, S.E., Arnold, R.K., Halton, P.M., Ivy, S.C., Kautz, W.H., Kroll, C.A., Levy, S., Mickley, R.B., Miller, P.D., Rainey, C.T., and Van Zandt, J.E.,** *Earthquake Prediction, Uncertainty, and Policies for the Future,* Stanford Research Institute, Menlo Park, Calif., 1977.

926. **International Association for Earthquake Engineering,** *Earthquake Resistant Regulations: A World List 1973,* Gakujutsu Bunken Fukyu-Kai, Tokyo, 1973.

927. **Dowrick, D.J.,** *Earthquake Resistant Design, A Manual for Engineers and Architects,* John Wiley & Sons, London, 1977.

928. **Committee on Seismology, Oliver, J.E., Chairman, Allen, C.A., Bolt, B.A., Hales, A.L., Herrin, E.T., Jr., Howell, B.F., Jr., Hudson, D.E., Kaufman, S., Kisslinger, C., Wilson, J.T., and Berg, J.W., Jr.,** Earthquake engineering, in *Seismology Responsibilities and Requirements of a Growing Science. Part II. Problems and Prospects,* National Academy of Sciences, Washington, D.C., 1969, 32.

929. **Clough, R.W.,** Earthquake response of structures, in *Earthquake Engineering,* Wiegel, R.L., Ed., Prentice-Hall, Englewood Cliffs, N.J., 1970, 307.

930. **Okamoto, S.,** *Introduction to Earthquake Engineering,* Halsted Press, New York, 1973.

931. **Newmark, N.M.,** Current trends in the seismic analysis and design of high-rise structures, in *Earthquake Engineering,* Wiegel, R. L., Ed., Prentice-Hall, Englewood Cliffs, N.J., 1970, 403.

932. **Degenkolb, H.J.,** Earthquake Forces on Tall Structures, Booklet 2717A, Bethlehem Steel Corporation, Bethlehem, Pa., 1977.

933. **Botsai, E.E., Goldberg, A., Fisher, J.L., Lagorio, H.J., and Wosser, T.D.,** *Architects and Earthquakes,* AIA Research Corporation, New York, 1977.

934. **Seismology Committee—Structural Engineers Association of California,** *Recommended Lateral Force Requirements and Commentary,* Structural Engineers Association of California, San Francisco, 1975.

935. **Clough, R.W. and Jenschke, V.A.,** The effect of diagonal bracing on the earthquake performance of a steel frame building, *Bull. Seismol. Soc. Am.,* 53, 389, 1963.

936. **Degenkolb, H.J. and Associates,** Tri-Cities structural hazards investigation, in *The Seismic Safety Study for the General Plan,* California Council on Intergovernmental Relations, Sacramento, 1973, 103.

937. **International Conference of Building Officials,** *Uniform Building Code 1976 Edition,* International Conference of Building Officials, Whittier, Calif., 1976.

938. Los Angeles Department of Building and Safety, *City of Los Angeles Building Code 1976 Edition,* Building News, Los Angeles, 1975.

939. Public Works Department, *City and County of San Francisco Building Code,* Building News, Los Angeles, 1975.

940. Joint Committee on Seismic Safety, Appendix B: history of earthquake code provisions and regulations in California, in Meeting the Earthquake Challenge, Joint Committee on Seismic Safety, California Legislature, Sacramento, 1974, 192.

941. **Smith, K.R.,** The application of seismic risk analysis to land use decisions, in *The Seismic Safety Study for the General Plan,* California Council on Intergovernmental Relations, Sacramento, 1973, 161.

942. Joint Committee of the San Francisco Section, American Society of Civil Engineers and Structural Engineers Association of Northern California, Lateral forces of earthquakes and wind, *Trans. Am. Soc. Civ. Eng.,* 117, 716, 1952.

943. **Moran, D.F. and Bockemohle, L.W.,** History and philosophy of California earthquake codes and elements of lateral force design, in San Fernando, California, Earthquake of February 9, 1971, Vol. 1 (Part A), Benfer, N.A., Coffman, J.L., and Dees, L.T., Eds., U.S. Government Printing Office, Washington, D.C., 1973, 23.

944. **Lord, J.,** Real-Time Simulated Earthquake Motion of High Rise Structures, Albert C. Martin and Associates, Los Angeles, no date.

945. **Lord, J.,** unpublished data, 1974.

946. **Muto, K.,** Earthquake-proof design gives rise to first Japanese skyscrapers, *Civ. Eng.,* 41, 49, 1971.

947. **Hauf, H.D.,** Architectural factors in earthquake resistance, in *The Great Alaska Earthquake of 1964—Engineering,* Wood, F.J., Ed., National Academy of Sciences, Washington, D.C., 1973, 340.

948. **Steinbrugge, K.V. and Bush, V.R.,** Review of earthquake damage in the western United States, 1933-1964, in *Earthquake Investigations in the Western United States* 1931-1964, Carter, D.S., Ed., U.S. Government Printing Office, Washington, D.C., 1964, 223.

949. **Meehan, J.F.,** Public school buildings, in San Fernando, California, Earthquake of February 9, 1971, Vol. 1 (Part B), Benfer, N.A., Coffman, J.L., and Dees, L.T., Eds., U.S. Government Printing Office, Washington, D.C., 1973, 667.

950. **Anon.,** 1,258 California schools don't meet quake standards, *Palo Alto Times,* February 6, 1974, 23.

951. **Meehan, J.F.,** Damage to public school buildings, in The San Fernando, California, Earthquake of February 9, 1971, Geological Survey Professional Paper 733, U.S. Government Printing Office, Washington, D.C., 1971, 209.

952. **Amimoto, P.Y.,** Review of new hospital sites for seismic safety, *Calif. Geol.,* 27, 110, 1974.

953. **Clark, W.B.,** Senate Bill 519, *Calif. Earthquake Sci. Newsl.,* 1, 2, 1973.

954. **Johnston, R.G. and Strand, D.R.,** Olive View Hospital, in San Fernando, California, Earthquake of February 9, 1971, Vol. 1 (Part A), Benfer, N.A., Coffman, J.L., and Dees, L.T., Eds., U.S. Government Printing Office, Washington, D.C., 1973, 255.

955. Structural Engineers Association of Southern California, *Report on Olive View Hospital, San Fernando, Calif., Earthquake of February 9, 1971,* Structural Engineers Association of Southern California, Los Angeles, 1971.

956. **Ewoldsen, H.M. and McNeill, R.C.,** Site studies Olive View Hospital, in San Fernando, California, Earthquake of February 9, 1971, Vol. 2 (Part A), Benfer, N.A., Coffman, J.L., and Dees, L.T., Eds., U.S. Government Printing Office, Washington, D.C., 1973, 293.

957. **Moran, D.F., Allen, C., Brugger, W.A., Crandall, L.L., Degenkolb, H.J., Duke, C.M., Meehan, J.F., Pinkham, C.W., and Wallace, R.E.,** Earthquake Engineering Research Institute Committee, Damage to Olive View Hospital buildings, in The San Fernando, California, Earthquake of February 9, 1971, U.S. Geological Survey Professional Paper 733, U.S. Government Printing Office, Washington, D.C., 1971, 188.

958. **Degenkolb, H.J.,** Preliminary structural lessons from the earthquake, in The San Fernando, California, Earthquake of February 9, 1971, U.S. Geological Survey Professional Paper 733, U.S. Government Printing Office, Washington, D.C., 1971, 133.

959. **Ropp, W.F.,** Damage to the Pacoima Memorial Lutheran Hospital, in The San Fernando, California, Earthquake of February 9, 1971, U.S. Geological Survey Professional Paper 733, U.S. Government Printing Office, Washington, D.C., 1971, 193.

960. **Degenkolb, H.J. and Fratessa, P.F.,** Pacoima Memorial Lutheran Hospital, in San Fernando, California, Earthquake of February 9, 1971, Vol. 1 (Part A), Benfer, N.A., Coffman, J.L., and Dees, L.T., Eds., U.S. Government Printing Office, Washington, D.C., 1973, 205.

961. **Barnes, S.B.,** Damage to the Holy Cross Hospital, in The San Fernando, California, Earthquake of February 9, 1971, U.S. Geological Survey Professional Paper 733, U.S. Government Printing Office, Washington, D.C., 1971, 195.

962. **Barnes, S.B. and Pinkham, C.W.,** Holy Cross Hospital, in San Fernando, California, Earthquake of February 9, 1971, Vol. 1 (Part A), Benfer, N.A., Coffman, J.L., and Dees, L.T., Eds., U.S. Government Printing Office, Washington, D.C., 1973, 235.

963. **Brugger, W.,** Damage to other buildings, in The San Fernando, California, Earthquake of February 9, 1971, U.S. Geological Survey Professional Paper 733, U.S. Government Printing Office, Washington, D.C., 1971, 213.

964. **Johnston, R.G.,** Veterans Administration Hospital, in San Fernando, California, Earthquake of February 9, 1971, Vol. 1 (Part B), Benfer, N.A., Coffman, J.L., and Dees, L.T., Eds., U.S. Government Printing Office, Washington, D.C., 1973, 655.

965. **Bolt, B.A., Johnston, R.G., Lefter, J., and Sozen, M.A.,** The study of earthquake questions related to Veterans Administration Hospital facilities, *Bull. Seismol. Soc. Am.,* 65, 937, 1975.

966. City Council of the City of Long Beach, Subdivision 80 — Earthquake Hazard Regulations for Rehabilation of Existing Structures within the City, City Council of the City of Long Beach, Long Beach, Calif., 1976.

967. **Anon.,** Subdivision 80 of the Building Regulations, Dept. of Planning and Building, Division of Building and Safety, City of Long Beach, Calif., 1977.

968. Department of Building and Safety, Specifications for Assessing the Capacity of Unreinforced Masonry Buildings, Department of Building and Safety, City of Long Beach, Calif., 1976.

969. **Woodward-McNeill and Associates,** Seismic Safety Element, The City Planning Commission, Long Beach, Calif., 1975.

970. **Abel, M.A.,** Unreinforced masonry buildings, in San Fernando, California, Earthquake of February 9, 1971, Vol. 1 (Part B), Benfer, N.A., Coffman, J. L., and Dees, L. T., Eds., U.S. Government Printing Office, Washington, D.C., 1973, 639.

971. Town of Portola Valley Council, Zoning Regulations for the Town of Portola Valley, Portola Valley, Calif., 1973.

972. U.S. Atomic Energy Commission, 1973 Annual Report to Congress, Volume 2, Regulatory Activities, U.S. Government Printing Office, Washington, D.C., 1974.

973. Office of the Federal Register, Code of Federal Regulations 10, Energy, Parts 0 to 199, U.S. Government Printing Office, Washington, D.C., 1977, 412.

974. **Anon.,** PG&E withdraws application to build Pt. Arena nuclear plant, *Calif. Geol.,* 26, 89, 1973.

975. **Carter, L.J.,** Nuclear initiative: impending vote stimulates legislative action, *Science,* 192, 975, 1976.

976. Committee on the Alaska Earthquake, *Toward Reduction of Losses from Earthquakes, Conclusions from the Great Alaska Earthquake of 1964,* National Academy of Sciences, Washington, D.C., 1969.

977. **Hodgson, J.H.,** *Earthquakes and Earth Structure,* Prentice-Hall, Englewood Cliffs, N.J., 1964.

978. **Berg, G.V. and Stratta, J.L.,** Anchorage and the Alaska earthquake of March 27, 1964, in *Earthquakes,* American Iron and Steel Institute, Washington, D.C., 1975, 145.

979. **Ayers, J.M.,** Damage to building equipment and contents, in The San Fernando, California, Earthquake of February 9, 1971, U.S. Geological Survey Professional Paper 733, U.S. Government Printing Office, Washington, D.C., 1971, 220.

980. **Iacopi, R.,** *Earthquake Country,* rev. ed., Lane Books, Menlo Park, Calif., 1971.

981. **Anon.,** Earthquake history of California, *Earthquake Inf. Bull.,* 3, 22, 1971.

982. **Steinbrugge, K.V. and Moran, D.F.,** An engineering study of the southern California earthquake of July 21, 1952, and its aftershocks, *Bull. Seismol. Soc. Am.,* 44, 199, 1954.

983. **Steinbrugge, K.V., Manning, J.H., Degenkob, H.J., Sardis, J.M., Biddison, C.M., Jr., Bush, V.R., Driskell, J.J., Holstein, J.J., Johnson, A.W., Kellam, H.S., Martin, J., McClure, F.E., McLaughlin, R.R., Moran, D.F., Stratta, J.L., and Zacher, E.G.,** Building damage in Anchorage, in The Prince William Sound, Alaska, Earthquake of 1964 and Aftershocks, Wood, F.J., Ed., Vol. 2 (Part A), U.S. Government Printing Office, Washington, D.C., 1967, 7.

984. **Steinbrugge, K.V.,** Introduction to the earthquake engineering of the 1964 Prince William Sound, Alaska, earthquake, in The Prince William Sound, Alaska, Earthquake of 1964 and Aftershocks, Wood, F.J., Ed., Vol. 2 (Part A), U.S. Government Printing Office, Washington, D.C., 1967, 1.

985. **Steinbrugge, K.V.,** Earthquake damage and structural performance in the United States, in *Earthquake Engineering,* Wiegel, R.L., Ed., Prentice-Hall, Englewood Cliffs, N.J., 1970, 167.

986. **George, W., Knowles, P., Allender, J.K., Sizemore, J.F., and Carson, D.E.,** Structures in Anchorage, in The Great Alaska Earthquake of 1964 — Engineering, National Academy of Sciences, Washington, D.C., 1973, 774.

987. **Ayers, J.M., Tseng-Yao, S., and Brown, F.R.,** Nonstructural damage to buildings, in The Great Alaska Earthquake of 1964 — Engineering, National Academy of Sciences, Washington, D.C., 1973, 346.

988. **Berg, G.V. and Stratta, J.L.,** *Anchorage and the Alaska Earthquake of March 27, 1964,* American Iron and Steel Institute, New York, 1964.

989. **Murphy, L.M., Steinbrugge, K.V., and Duke, C.M.,** The San Fernando earthquake, in San Fernando, California, Earthquake of February 9, 1971, Vol. 1 (Part A), Benfer, N.A., Coffman, J.L., and Dees, L.T., Eds., U.S. Government Printing Office, Washington, D.C., 1973, 3.

990. **Kachadoorian, R.,** An estimate of the damage, in The San Fernando, California, Earthquake of February 9, 1971, U.S. Geological Survey Professional Paper 733, U.S. Government Printing Office, Washington, D.C., 1971, 5.

991. **Grantz, A.,** Introduction, in The San Fernando, California, Earthquake of February 9, 1971, U.S. Geological Survey Professional Paper 733, U.S. Government Printing Office, Washington, D.C., 1971, 1.

992. **Housner, G.W. and Hudson, D.E.,** Preliminary report on the engineering aspects of the San Fernando, California, earthquake of February 9, 1971, *Bull. Seismol. Soc. Am.,* 61, 499, 1971.

993. **Greensfelder, R.,** Seismologic and crustal movement investigations of the San Fernando earthquake, *Calif. Geol.,* 24, 62, 1971.

994. **Steinbrugge, K.V. and Schader, E.E.,** Earthquake damage and related statistics, in San Fernando, California, Earthquake of February 9, 1971, Vol. 1 (Part B), Benfer, N.A., Coffman, J.L., and Dees, L.T., Eds., U.S. Government Printing Office, Washington, D.C., 1973, 691.

995. Subcommittee on Buildings, NOAA/EERI Earthquake Investigation Committee, Low-rise industrial and commercial buildings, in San Fernando, California, Earthquake of February 9, 1971, Vol. 1 (Part A), Benfer, N.A., Coffman, J.L., and Dees, L.T., Eds., U.S. Government Printing Office, Washington, D.C., 1973, 41.

996. Subcommittee on Buildings, NOAA/EERI Earthquake Investigation Committee, Sylmar Industrial Tract, in San Fernando, California, Earthquake of February 9, 1971, Vol. 1 (Part A), Benfer, N.A., Coffman, J.L., and Dees, L.T., Eds., U.S. Government Printing Office, Washington, D.C., 1973, 43.

997. **Wheeler & Gray Consulting Engineers,** Stone's Liquor Store, in San Fernando, California, Earthquake of February 9, 1971, Vol. 1 (Part A), Benfer, N.A., Coffman, J.L., and Dees, L.T., Eds., U.S. Government Printing Office, Washington, D.C., 1973, 45.

998. **Wheeler & Gray Consulting Engineers,** Warehouse Building, in San Fernando, California, Earthquake of February 9, 1971, Vol. 1 (Part A), Benfer, N.A., Coffman, J.L., and Dees, L.T., Eds., U.S. Government Printing Office, Washington, D.C., 1973, 51.

999. **Wheeler & Gray Consulting Engineers,** All Phase Color, in San Fernando, California, Earthquake of February 9, 1971, Vol. 1 (Part A), Benfer, N.A., Coffman, J.L., and Dees, L.T., Eds., U.S. Government Printing Office, Washington, D.C., 1973, 55.

1000. Subcommittee on Buildings, NOAA/EERI Earthquake Investigation Committee, San Fernando Industrial Tract, in San Fernando, California, Earthquake of February 9, 1971, Vol. 1 (Part A), Benfer, N.A., Coffman, J.L., and Dees, L.T., Eds., U.S. Government Printing Office, Washington, D.C., 1973, 61.

1001. **Wheeler & Gray Consulting Engineers,** Bell Metrics, in San Fernando, California, Earthquake of February 9, 1971, Vol. 1 (Part A), Benfer, N.A., Coffman, J.L., and Dees, L.T., Eds., U.S. Government Printing Office, Washington, D.C., 1973, 63.

1002. **Wheeler & Gray Consulting Engineers,** M & L Machine Shop, in San Fernando, California, Earthquake of February 9, 1971, Vol. 1 (Part A), Benfer, N.A., Coffman, J.L., and Dees, L.T., Eds., U.S. Government Printing Office, Washington, D.C., 1973, 69.

1003. **Wheeler & Gray Consulting Engineers,** Vector Electronics, in San Fernando, California, Earthquake of February 9, 1971, Vol. 1 (Part A), Benfer, N.A., Coffman, J.L., and Dees, L.T., Eds., U.S. Government Printing Office, Washington, D.C., 1973, 75.

1004. **Wheeler & Gray Consulting Engineers,** Wendell Machine Shop, in San Fernando, California, Earthquake of February 9, 1971, Vol. 1 (Part A), Benfer, N.A., Coffman, J.L., and Dees, L.T., Eds., U.S. Government Printing Office, Washington, D.C., 1973, 83.

1005. **Thompson, J.H.,** Bennett Industries, in San Fernando, California, Earthquake of February 9, 1971, Vol. 1 (Part A), Benfer, N.A., Coffman, J.L., and Dees, L.T., Eds., U.S. Government Printing Office, Washington, D.C., 1973, 87.

1006. **Thompson, J.H.,** Thriftimart Market, in San Fernando, California, Earthquake of February 9, 1971, Vol. 1 (Part A), Benfer, N.A., Coffman, J.L., and Dees, L.T., Eds., U.S. Government Printing Office, Washington, D.C., 1973, 91.

1007. **Thompson, J.H.,** W.T. Grants, in San Fernando, California, Earthquake of February 9, 1971, Vol. 1 (Part A), Benfer, N.A., Coffman, J.L., and Dees, L.T., Eds., U.S. Government Printing Office, Washington, D.C., 1973, 95.

1008. **Thompson, J.H.,** Builder's Emporium, in San Fernando, California, Earthquake of February 9, 1971, Vol. 1 (Part A), Benfer, N.A., Coffman, J.L., and Dees, L.T., Eds., U.S. Government Printing Office, Washington, D.C., 1973, 99.

1009. **Wheeler & Gray Consulting Engineers,** Alpha Beta Market, in San Fernando, California, Earthquake of February 9, 1971, Vol. 1 (Part A), Benfer, N.A., Coffman, J.L., and Dees, L.T., Eds., U.S. Government Printing Office, Washington, D.C., 1973, 103.

1010. **Wheeler & Gray Consulting Engineers,** Boys Market, in San Fernando, California, Earthquake of February 9, 1971, Vol. 1 (Part A), Benfer, J.L., Coffman, and Dees, L.T., Eds., U.S. Government Printing Office, Washington, D.C., 1973, 113.

1011. Subcommittee on Buildings, NOAA/EERI Earthquake Investigation Committee, Wood roof and masonry wall buildings—summary, conclusions and recommendations, in San Fernando, California, Earthquake of February 9, 1971, Vol. 1 (Part A), Benfer, N.A., Coffman, J.L., and Dees, L.T., Eds., U.S. Government Printing Office, Washington, D.C., 1973, 127.

1012. **Briasco, E.,** Behavior of joist anchors versus wood ledgers, in San Fernando, California, Earthquake of February 9, 1971, Vol. 1 (Part A), Benfer, N.A., Coffman, J.L., and Dees, L.T., Eds., U.S. Government Printing Office, Washington, D.C., 1973, 121.

1013. **Kesler, J.J.,** Hospitals and medical facilities, in San Fernando, California, Earthquake of February 9, 1971, Vol. 1 (Part A), Benfer, N.A., Coffman, J.L., and Dees, L.T., Eds., U.S. Government Printing Office, Washington, D.C., 1973, 175.

1014. **Pinkham, C.W.,** Summary of conclusions and recommendations, in San Fernando, California, Earthquake of February 9, 1971, Vol. 1 (Part B), Benfer, N.A., Coffman, J.L., and Dees, L.T., Eds., U.S. Government Printing Office, Washington, D.C., 1973, 777.

1015. **John A. Blume & Associates,** Engineers, Bank of California, in San Fernando, California, Earthquake of February 9, 1971, Vol. 1 (Part A), Benfer, N.A., Coffman, J.L., and Dees, L.T., Eds., U.S. Government Printing Office, Washington, D.C., 1973, 327.

1016. **John A. Blume & Associates,** Engineers, High-rise building — not instrumented Union Bank, in San Fernando, California, Earthquake of February 9, 1971, Vol. 1 (Part B), Benfer, N.A., Coffman, J.L., and Dees, L.T., Eds., U.S. Government Printing Office, Washington, D.C., 1973, 629.

1017. **Hudson, D.E. and Jephcott, D.K.,** The San Fernando earthquake and public school safety, *Bull. Seismol. Soc. Am.,* 64, 1653, 1974.

1018. **Yanev, P.,** Earthquake safety for homes, *Calif. Geol.,* 30, 272, 1977.

1019. **Foth, U.A.,** Earthquake damage repair techniques, in San Fernando, California, Earthquake of February 9, 1971, Vol. 1 (Part B), Benfer, N.A., Coffman, J.L., and Dees, L.T., Eds., U.S. Government Printing Office, Washington, D.C., 1973, 685.

1020. **Slosson, J.E. and Krohn, J.P.,** Effective building codes, *Calif. Geol.,* 30, 136, 1977.

1021. **Yanev, P.,** *Peace of Mind in Earthquake Country — How to Save Your Home and Life,* Chronicle Books, San Francisco, 1974.

1022. **Merz, K.L. and Ayers, J.M.,** Nonstructural component considerations, in *Architects and Earthquakes: Research Needs,* AIA Research Corporation, Washington, D.C., 1976, 32.

1023. **John A. Blume & Associates,** Holiday Inn, in San Fernando, California, Earthquake of February 9, 1971, Vol. 1 (Part A), Benfer, N.A., Coffman, J.L., and Dees, L.T., Eds., U.S. Government Printing Office, Washington, D.C., 1973, 359.

1024. **John A. Blume & Associates,** Holiday Inn, in San Fernando, California, Earthquake of February 9, 1971, Vol. 1 (Part A), Benfer, N.A., Coffman, J.L., and Dees, L.T., Eds., U.S. Government Printing Office, Washington, D.C., 1973, 395.

1025. **Ayers, J.M. and Sun, T.Y.,** Nonstructural damage, in San Fernando, California, Earthquake of February 9, 1971, Vol. 1 (Part B), Benfer, N.A., Coffman, J.L., and Dees, L.T., Eds., U.S. Government Printing Office, Washington, D.C., 1973, 735.

1026. **Dewey, J.W. and Grantz, A.,** The Ghir earthquake of April 10, 1972 in the Zagros Mountains of southern Iran: seismotectonic aspects and some results of a field reconnaissance, *Bull. Seismol. Soc. Am.,* 63, 2071, 1973.

1027. **McEvilly, T.V. and Razani, R.,** A preliminary report: the Ghir, Iran earthquake of April 10, 1972, *Bull. Seismol. Soc. Am.,* 63, 339, 1973.

1028. **Dewey, J.W.,** Field observations of the Fars, Iran, earthquake of April 10, 1972, *Earthquake Inf. Bull.,* 4, 22, 1972.

1029. **Anon.,** Another killer earthquake strikes Iran, *Earthquake Inf. Bull.,* 4, 8, 1972.

1030. **Coffman, J.L. and Stover, C.W., Eds.,** *United States Earthquakes, 1975,* U.S. Government Printing Office, Washington, D.C., 1977.

1031. **Yanev, P.I.,** *The Lice, Turkey, Earthquake of September 6, 1975 (Reconnaisance Report),* URS/John A. Blume & Associates, San Francisco, 1975.

1032. **Yanev, P.I.,** The Lice, Turkey, earthquake of September 6, 1975: a preliminary engineering investigation, *Earthquake Inf. Bull.,* 8, 4, 1976.

1033. **Stratta, J.L. and Feldman, J.,** Interaction of infill walls and concrete frames during earthquakes, *Bull. Seismol. Soc. Am.,* 61, 609, 1971.

1034. **Uzsoy, S.Z. and Ersoy, U.,** Damage to reinforced concrete buildings caused by the July 22, 1967 earthquake in Turkey, *Bull. Seismol. Soc. Am.,* 59, 631, 1969.

1035. **Meehan, J.F.,** Performance of school buildings in the Peru earthquake of May 31, 1970, *Bull. Seismol. Soc. Am.,* 61, 591, 1971.

1036. **Uzsoy, S.Z. and Citipitioglu, E.,** Influence of infill walls on building frames: an example from the May 12, 1971 earthquake in Turkey, *Bull. Seismol. Soc. Am.,* 62, 1113, 1972.

1037. **Keightley, W.O.,** Destructive Earthquakes in Burdur and Bingöl, Turkey—May 1971, National Academy of Sciences, Washington, D.C., 1975.

1038. **Duke, C.M. and Moran, D.F.,** personal communication, 1978.

1039. **Duke, C.M. and Mathiesen, R.B.,** Earthquakes, lifelines and ASCE, *Civil Eng.,* 43, 65, 1973.

1040. **Taleb-Agha, G.,** Seismic risk analysis of lifeline networks, *Bull. Seismol. Soc. Am.,* 67, 1625, 1977.

1041. Subcommittee on Energy and Communication Systems-NOAA/EERI Earthquake Investigation Committee (West, P.J.—Chairman, Hamilton, P., Jr., Cree, D.E., McNorgan, J., Simon, E., Schoustra, J.J., Wong, P.P., and Robb, H.), Preface to energy and communication systems, in San Fernando, California, Earthquake of February 9, 1971, Vol. 2, Benfer, N.A., Coffman, J.L., and Dees, L.T., Eds., U.S. Government Printing Office, Washington, D.C., 1973, 7.

1042. Subcommittee on Water and Sewerage Systems-NOAA/EERI Earthquake Investigation Committee (Phillips, R.V.—Chairman, Georgeson, D.L.—Vice Chairman, Anderson, L.L., James, L.B., Steinwert, D.C., Kruse, G.H., Koehm, E., Bird, J.M., Bruington, A.E., Bradley, N., Bargman, R.D., Jones, G.W., Lund, L.V., Abbott, W.O., Jr., Marynick, S.M., Wool, J.M., Braxton, E.G., Anderson, M.M., Riggs, E.H., King, P.V., Comstock, J., and Putnam, K.R.), Preface to water and sewerage systems, in San Fernando, California, Earthquake of February 9, 1971, Vol. 2, Benfer, N.A., Coffman, J.L., and Dees, L.T., Eds., U.S. Government Printing Office, Washington, D.C., 1973, 73.

1043. Subcommittee on Transportation Systems-NOAA/EERI Earthquake Investigation Committee (Meehan, J.F.—Chairman, Hare, J.F., Jr.—Vice Chairman, Elliot, A. L., Nagai, I., Wilson, M. R., Newby, J. E., Beaton, J.L., Ayanian, H., Himelhock, A.L., Drosendahl, R., Smith, T., Durr, D.L., Sr., Prysock, R.H., and Egan, J.P.), Preface to transportation systems, in San Fernando, California, Earthquake of February 9, 1971, Vol. 2, Benfer, N.A., Coffman, J.L., and Dees, L.T., Eds., U.S. Government Printing Office, Washington, D.C., 1973, 197.

1044. **Benfer, N.A., Coffman, J.L., and Dees, L.T., Eds.,** Utilities, transportation, and sociological aspects, in San Fernando, California, Earthquake of February 9, 1971, Vol. 2, U.S. Government Printing Office, Washington, D.C., 1973.

1045. **Wong, P.P.,** Earthquake effects on power system facilities of the City of Los Angeles, in San Fernando, California, Earthquake of February 9, 1971, Vol. 2, Benfer, N.A., Coffman, J.L., and Dees, L.T., Eds., U.S. Government Printing Office, Washington, D.C., 1973, 9.

1046. Southern California Edison Company, Earthquake damage to Southern California Edison Company power facilities, in San Fernando, California, Earthquake of February 9, 1971, Vol. 2, Benfer, N.A., Coffman, J.L., and Dees, L.T., Eds., U.S. Government Printing Office, Washington, D.C., 1973, 27.

1047. **Moran, D.F.,** Damage to energy and communication systems, in The San Fernando, California, Earthquake of February 9, 1971, U.S. Geological Survey Professional Paper 733, U.S. Government Printing Office, Washington, D.C., 1971, 245.

1048. Southern California Gas Company, Earthquake effects on Southern California Gas Company facilities, in San Fernando, California, Earthquake of February 9, 1971, Vol. 2, Benfer, N.A., Coffman, J.L., and Dees, L.T., Eds., U.S. Government Printing Office, Washington, D.C., 1973, 59.

1049. General Telephone Company of California, Earthquake damage to General Telephone Company of California facilities, in San Fernando, California, Earthquake of February 9, 1971, Vol. 2, Benfer, N.A., Coffman, J.L., and Dees, L.T., Eds., U.S. Government Printing Office, Washington, D.C., 1973, 47.

1050. **Waananen, A.O. and Moyle, W.R., Jr.,** Water-resources aspects, in The San Fernando, California, Earthquake of February 9, 1971, U.S. Geological Survey Professional Paper 733, U.S. Government Printing Office, Washington, D.C., 1971, 119.

1051. **Duke, C.M.,** Damage to water supply systems, in The San Fernando, California, Earthquake of February 9, 1971, U.S. Geological Survey Professional Paper 733, U.S. Government Printing Office, Washington, D.C., 1971, 225.

1052. Subcommittee on Water and Sewerage Systems-NOAA/EERI Earthquake Investigation Committee, Earthquake damage to water and sewerage facilities, in San Fernando, California, Earthquake of February 9, 1971, Vol. 2, Benfer, N.A., Coffman, J.L., and Dees, L.T., Eds., U.S. Government Printing Office, Washington, D.C., 1973, 75.

1053. Joint Committee on Seismic Safety, The San Fernando earthquake study, in Meeting the Earthquake Challenge, California Legislature, Sacramento, 1974, 215.

1054. Earthquake Engineering Research Institute Committee, Sociological aspects of the earthquake, in The San Fernando, California, Earthquake of February 9, 1971, U.S. Geological Survey Professional Paper 733, U.S. Government Printing Office, Washington, D.C., 1971, 251.

1055. **Dukleth, G.W. and Stroppini, E.W.**, Seismic safety for California dams, *Calif. Geol.*, 29, 243, 1976.

1056. **Elliott, A. and Nagai, I.**, Earthquake damage to freeway bridges, in San Fernando, California, Earthquake of February 9, 1971, Vol. 2, Benfer, N.A., Coffman, J.L., and Dees, L.T., Eds., U.S. Government Printing Office, Washington, D.C., 1973, 201.

1057. Foundation Section, California Division of Highways, Earthquake damage to California highways, in San Fernando, California, Earthquake of February 9, 1971, Vol. 2, Benfer, N.A., Coffman, J.L., and Dees, L.T., Eds., U.S. Government Printing Office, Washington, D.C., 1973, 235.

1058. **Meehan, J.F.**, Damage to transportation systems, in The San Fernando, California, Earthquake of February 9, 1971, U.S. Geological Survey Professional Paper 733, U.S. Government Printing Office, Washington, D.C., 1971, 241.

1059. **Rojahn, C.**, Earthquake damage to Los Angeles County roads, in San Fernando, California, Earthquake of February 9, 1971, Vol. 2, Benfer, N.A., Coffman, J.L., and Dees, L.T., Eds., U.S. Government Printing Office, Washington, D.C., 1973, 247.

1060. **Newby, J.E.**, Earthquake damage to railroads, in San Fernando, California, Earthquake of February 9, 1971, Vol. 2, Benfer, N.A., Coffman, J.L., and Dees, L.T., Eds., U.S. Government Printing Office, Washington, D.C., 1973, 253.

1061. **Wilson, M.R.**, Earthquake effects on airports, in San Fernando, California, Earthquake of February 9, 1971, Vol. 2, Benfer, N.A., Coffman, J.L., and Dees, L.T., Eds., U.S. Government Printing Office, Washington, D.C., 1973, 199.

1062. **Duke, C.M. and Moran, D.F.**, Lifelines, in *Architects and Earthquakes: Research Needs*, AIA Research Corp., Washington, D.C., 1976, 86.

1063. Earthquake Engineering Research Institute, Learning from Earthquakes, Earthquake Engineering Research Institute, Oakland, Calif., 1977.

1064. American Society of Civil Engineers, *The Current State of Knowledge of Lifeline Earthquake Engineering*, American Society of Civil Engineers, New York, 1977.

1065. **Duke, C.M. and Moran, D.F.**, Earthquakes and city lifelines, in San Fernando Earthquake of February 9, 1971 and Public Policy, Joint Committee on Seismic Safety, California Legislature, Sacramento, 1972, 53.

1066. **Way, G.L.**, Report of Subcommittee on Earthquake Resistance of Public Utility Systems, Governor's Interagency Earthquake Committee, Public Utilities Commission, San Francisco, 1974.

1067. **Will, A.G.**, Implementation of Recommendations Developed from the Reports of the Earthquake Commission and the Earthquake Task Force, Los Angeles County, 1973.

1068. **Hudson, D.E.**, Dynamic tests of full-scale structures, in *Earthquake Engineering*, Wiegel, R.L., Ed., Prentice-Hall, Englewood Cliffs, N.J., 1970, 127.

1069. **Hudson, D.E.**, Ground motion measurements, in *Earthquake Engineering*, Wiegel, R.L., Ed., Prentice-Hall, Englewood Cliffs, N.J., 1970, 107.

1070. **Gates, W.E.**, High-rise buildings with strong-motion instruments — dynamic analyses, in San Fernando, California, Earthquake of February 9, 1971, Vol. I (Part A), Benfer, N.A., Coffman, J.L., and Dees, L.T., Eds., U.S. Government Printing Office, Washington, D.C., 1973, 297.

1071. **Jennings, P.C. and Bielak, J.**, Dynamics of building-soil interaction, *Bull. Seismol. Soc. Am.*, 63, 9, 1973.

1072. **Hudson, D.E., Brady, A.G., and Trifunac, M.D.**, Strong-Motion Earthquake Accelerograms, Digitized and Plotted Data, Vol. IA, EERL 69-20, Earthquake Engineering Research Laboratory, California Institute of Technology, Pasadena, 1969.

1073. **Trifunac, M.D.**, Low-Frequency Digitization Errors and a New Method for Zero Base-Line Correction of Strong-Motion Accelerograms, EERL 70-07, Earthquake Engineering Research Laboratory, California Institute of Technology, Pasadena, 1970.

1074. **Trifunac, M. D.**, Zero base line correction of strong-motion accelerograms, *Bull. Seismol. Soc. Am.*, 61, 1201, 1971.

1075. **Trifunac, M.D.**, A note on correction of strong-motion accelerograms for instrument response, *Bull. Seismol. Soc. Am.*, 62, 401, 1972.

1076. **Hudson, D.E. and Brady, A.G.**, Strong-Motion Earthquake Accelerograms, Digitized and Plotted Data, Vol. IB, EERL 70-21, Earthquake Engineering Research Laboratory, California Institute of Technology, Pasadena, 1970.

1077. **Hudson, D.E., Brady, A.G., Trifunac, M.D., and Vijayaraghavan, A.**, Strong-Motion Earthquake Accelerograms — Corrected Accelerograms and Integrated Velocity and Displacement Curves, Vol. 2 (Part A), EERL 71-51, Earthquake Engineering Research Laboratory, California Institute of Technology, Pasadena, 1971.

1078. **Trifunac, M.D., Brady, A.G., Hudson, D.E., and Hanks, T.C.**, Strong-Motion Earthquake Accelerograms — Corrected Accelerograms and Integrated Velocity and Displacement Curves, Vol. 2 (Part G), EERL 73-52, Earthquake Engineering Research Laboratory, California Institute of Technology, Pasadena, 1973.

1079. **Trifunac, M.D. and Lee, V.W.,** Routine Computer Processing of Strong-Motion Accelerograms, EERL 73-03, Earthquake Engineering Research Laboratory, California Institute of Technology, Pasadena, 1973.
1080. **Trifunac, M.D., Udwadia, F.E., and Brady, A.G.,** Analysis of errors in digitized strong-motion accelerograms, *Bull. Seismol. Soc. Am.,* 63, 157, 1973.
1081. **Trifunac, M.D. and Lee, V.W.,** A note on the accuracy of computed ground displacements from strong-motion accelerographs, *Bull. Seismol. Soc. Am.,* 64, 1209, 1974.
1082. **U.S. Geological Survey,** The national strong-motion instrumentation network, *Earthquake Inf. Bull.,* 7, 16, 1975.
1083. **Veletsos, A.S.,** Site and soil-structure interaction effects, in *Architects and Earthquakes: Research Needs,* AIA Research Corp., Washington, D.C., 1976, 18.
1084. **Mohraz, B.,** A study of earthquake response spectra for different geological conditions, *Bull. Seismol. Soc. Am.,* 66, 915, 1976.
1085. **Hays, W.W.,** A note on the duration of earthquake and nuclear-explosion ground motions, *Bull. Seismol. Soc. Am.,* 65, 875, 1975.
1086. **Housner, G.W.,** Design spectrum, in *Earthquake Engineering,* Wiegel, R.L., Ed., Prentice-Hall, Englewood Cliffs, N.J., 1970, 93.
1087. **Hays, W.W.,** The design earthquake and earthquake response spectra, *Earthquake Inf. Bull.,* 6, 18, 1974.
1088. **Hudson, D.E.,** Ground motion measurements in earthquake engineering, in *Proceedings, Symposium on Earthquake Engineering,* University of British Columbia, Vancouver, 1965.
1089. **Hudson, D.E.,** The measurement of ground motion of destructive earthquakes, *Bull. Seismol. Soc. Am.,* 53, 419, 1963.
1090. **Anon.,** Recent strong-motion records, in Seismic Engineering Program Report January-April 1977, U.S. Geological Survey Circular 762-A, 1977, 1.
1091. **Porcella, R.L.,** Recent strong-motion records, in Seismic Engineering Program Report October-December 1976, U.S. Geological Survey Circular 736-D, 1977, 1.
1092. **Anon.,** Seismic estuarine studies in the Puget Sound area, *NOAA,* 2, 60, 1972.
1093. **Anon.,** A new instrument net design aid estimates of Puget Sound earthquake hazard, *Earthquake Inf. Bull.,* 4, 20, 1972.
1094. **Morrill, B.J.,** The APEEL Array: A Site Study, NOAA Technical Report ERL 245-ESL, U.S. Government Printing Office, Washington, D.C., 1972.
1095. **Matthiesen, R.B.,** Planning and design of strong-motion instrument networks, in Seismic Engineering Program Report April—June 1976, U.S. Geological Survey Circular 736-B, 1976, 9.
1096. **Hill, M.R.,** Facts about California's new strong-motion instrumentation program, *Calif. Geol.,* 25, 162, 1972.
1097. **Anon.,** First earthquake devices installed, *Calif. Geol.,* 26, 12, 1973.
1098. **Rojahn, C.,** California building strong-motion earthquake instrumentation program, in Seismic Engineering Program Report January-March 1976, U.S. Geological Survey Circular 736-A, 1976, 3.
1099. **Rojahn, C.,** Building strong-motion earthquake instrumentation, in Seismic Engineering Program Report April—June 1976, U.S. Geological Survey Circular 736-B, 1976, 10.
1100. **Wootton, T.M.,** Earthquake ground motion records, *Calif. Geol.,* 30, 86, 1977.
1101. **California Division of Mines and Geology,** Second report on the strong-motion instrumentation program, in Seismic Engineering Program Report July—September 1976, U.S. Geological Survey Circular 736-C, 1976, 6.
1102. **Power, J.H. and Real, C.R.,** Shear wave velocity, *Calif. Geol.,* 29, 27, 1976.
1103. **Dallaire, G.,** San Francisco ce's spark code improvements, *Civil Eng.,* 43, 36, 1973.
1104. **Anon.,** Strong-motion instruments for VA hospitals, *Earthquake Inf. Bull.,* 4, 10, 1972.
1105. **Anon.,** Strong-motion instruments for California's highways in the sky, *Earthquake Inf. Bull.,* 4, 18, 1972.
1106. **Gates, J.,** California Division of Highways strong-motion studies, *Calif. Earthquake Sci. Newsl.,* 1, 3, 1973.
1107. **Maley, R.P. and Cloud, W.K.,** Strong-motion accelerograph records, in San Fernando, California, Earthquake of February 9, 1971, Vol. 3, Benfer, N.A., Coffman, J.L., Bernick, J.R., and Dees, L. T., Eds., U.S. Government Printing Office, Washington, D.C., 1973, 325.
1108. **Murphy, J.R. and O'Brien, L.J.,** The correlation of peak ground acceleration amplitude with seismic intensity and other physical parameters, *Bull. Seismol. Soc. Am.,* 67, 877, 1977.
1109. **Herrmann, R.B., Fischer, G.W., and Zollweg, J.E.,** The June 13, 1975 earthquake and its relationship to the New Madrid seismic zone, *Bull. Seismol. Soc. Am.,* 67, 209, 1977.
1110. **Blume, J.A.,** Response of highrise buildings to ground-motion from underground nuclear detonations, *Bull. Seismol. Soc. Am.,* 59, 2343, 1969.

1111. **Borcherdt, R.D.,** Effects of local geology on ground motion near San Francisco Bay, *Bull. Seismol. Soc. Am.,* 60, 29, 1970.

1112. **Blume, J.A.,** The motion and damping of buildings relative to seismic response spectra, *Bull. Seismol. Soc. Am.,* 60, 231, 1970.

1113. **Blume, J. A.,** Highrise building characteristics and responses determined from nuclear seismology, *Bull. Seismol. Soc. Am.,* 62, 519, 1972.

1114. **Borcherdt, R.D. and Gibbs, J.F.,** Effects of local geological conditions in the San Francisco Bay region on ground motions and the intensities of the 1906 earthquake, *Bull. Seismol. Soc. Am.,* 66, 467, 1976.

1115. **Anon.,** The building shakers, *Earthquake Inf. Bull.,* 4, 3, 1972.

1116. **Blume, J.A.,** The building vibrator, in *Earthquake Investigations in the Western United States 1931-1964,* Carder, D.S., Ed., U.S. Government Printing Office, Washington, D.C., 1964, 177.

1117. **Mader, G.G.,** Land use planning, in *Architects and Earthquakes: Research Needs,* AIA Research Corporation, Washington, D.C., 1976, 11.

1118. **Mader, G.G.,** Land use planning for seismic safety, in *Summer Seismic Institute for Architectural Faculty,* AIA Research Corporation, Washington, D.C., 1977, 27.

1119. Tri-Cities Citizens Advisory Committee on Seismic Safety, The seismic safety program, in The Seismic Safety Study for the General Plan, California Council on Intergovernmental Relations, Sacramento, 1973, 1.

1120. Tri-Cities Citizens Advisory Committee on Seismic Safety, The Seismic Safety Study for the General Plan, California Council on Intergovernmental Relations, Sacramento, 1973.

1121. Grading Codes Advisory Board and Building Code Committee, Southern California Section, Association of Engineering Geologists, Guidelines to Geologic/Seismic Reports, CDMG Note No. 37, California Division of Mines and Geology, Sacramento, 1973: also printed in *Calif. Geol.,* 27, 111, 1974.

1122. **Anon.,** Alquist-Priolo Geologic Hazard Zones Act, *Calif. Geol.,* 27, 7, 1974.

1123. **Hart, E.W.,** Zoning for surface fault hazards in California: the new special studies zones maps, *Calif. Geol.,* 27, 227, 1974.

1124. State Geologist, Explanation of Special Studies Zones Maps, Compiled by the State Geologist, California Division of Mines and Geology, Sacramento, no date.

1125. **Hart, E.W.,** New special studies zones maps, *Calif. Geol.,* 31, 55, 1978.

1126. California Division of Mines and Geology, Recommended Guidelines for Determining the Maximum Credible and the Maximum Probable Earthquakes, CDMG Note No. 43, California Division of Mines and Geology, Sacramento, 1975; also printed in *Calif. Geol.,* 30, 227, 1977.

1127. California Division of Mines and Geology, Checklists for the Review of Geologic/Seismic Reports, CDMG Note No. 48, California Division of Mines and Geology, Sacramento, 1975; also printed in *Calif. Geol.,* 30, 226, 1977.

1128. **Murphy, G.H.,** Geologists and environmental impact reports, *Calif. Geol.,* 29, 22, 1976.

1129. **Stewart, R.M., Hart, E.W., and Amimoto, P.Y.,** The review process and the adequacy of geologic reports, *Calif. Geol.,* 30, 224, 1977.

1130. **Amimoto, P.Y. and Slosson, J.E.,** Guidelines for Geologic/Seismic Considerations in Environmental Impact Reports, CDMG Note No. 46, California Division of Mines and Geology, Sacramento, 1975; also printed in *Calif. Geol.,* 30, 13, 1977.

1131. **Amimoto, P.Y.,** Environmental impact reports, *Calif. Geol.,* 30, 12, 1977.

1132. **Nichols, D.R. and Buchannan-Banks, J.M.,** Seismic Hazards and Land Use Planning, U.S. Geological Survey Circular 690, 1974.

1133. **Linville, J., Jr.,** Land-use planning to mitigate earthquake disasters, paper presented at the 2nd Annu. Natural Hazards Research Workshop, Institute of Behavioral Science, University of Colorado, Boulder, 1977.

1134. **Borcherdt, R.D.,** Foreword, in Studies for Seismic Zonation of the San Francisco Bay Region, Borcherdt, R.D., Ed., U.S. Geological Survey Professional Paper 941-A, U.S. Government Printing Office, Washington, D.C., 1975, iii.

1135. **Wallace, R.E.** (Project Director), San Francisco Bay Region Environment and Resources Planning Study, U.S. Department of Interior and U.S. Department of Housing and Urban Development, Washington, D.C., 1972.

1136. **Anon.,** San Francisco Bay study, *Calif. Geol.,* 30, 66, 1977.

1137. **Adams, V.W.,** Earth Science Data in Urban and Regional Information Systems — a Review, U.S. Geological Survey Circular 712, 1975.

1138. **Borcherdt, R.D., Ed.,** Studies for Seismic Zonation of the San Francisco Bay Region, U.S. Geological Survey Professional Paper 941-A, U.S. Government Printing Office, Washington, D.C., 1975.

1139. **Lajoie, K.R. and Helley, E.J.,** Differentiation of sedimentary deposits for purposes of seismic zonation, in Studies for Seismic Zonation of the San Francisco Bay Region, Borcherdt, R.D., Ed., U.S. Geological Survey Professional Paper 941-A, U.S. Government Printing Office, Washington, D.C., 1975, 39.

1140. **Ritter, J.R. and Dupre, W.R.,** Map Showing Areas of Potential Inundation by Tsunamis in the San Francisco Bay Region, California, U.S. Geological Survey Miscellaneous Field Investigation Map MF-480, 1972.

1141. **Youd, T.L., Nichols, D.R., Helley, E.J., and Lajoie, K.R.,** Liquefaction potential, in Studies for Seismic Zonation of the San Francisco Bay Region, Borcherdt, R.D., Ed., U.S. Geological Survey Professional Paper 941-A, U.S. Government Printing Office, Washington, D.C., 1975, 68.

1142. **Sabins, F.F., Jr.,** *Remote Sensing Principles and Interpretation,* W.H. Freeman & Co., San Francisco, 1978.

1143. **Estes, J.E. and Senger, L.W., Eds.,** *Remote Sensing—Techniques for Environmental Analysis,* Hamilton Publishing Co., Santa Barbara, Calif., 1974.

1144. **Rudd, R.D.,** *Remote Sensing: A Better View,* Duxbury Press, North Scituate, Mass., 1974.

1145. **Reeves, R.G., Ed.,** *Manual of Remote Sensing,* vol. I and II, American Society of Photogrammetry, Falls Church, Va., 1975.

1146. **Schanda, E., Ed.,** *Remote Sensing for Environmental Sciences,* Springer-Verlag, Berlin, 1976.

1147. **Lintz, J. and Simonett, D.S.,** *Remote Sensing of Environment,* Addison-Wesley Publishing, Reading, Mass., 1976.

1148. **Vestappen, H.T.,** *Remote Sensing in Geomorphology,* Elsevier, Amsterdam, 1977.

1149. **Avery, T.E.,** *Interpretation of Aerial Photographs,* 3rd ed., Burgess, Minneapolis, 1977.

1150. **O'Leary, D.W., Friedman, J.D., and Pohn, H.A.,** Lineament, linear, lineation: some proposed new standards for old terms, *Geol. Soc. Am. Bull.,* 87, 1463, 1976.

1151. **Lueder, D.R.,** *Aerial Photographic Interpretation,* McGraw-Hill, New York, 1959.

1152. **Ray, R.G.,** Aerial Photographs in Geologic Mapping and Interpretation, U.S. Geological Survey Professional Paper 373, U.S. Government Printing Office, Washington, D.C., 1960.

1153. **Colwell, R.N., Ed.,** *Manual of Photographic Interpretation,* American Society of Photogrammetry, Falls Church, Va., 1960.

1154. **Mollard, J.D.,** *Air Photo Analysis and Interpretation,* Bellhaven House, Scarborough, Ontario, 1960.

1155. **Miller, V.C.,** *Photogeology,* McGraw-Hill, New York, 1961.

1156. **von Bandat, H.F.,** *Aerogeology,* Gulf Publishing, Houston, Tex., 1962.

1157. **Lattman, L.H. and Ray, R.G.,** *Aerial Photographs in Field Geology,* Holt, Rinehart & Winston, New York, 1965.

1158. **Strandberg, C.H.,** *Aerial Discovery Manual,* John Wiley & Sons, New York, 1967.

1159. **Smith, J.T. and Anson, H., Eds.,** *Manual of Color Aerial Photography,* American Society of Photogrammetry, Falls Church, Va., 1968.

1160. **Way, D.S.,** *Terrain Analysis,* Dowden, Hutchinson, & Ross, Stroudsburg, Pa., 1973.

1161. **Clark, M.M.,** Geologic utility of small-scale airphotos, in Second Annual Earth Resources Aircraft Program, Status Review, Vol. I—Geology and Geography, NASA Manned Spacecraft Center, Houston, Tex., 1969, 8—1.

1162. **Cluff, L.S. and Slemmons, D.B.,** Wasatch fault zone—features defined by low-sun-angle photography, in *Environmental Geology of the Wasatch Front, 1971,* Utah Geological Association, Salt Lake City, 1972, Gl.

1163. **Clanton, U.S. and Amsbury, D.L.,** Active faults in southeastern Harris County, Texas, *Environ. Geol.,* 1, 149, 1975.

1164. **Nicks, O.W., Ed.,** This Island Earth, NASA SP-250, U.S. Government Printing Office, Washington, D.C., 1970.

1165. **Lowman, P.D. and Tiedeman, H.A.,** Terrain Photography from Gemini Spacecraft—Final Geologic Report, NASA Goddard Space Flight Center, Greenbelt, Md., 1971.

1166. **Bodechtel, J. and Gierloff-Emden, H.-G.,** *The Earth from Space,* Arco Publishing, New York, 1974.

1167. National Aeronautics and Space Administration, Skylab Earth Resources Data Catalog, U.S. Government Printing Office, Washington, D.C., 1974.

1168. National Aeronautics and Space Administration, Landsat Users Handbook, Goddard Space Flight Center, Greenbelt, Md., 1976.

1169. **Short, N.M., Lowman, P.D., Freden, S.C., and Finch, W.A.,** Mission to Earth, Landsat Views the World, NASA SP-360, U.S. Government Printing Office, Washington, D.C., 1976.

1170. **Williams, R.S. and Carter, W.D., Eds.,** ERTS-1, A New Window on Our Planet, U.S. Geological Survey Professional Paper 929, U.S. Government Printing Office, Washington, D.C., 1976.

1171. **Otterman, J., Lowman, P.D., and Salmonson, V.V.,** Surveying earth resources by remote sensing from satellites, *Geophys. Surv.,* 2, 431, 1976.

1172. National Aeronautics and Space Administration, Skylab Explores the Earth, NASA SP-380, U.S. Government Printing Office, Washington, D.C., 1977.

1173. **Marshall, G.C.,** Skylab, Our First Space Station, NASA SP-400, U.S. Government Printing Office, Washington, D.C., 1977.

1174. **Summerlin, L.B., Ed.,** Skylab, Classroom in Space, NASA SP-401, U.S. Government Printing Office, Washington, D.C., 1977.

1175. **Kent, M.I., Stuhlinger, E., and Wu, S.T., Eds.,** *Scientific Investigations on the Skylab Satellite,* American Institute of Aeronautics and Astronautics, New York, 1977.

1176. **Sabins, F.F., Jr.,** Geologic investigation of radar and space imagery of California, *Am. Assoc. Petrol. Geol. Bull.,* 57, 802, 1973.

1177. **Isachsen, Y.W., Fakundiny, R.H., and Forster, S.W.,** Evaluation of ERTS-1 imagery for geological sensing over the diverse geological terranes of New York State, in Symposium on Significant Results Obtained from the Earth Resources Technology Satellite-1, Vol. 1, Freden, S.C., Mercanti, E.P., and Becker, M.A., Eds., NASA SP-327, U.S. Government Printing Office, Washington, D.C., 1973, 223.

1178. **Lathram, E.H., Tailleur, I.L., Patton, W.W., Jr., and Fischer, W.A.,** Preliminary geologic application of ERTS-1 imagery in Alaska, in Symposium on Significant Results Obtained from the Earth Resources Technology Satellite-1, Vol. 1, Freden, S.C., Mercanti, E.P., and Becker, M.A., Eds., NASA SP-327, U.S. Government Printing Office, Washington, D.C., 1973, 257.

1179. **Allen, W.H., Martin, J.A., and Rath, D.L.,** First-look analysis of geologic ground patterns on ERTS-1 imagery of Missouri, in Symposium on Significant Results Obtained from the Earth Resources Technology Satellite-1, Vol. 1, Freden, S.C., Mercanti, E.P., and Becker, M.A., Eds., NASA SP-327, U.S. Government Printing Office, Washington, D.C., 1973, 371.

1180. **Goetz, A.F.H., Billingsley, F.C., Elston, D., Lucchitta, I., and Shoemaker, E.M.,** Preliminary geologic investigations in the Colorado Plateau using enhanced ERTS images, in Symposium on Significant Results Obtained from the Earth Resources Technology Satellite-1, Vol. 1, Freden, S.C., Mercanti, E.P., and Becker, M.A., Eds., NASA SP-327, U.S. Government Printing Office, Washington, D.C., 1973, 403.

1181. **Rowan, L.C. and Wetlaufer, P.H.,** Structural geologic analysis in Nevada using ERTS-1 images: a preliminary report, in Symposium on Significant Results Obtained from the Earth Resources Technology Satellite-1, Vol. 1, Freden, S.C., Mercanti, E.P., and Becker, M.A., Eds., NASA SP-327, U.S. Government Printing Office, Washington, D.C., 1973, 413.

1182. **Bechtold, I.C., Liggett, M.A., and Childs, J.F.,** Regional tectonic control of Tertiary mineralization and Recent faulting in the southern Basin-Range Province, an application of ERTS-1 data, in Symposium on Significant Results Obtained from the Earth Resources Technology Satellite-1, Vol. 1, Freden, S.C., Mercanti, E.P., and Becker, M.A., Eds., NASA SP-327, U.S. Government Printing Office, Washington, D.C., 1973, 425.

1183. **Abdel-Gawad, M. and Silverstein, J.,** ERTS application in earthquake research and mineral exploration in California, in Symposium on Significant Results Obtained from the Earth Resources Technology Satellite-1, Vol. 1, Freden, S.C., Mercanti, E.P., and Becker, M.A., Eds., NASA SP-327, U.S. Government Printing Office, Washington, D.C., 1973, 433.

1184. **Gedney, L.D. and Van Wormer, J.D.,** Some aspects of active tectonism in Alaska as seen on ERTS-1 imagery, in Symposium on Significant Results Obtained from the Earth Resources Technology Satellite-1, Vol. 1, Freden, S.C., Mercanti, E.P., and Becker, M.A., Eds., NASA SP-327, U.S. Government Printing Office, Washington, D.C., 1973, 451.

1185. **Carter, W.D. and Eaton, G.P.,** ERTS-1 image contributes to understanding of geologic structures related to Managua earthquake, 1972, in Symposium on Significant Results Obtained from the Earth Resources Technology Satellite-1, Vol. 1, Freden, S.C., Mercanti, E.P., and Becker, M.A., Eds., NASA SP-327, U.S. Government Printing Office, Washington, D.C., 1973, 459.

1186. **Bodechtel, J. and Lammerer, B.,** New aspects on the tectonics of the Alps and the Apennines revealed by ERTS-1 data, in Symposium on Significant Results Obtained from the Earth Resources Technology Satellite-1, Vol. 1, Freden, S.C., Mercanti, E.P., and Becker, M.A., Eds., NASA SP-327, U.S. Government Printing Office, Washington, D.C., 1973, 493.

1187. **Steffensen, R.,** Structural lineaments of Gaspe from ERTS imagery, in Symposium on Significant Results Obtained from the Earth Resources Technology Satellite-1, Vol. 1, Freden, S.C., Mercanti, E.P., and Becker, M.A., Eds., NASA SP-327, U.S. Government Printing Office, Washington, D.C., 1973, 501.

1188. **Withington, C.F.,** Lineaments in Coastal Plain sediments as seen in ERTS imagery, in Symposium on Significant Results Obtained from the Earth Resources Technology Satellite-1, Vol. 1, Freden, S.C., Mercanti, E.P., and Becker, M.A., Eds., NASA SP-327, U.S. Government Printing Office, Washington, D.C., 1973, 517.

1189. **Hoppin, R.A.,** Structural interpretations based on ERTS-1 imagery, Bighorn region, Wyoming-Montana, in Symposium on Significant Results Obtained from the Earth Resources Technology Satellite-1, Vol. 1, Freden, S.C., Mercanti, E.P., and Becker, M.A., Eds., NASA SP-327, U.S. Government Printing Office, Washington, D.C., 1973, 531.

1190. **Weidman, R.M., Alt, D.D., Flood, R.E., Jr., Hawley, K.T., Wackwitz, L.K., Berg, R.B., and Johns, W.M.,** Applicability of ERTS-1 to lineament and photogeologic mapping in Montana — preliminary report, in Symposium on Significant Results Obtained from the Earth Resources Technology Satellite-1, Vol. 1, Freden, S.C., Mercanti, E.P., and Becker, M.A., Eds., NASA SP-327, U.S. Government Printing Office, Washington, D.C., 1973, 539.

1191. **Pease, R.W. and Johnson, C.W.,** A new fault lineament in southern California, in Symposium on Significant Results Obtained from the Earth Resources Technology Satellite-1, Vol. 1, Freden, S.C., Mercanti, E.P., and Becker, M.A., Eds., NASA SP-327, U.S. Government Printing Office, Washington, D.C., 1973, 547.

1192. **List, F.K., Roland, N.W., and Helmcke, D.,** Comparison of geological information from satellite imagery, aerial photography, and ground investigations in the Tibesti Mountains, Chad, in *Proc. Symp. Remote Sensing and Photo Interpretation,* Vol. 2, Canadian Institute of Surveying, Ottawa, 1974, 543.

1193. **el Shazly, E.M., Abdel-Hady, M.A., el Gxawaby, M.A., and el Kassas, I.A.,** Geologic interpretation of ERTS-1 satellite images for east Aswan area, Egypt, in *Proc. 9th Int. Symp. Remote Sensing of Environment,* Vol. 1, Environmental Research Institute of Michigan, Ann Arbor, 1974, 105.

1194. **el Shazly, E.M., Abdel-Hady, M.A., el Gxawaby, M.A., and el Kassas, I.A.,** Geologic interpretation of ERTS-1 satellite images for west Aswan area, Egypt, in *Proc. 9th Int. Symp. Remote Sensing of Environment,* Vol. 1, Environmental Research Institute of Michigan, Ann Arbor, 1974, 119.

1195. **Makarov, V.I., Skobelev, S.F., Trifonov, V.G., Florenskiy, P.V., and Shchukin, Y.K.,** Plutonic structure of the earth's crust on space images, in *Proc. 9th Int. Symp. Remote Sensing of Environment,* Vol. 1, Environmental Research Institute of Michigan, Ann Arbor, 1974, 369.

1196. **Miller, J.M. and Belon, A.E.,** A summary of ERTS data applications in Alaska, in *Proc. 9th Int. Symp. Remote Sensing of Environment,* Vol. 3, Environmental Research Institute of Michigan, Ann Arbor, 1974, 2113.

1197. **Gedney, L. and van Wormer, J.,** In Alaska: remote sensing of seismic hazards, *Geotimes,* 19, 15, 1974.

1198. **Lamar, D.L. and Merifield, P.M.,** Application of Skylab and ERTS Imagery to Fault Tectonics and Earthquake Hazards of Peninsular Ranges, Southwestern California, Technical Report 75-2, California Earth Science Corporation, Santa Monica, 1975.

1199. **O'Leary, D. and Simpson, S.,** Lineaments and tectonism in the northern part of the Mississippi embayment, in *Proc. 10th Int. Symp. Remote Sensing of Environment,* Vol. 2, Environmental Research Institute of Michigan, Ann Arbor, 1975, 965.

1200. **Baker, R.N.,** LANDSAT data: a new perspective for geology, *Photogramm. Eng. Remote Sensing,* 41, 1233, 1975.

1201. **Allen, C.R.,** Geological criteria for evaluating seismicity, *Geol. Soc. Am. Bull.,* 86, 1041, 1975.

1202. **Muehlberger, W. R. and Ritchie, A. W.,** Caribbean-Americas plate boundary in Guatemala and southern Mexico as seen on Skylab IV orbital photography, *Geology,* 3, 232, 1975.

1203. **Merifield, P. M. and Lamar, D. L.,** Active and inactive faults in southern California viewed from Skylab, in NASA Earth Resources Survey Symposium, Vol. 1B, NASA TMX-58168, U.S. Government Printing Office, Washington, D.C., 1975, 779.

1204. **Eggenberger, A. J., Rowlands, D., and Rizzo, P. C.,** The utilization of Landsat imagery in nuclear power plant siting, in NASA Earth Resources Survey Symposium, Vol. 1B, NASA TMX-58168, U.S. Government Printing Office, Washington, D.C., 1975, 799.

1205. **Eggenberger, A. J., Rowlands, D., and Rizzo, P. C.,** The utilization of Landsat imagery in nuclear power plant siting, in Abstracts of the NASA Earth Resources Survey Symposium, Lyndon B. Johnson Space Center, Houston, Tex., 1975, 130.

1206. **Anon.,** Lineaments linked with earthquakes in Mississippi embayment, *Earthquake Inf. Bull.,* 8, 21, 1976.

1207. **Berlin, G. L., Chavez, P. S., Jr., Grow, T. E., and Soderblom, L. A.,** Preliminary geologic analysis of southwest Jordan from computer enhanced Landsat 1 image data, in *Proc. Am. Soc. Photogrammetry 36th Annu. Meet.,* American Society of Photogrammetry, Falls Church, Va., 1976, 545.

1208. **Lowman, P. D., Jr.,** Geologic structure in California: three studies with Landsat-1 imagery, *Calif. Geol.,* 29, 75, 1976.

1209. **Clarke, A. O.,** Structural features of the Salton Trough as seen from Skylab, *Calif. Geol.,* 29, 187, 1976.

1210. **Longshaw, T. G., Viljoen, R. P., and Hodson, M. C.,** Photogeologic display of Landsat-1 CCT images for improved geological definition, *IEEE Trans. Geosci. Electron.,* 14, 66, 1976.

1211. **Gregory, A. F. and Moore, H. D.,** Recent advances in geologic applications of remote sensing from space, in *Proc. 24th Int. Astronautical Congress,* Pergamon Press, New York, 1976, 153.

1212. **Eichen, L. and Pascucci, R. F.,** Applications of remote-sensing technology for powerplant siting, in Proc. 1st Annu. William T. Pecora Memorial Symp., U.S. Geological Survey Professional Paper 1015, Woll, P.W. and Fischer, W.A., Eds., U.S. Government Printing Office, Washington, D.C., 1977, 133.

1213. **Cardamone, P., Lechi, G. M., Cavallin, A., Marino, C. M., and Zanferrari, A.,** Application of conventional and advanced techniques for the interpretation of Landsat 2 images for the study of linears in the Friuli earthquake area, in *Proc. 11th Int. Symp. Remote Sensing of Environment,* Vol. 2, Environmental Research Institute of Michigan, Ann Arbor, 1977, 1337.

1214. **Iranpanah, A.,** Geologic applications of Landsat imagery, *Photogramm. Eng. Remote Sensing,* 43, 1037, 1977.

1215. **O'Leary, D.,** Remote sensing for lineaments in the Mississippi embayment, *Earthquake Inf. Bull.,* 9, 14, 1977.

1216. **Chavez, P. S., Jr., Berlin, G. L., and Acosta, A. V.,** Computer processing of Landsat MSS digital data for linear enhancements, in *Proc. 2nd Annu. William T. Precora Memorial Symp.,* U.S. Geological Survey and American Society of Photogrammetry, Falls Church, Va., 1977, 235.

1217. **O'Leary, D. W. and Simpson, S. L.,** Remote sensor applications to tectonism and seismicity in the northern part of the Mississippi embayment, *Geophysics,* 42, 542, 1977.

1218. **Muehlberger, W. R., Gucwa, P. R., Ritchie, A. W., and Swanson, E. R.,** Global tectonics: some geologic analyses of observations and photographs from Skylab, in Skylab Explores the Earth, NASA SP-380, U.S. Government Printing Office, Washington, D.C., 1977, 49.

1219. **Silver, L. T., Anderson, T. H., Conway, C. M., Murray, J. D., and Powell, R. E.,** Geological features of southwestern North America, in Skylab Explores the Earth, NASA SP-380, U.S. Government Printing Office, Washington, D.C., 1977, 89.

1220. **Molnar, P. and Tapponnier, P.,** The collision between India and Eurasia, *Sci. Am.,* 236, 30, 1977.

1221. **Burkart, B.,** Offset across the Polochic fault of Guatemala and Chiapas, Mexico, *Geology,* 6, 328, 1978.

1222. **Longwell, C. R.,** Reconnaissance geology between Lake Mead and Davis Dam, Arizona-Nevada, in Shorter Contributions to General Geology 1960, U.S. Geological Survey Professional Paper 374, U.S. Government Printing Office, Washington, D.C., 1963, E-1.

1223. **Skolnik, M. I., Ed.,** *Radar Handbook,* McGraw-Hill, New York, 1970.

1224. **Matthews, R. E., Ed.,** Active Microwave Workshop Report, NASA SP-376,U.S. Government Printing Office, Washington, D.C., 1975.

1225. **Long, M. W.,** *Radar Reflectivity of Land and Sea,* D.C. Heath & Co., Lexington, Mass., 1975.

1226. **Jensen, H., Graham, L. C., Porcello, L. J., and Leith, E. N.,** Side-looking airborne radar, *Sci. Am.,* 237, 84, 1977.

1227. **Sabins, F. F., Jr.,** Engineering geology applications of remote sensing, in *Geology, Seismicity, and Environmental Impact,* Moran, D.E., Ed., Association of Engineering Geologists, Los Angeles, 1973, 141.

1228. **Cameron, H. L.,** Radar as a surveying instrument in hydrology and geology, in *Proc. 3rd Symp. Remote Sensing of Environment,* University of Michigan, Ann Arbor, 1964, 441.

1229. **Hilpert, L. S.,** Geological Evaluation of Radar Imagery, Southwestern and Central Utah, Technical Letter NASA-38, U.S. Geological Survey, Washington, D.C., 1966.

1230. **Hackman, R. J.,** Geologic Evaluation of Radar Imagery in Southern Utah, Technical Letter NASA-58, U.S. Geological Survey, Washington, D.C., 1966.

1231. **Roberts, R. J.,** Geological Evaluation of K-Band Radar Imagery, North-Central Nevada, Technical Letter NASA-49, U.S. Geological Survey, Washington, D.C., 1966.

1232. **Dellwig, L. F., Kirk, J. N., and Walters, R. L.,** The potential of low-resolution radar imagery in regional geologic studies, *J. Geophys. Res.,* 71, 4995, 1966.

1233. **Snavely, P. D., Jr. and Wagner, H. C.,** Geological Evaluation of Radar Imagery, Oregon Coast, Technical Letter NASA-16, U.S. Geological Survey, Washington, D.C., 1966.

1234. **MacDonald, H. C., Brennan, P. A., and Dellwig, L. F.,** Geologic evaluation by radar of NASA sedimentary test site, *IEEE Trans. Geosci. Electron.,* 5, 72 1967.

1235. **Hackman, R. J.,** Geologic evaluation of radar imagery in southern Utah, in Geological Survey Research 1967, U.S. Geological Survey Professional Paper 575D, U.S. Government Printing Office, Washington, D.C., 1967, 135.

1236. **Kirk, J. N. and Walters, R. L.,** Preliminary report on radar lineaments in the Boston Mountains of Arkansas, *Compass Sigma Gamma Epsilon,* 45, 122, 1968.

1237. **Reeves, R. G.,** Structural Geologic Interpretation from Radar Imagery, Technical Letter NASA-102, U.S. Geological Survey, Washington, D.C., 1968.

1238. **Dellwig, L. F., MacDonald, H.C., and Kirk, J.N.,** The potential of radar in geological exploration, in *Proc. 5th Symp. Remote Sensing of Environment,* University of Michigan, Ann Arbor, 1968, 747.

1239. **Snavely, P. D., Jr. and MacLeod, N. S.,** Preliminary Evaluation of Infrared and Radar Imagery, Washington and Oregon Coasts, U.S. Geological Survey Open File Report No. 118, U.S. Geological Survey, Washington, D.C., 1968.

1240. **Jefferis, L. H.,** Lineaments in the Grand Canyon Area, Northern Arizona — a Radar Analysis, Report 118—9, University of Kansas, Lawrence, 1969.

1241. **Jefferis, L. H.,** An Evaluation of Radar Imagery for Structural Analysis in Gently Deformed Strata: A Study in Northeast Kansas, CRES Report 118—16, University of Kansas, Lawrence, 1969.

1242. **MacDonald, H. C.,** Geologic evaluation of radar imagery from Darien Province, Panama, *Mod. Geol.,* 1, 1, 1969.

1243. **Rowan, L. C. and Cannon, P. J.,** Remote-sensing investigations near Mill Creek, Oklahoma, *Okla. Geol. Notes,* 30, 127, 1970.

1244. **Wing, R. S.,** Cholame area-San Andreas fault zone-California, a study in SLAR, *Mod. Geol.,* 1, 173, 1970.

1245. **Gillerman, E.,** Roselle lineament of southeast Missouri, *Geol. Soc. Am. Bull.,* 81, 975, 1970.

1246. **Wing, R.S.,** Structural analysis from radar imagery of the eastern Panamanian Isthmus. Part I, *Mod. Geol.,* 2, 1, 1971.

1247. **Wing, R. S.,** Structural analysis from radar imagery of the eastern Panamanian Isthmus. Part II, *Mod. Geol.,* 2, 75, 1971.

1248. **Walker, A.S.,** Geological evaluation of remote sensing imagery of Mesabi Range, Minnesota, in *Proc. 8th Int. Symp. Remote Sensing of Environment,* Vol. 2, Environmental Research Institute of Michigan, Ann Arbor, 1972, 1137.

1249. **Wing, R.S. and MacDonald, H.C.,** Radar imagery identifies hidden jungle structures, *World Oil,* 176, 67, 1973.

1250. **Gelnett, R.H.,** Airborne Remote Sensors Applied to Engineering Geology and Civil Works Design Investigations, Technical Report TR-17621, Motorola Aerial Remote Sensing, Inc., Phoenix, Ariz., 1975.

1251. **Banks, P.T.,** A Geologic Analysis of the Side-Looking Airborne Radar Imagery of Southern New England, U.S. Geological Survey Open File Report 75-207, U.S. Geological Survey, Reston, Va., 1975.

1252. **Sabins, F.F., Jr.,** Infrared imagery and geologic aspects, *Photogramm. Eng.,* 29, 83, 1967.

1253. **Wallace, R.E. and Moxham, R.M.,** Use of infrared imagery in study of the San Andreas fault system, California, in Geological Survey Research 1967, U.S. Geological Survey Professional Paper 575-D, U.S. Government Printing Office, Washington, D.C., 1967, 147.

1254. **Sabins, F.F., Jr.,** Thermal infrared imagery and its application to structural mapping in southern California, *Geol. Soc. Am. Bull.,* 80, 397, 1969.

1255. **Brown, R.D., Jr.,** Post-Earthquake Infrared Imagery, Parkfield-Cholame, California Earthquakes of June-August, 1966, U.S. Geological Survey Open File Report 1167, U.S. Geological Survey, Washington, D.C., 1969.

1256. **Rowan, L.C., Offield, T.W., Watson, K., and Watson, R.D.,** Thermal infrared investigations, Mill Creek area, Oklahoma, in 2nd Annu. Earth Resources Aircraft Program, Status Review, Vol. 1, NASA Manned Spacecraft Center, Houston, Tex., 1969, 5-1.

1257. **Offield, T.W., Rowan, L.C. and Watson, R.D.,** Linear geologic structure and mafic rock discrimination as determined from infrared data, in 3rd Annu. Earth Resources Program Review, Vol. 1, NASA Manned Spacecraft Center, Houston, Tex., 1970, 11-1.

1258. **Rowan, L.C., Offield, T.W., Watson, K., Cannon, P.J., and Watson, R.D.,** Thermal infrared investigations, Arbuckle Mountains, Oklahoma, *Geol. Soc. Am. Bull.,* 81, 3549, 1970.

1259. **Morris, R.H., Carter, W.D., and Orkild, P.P.,** Preliminary Analysis of Infrared Imagery Pahute Mesa Area, Nevada Test Site, NASA Site 52, U.S. Geological Survey Open File Report 1466, U.S. Geological Survey, Washington, D.C., 1970.

1260. **Kilinc, I.A. and Lyon, R.J.P.,** Geologic Interpretation of Airborne Infrared Thermal Imagery of Goldfield, Nevada, Report 70-3, Remote Sensing Laboratory, Stanford University, Palo Alto, Calif., 1970.

1261. **Offield, T.W.,** Thermal-infrared images as a basis for structure mapping, Front Range and adjacent plains in Colorado, *Geol. Soc. Am. Bull.,* 86, 495, 1975.

1262. **Sabins, F.F., Jr.,** Geologic applications of remote sensing, in *Remote Sensing of Environment,* Lintz, J., Jr. and Simonett, D.S., Eds., Addison-Wesley Publishing, Reading, Mass., 1976, 508.

1263. **Lee, W.H.K. and Eaton, J.S.,** A detailed study of the Danville, California, earthquakes of May to July, 1970, *EOS Trans. Am. Geophys. Union,* 51, 779, 1970.

1264. **Lee, W.H.K., Eaton, J.S., and Brabb, E.E.,** The earthquake sequence near Danville, California, 1970, *Bull. Seismol. Soc. Am.,* 61, 1771, 1971.

1265. **Dobrin, M.B.,** *Introduction to Geophysical Prospecting,* 2nd ed., McGraw-Hill, New York, 1960.

1266. **Griffiths, D.H. and King, R.F.,** *Applied Geophysics for Engineers and Geologists,* Pergamon Press, New York, 1965.

1267. **Parasnis, D.S.,** *Mining Geophysics,* Elsevier, 1973.
1268. **Anon.,** New method tested in geophysical study, *Calif. Geol.,* 27, 135, 1974.
1269. **Bailey, A.D.,** Near surface fault detection by magnetometer, *Calif. Geol.,* 27, 274, 1974.
1270. **Chase, G.W. and Chapman, R.H.,** Black-box geology—uses and misuses of geophysics in engineering geology, *Calif. Geol.,* 29, 8, 1976.
1271. **Bryant, W.A.,** The Raymond Hill fault, an urban geological investigation, *Calif. Geol.,* 31, 127, 1978.
1272. **Bonilla, M.G., Alt, J.N., and Hodgen, L.D.,** Trenches across the 1906 trace of the San Andreas fault in northern San Mateo County, California, *J. Res. U.S. Geol. Surv.,* 6, 347, 1978.
1273. **Brabb, E.E.,** Finding faults in the San Francisco Bay region, *Earthquake Inf. Bull.,* 5, 10, 1973.
1274. **Carder, D.S.,** Seismic investigations in the Boulder Dam area, 1940—1944, and the influence of reservoir loading and local earthquake activity, *Bull. Seismol. Soc. Am.,* 35, 175, 1945.
1275. **Carder, D.S. and Small, J.B.,** Level divergencies, seismic activity, and reservoir loading in the Lake Mead area, Nevada and Arizona, *EOS Trans. Am. Geophys. Union,* 29, 767, 1948.
1276. **Carder, D.S.,** Reservoir loading and local earthquakes, in *Engineering Seismology—The Works of Man,* Adams, W.M., Ed., Geological Society of America, Boulder, Colo., 1970, 51.
1277. **Mickey, W.V. and Dunphy, G.J.,** Lake Mead seismicity, *EOS Trans. Am. Geophys. Union,* 52, 278, 1971.
1278. **Mickey, W.V.,** Reservoir seismic effects, in *Man-Made Lakes, Their Problems and Environmental Effects,* Ackerman, W.C., White, G.F., and Worthington, E.B., Eds., American Geophysical Union, Washington, D.C., 1973, 472.
1279. **Anderson, R.E. and Laney, R.L.,** The influence of late Cenozoic stratigraphy on distribution of impoundment-related seismicity at Lake Mead, Nevada-Arizona, *J. Res. U.S. Geol. Surv.,* 3, 337, 1975.
1280. **Schleicher, D.,** A model for earthquakes near Palisades Reservoir, southeast Idaho, *J. Res. U.S. Geol. Surv.,* 3, 393, 1975.
1281. **Rogers, A.M. and Lee, W.H.K.,** Seismic study of earthquakes in the Lake Mead, Nevada-Arizona region, *Bull. Seismol. Soc. Am.,* 66, 1657, 1976.
1282. **Anon.,** Seismicity near Lake Mead, Nevada, and Palisades Reservoir, Idaho, *Earthquake Inf. Bull.,* 8, 16, 1976.
1283. **Talwani, P.,** Earthquakes associated with Clark Hill Reservoir, South Carolina — A case of induced seismicity, *Eng. Geol.,* 10, 239, 1976.
1284. **Fogle, G.H., White, R.M., Benson, A.F., Long, L.T., and Sowers, G.F.,** Reservoir induced seismicity at Lake Jocassee, northwestern South Carolina, *EOS Trans. Am. Geophys. Union,* 57, 759, 1976.
1285. **Leblanc, G.,** Induced seismicity at the Manic 3 Reservoir, Province of Quebec, *EOS Trans. Am. Geophys. Union,* 57, 759, 1976.
1286. **Milne, W.G. and Berry, M.J.,** Induced seismicity in Canada, *Eng. Geol.,* 10, 219, 1976.
1287. **Hagiwara, T. and Ohtake, M.,** Seismic activity associated with the filling of the reservoir behind the Kurobe Dam, Japan, 1963—1970, *Tectonophysics,* 15, 241, 1972.
1288. **Gupta, H. and Rastogi, B.K.,** *Dams and Earthquakes,* Elsevier, Amsterdam, 1976.
1289. **Adams, R.D.,** Statistical studies of earthquakes associated with Lake Benmore, New Zealand, *Eng. Geol.,* 8, 155, 1974.
1290. **Chung-kang, S., Hou-chun, C., Li-shing, H., Tzu-chiang, L., Chen-yung, Y., Ta-chun, W., and Hsueh-hai, L.,** Earthquakes induced by reservoir impounding and their effect on the Hsinfengkiang Reservoir, *Sci. Sinica,* 17, 239, 1974.
1291. **Miao-yueh, W., Mao-yuan, Y., Yu-liang, H., Tzu-chiang, L., Yun-tai, C., Yen, C., and Jui, F.,** Mechanism of the reservoir impounding earthquakes at Hsinfengkiang and a preliminary endeavour to discuss their cause, *Eng. Geol.,* 10, 331, 1976.
1292. **Tsung-ho, H., Hsueh-hai, L., Tu-hsin, H., and Cheng-jung, Y.,** Strong-motion observation of water-induced earthquakes at Hsinfengkiang Reservoir in China, *Eng. Geol.,* 10, 315, 1976.
1293. **Narain, H. and Gupta, H.K.,** Koyna earthquake, *Nature (London),* 217, 1138, 1968.
1294. **Rothe, J. P.,** Fill a lake, start an earthquake, *New Sci.,* 11, 75, 1968.
1295. **Gupta, H.K., Narain, H., Rastogi, B.K., and Mohan, I.,** A study of the Koyna earthquake of December 10, 1967, *Bull. Seismol. Soc. Am.,* 59, 1149, 1969.
1296. **Gupta, H.K., Rastogi, B.K., and Narain, H.,** The Koyna earthquake of December 10: a multiple seismic event, *Bull. Seismol. Soc. Am.,* 61, 167, 1971.
1297. **Chopra, A.K. and Chakrabarti, P.,** The Koyna earthquake and the damage to Koyna Dam, *Bull. Seismol. Soc. Am.,* 63, 381, 1973.
1298. **Gupta, H.K., Rastogi, B.K., and Narain, H.,** Earthquakes in the Koyna region and common features of the reservoir-associated seismicity, in *Man-Made Lakes, Their Problems and Environmental Effects,* Ackerman, W.C., White, G.F., and Worthington, E.B., Eds., American Geophysical Union, Washington, D.C., 1973, 455.

1299. **Guha, S.K., Gosavi, P.D., Agarwal, B.N.P., Padale, J.G., and Marwadi, S.C.,** Case histories of some artificial crustal disturbances, *Eng. Geol.,* 8, 59, 1974.

1300. **Gupta, H.K. and Combs, J.,** Continued seismic activity at the Koyna Reservoir site, India, *Eng. Geol.,* 10, 307, 1976.

1301. **Adams, R.D. and Ahmed, A.,** Seismic effects at Mangla Dam, Pakistan, *Nature (London),* 222, 1153, 1969.

1302. **Nikolaev, N.I.,** The first case of induced earthquakes during construction of a hydro-electric power-station in the U.S.S.R., *Eng. Geol.,* 8, 107, 1974.

1303. **Soboleva, O.V., Mamadaliev, U.A., and Simpson, D.W.,** Seismicity changes during the first stage of filling of Nurek Reservoir, Tadzikistan U.S.S.R., *EOS Trans. Am. Geophys. Union,* 56, 1025, 1975.

1304. **Simpson, D.W., Soboleva, O.V., Koczynski, T.A., and Abramov, O.,** Investigations of the seismic regime during filling of Nurek Reservoir, Tadzikistan U.S.S.R., *EOS Trans. Am. Geophys. Union,* 56, 1025, 1975.

1305. **Soboleva, O.V. and Mamadaliev, U.A.,** The influence of the Nurek Reservoir on local earthquake activity, *Eng. Geol.,* 10, 293, 1976.

1306. **Simpson, D.W. and Soboleva, O.V.,** Water level variations and reservoir-induced seismicity at Nurek, U.S.S.R., *EOS Trans. Am. Geophys. Union,* 58, 1196, 1977.

1307. **Roksandic, M.M.,** Influence de la charge d'un reservoir sur l'activité séismique, *Proc. 2nd Congr. Int. Soc. Rock Mech.,* 3, 8, 1970.

1308. **Božović, A.,** Review and appraisal of case histories related to seismic effects of reservoir impounding, *Eng. Geol.,* 8, 9, 1974.

1309. **Galanopoulos, A.G.,** The influence of the fluctuation of Marathon Lake elevation on local earthquake activity in the Attica Basin area, *Ann. Geol. Hell.,* 18, 281, 1967.

1310. **Comminakis, P., Drakopoulos, J., Moumoulidis, G., and Papazachos, B.,** Foreshock sequences of the Kremasta earthquake and their relation to the water loading of the Kremasta artificial lake, *Ann. Geofis.,* 21, 39, 1968.

1311. **Therianos, A.D.,** The seismic activity of the Kremasta area—Greece—between 1967 and 1972, *Eng. Geol.,* 8, 49, 1974.

1312. **Susstrunk, A.,** Seismic shocks in the Verzasca Valley during the filling of the Vogorno Reservoir, *Verh. Schweiz. Naturforsch. Ges.,* 148, 89, 1968.

1313. **Caloi, P.,** La geodinamica el servizio delle grandi dighe, *Ann. Geofis.,* 17, 449, 1964.

1314. **Caloi, P.,** L'enento del Vajont nei suoi aspetti geodinamici, *Ann. Geofis.,* 19, 1, 1966.

1315. **Rothé, J.P.,** Earthquakes and reservoir loadings, in *Proc. Assoc. Chilena Seisologia Ing. Antisimica,* Santiago, 1969, 28.

1316. **Rothé, J.P.,** Seismes artificiels, *Tectonophysics,* 9, 215, 1970.

1317. **Gourinard, Y.,** La géologie et les problemes de l'eau en Algerie. I. Le barrage de l'Qued Fodda, *19th Congr. Geol. Int.,* 1, 155, 1952.

1318. **Gough, D.I. and Gough, W.I.,** Stress, strain and seismicity at Lake Kariba, *EOS Trans. Am. Geophys. Union,* 51, 351, 1970.

1319. **Gough, D.I. and Gough, W.I.,** Stress and deflection in the lithosphere near Lake Kariba. I, *Geophys. J. R. Astron. Soc.,* 21, 65, 1970.

1320. **Gough, D.I. and Gough, W.I.,** Load-induced earthquakes at Lake Kariba. II, *Geophys. J.R. Astron. Soc.,* 21, 79, 1970.

1321. **Gough, D.I. and Gough, W.I.,** Time dependence and trigger mechanisms for the Kariba (Rhodesia) earthquakes, *Eng. Geol.,* 10, 211, 1976.

1322. UNESCO Working Group on Seismic Phenomena Associated with Large Reservoirs, Report of the First Meeting, United Nations Educational, Scientific, and Cultural Organization, Paris, 1970.

1323. **Gupta, H.K., Rastogi, B.K., and Narain, H.,** Common features of the reservoir-associated seismic activities, *Bull. Seismol. Soc. Am.,* 62, 481, 1972.

1324. **Gupta, H.K., Rastogi, B.K., and Narain, H.,** Some discriminatory characteristics of earthquakes near the Kariba, Kremasta and Koyna artificial lakes, *Bull. Seismol. Soc. Am.,* 62, 493, 1972.

1325. Joint Panel on Problems Concerning Seismology and Rock Mechanics, *Earthquakes Related to Reservoir Filling,* National Academy of Sciences, Washington, D.C., 1972.

1326. **Rothé, J.P.,** Summary: geophysics report, in *Man-Made Lakes, Their Problems and Environmental Effects,* Ackerman, W.C., White, G.F., and Worthington, E.B., Eds., American Geophysical Union, Washington, D.C., 1973, 441.

1327. **Anon.,** Large reservoirs and induced earthquakes, *Earthquake Inf. Bull.,* 7, 9, 1975.

1328. **Kanasewich, E.R.,** Instrumentation for observation of induced seismicity, *Eng. Geol.,* 11, 139, 1977.

1329. **Kisslinger, C.,** A review of theories of mechanics of induced seismicity, *Eng. Geol.,* 10, 85, 1976.

1330. **Simpson, D.W.,** Seismicity changes associated with reservoir loading, *Eng. Geol.,* 10, 123, 1976.

1331. **Klein, F.W.,** Tidal triggering of reservoir-associated earthquakes, *Eng. Geol.,* 10, 197, 1976.

1332. **Handin, J. and Nelson, R.A.,** Why is Lake Powell aseismic?, in *Int. Colloq. Seismic Effects of Reservoir Impounding,* The Royal Society, London, 1973, 53.

1333. **Lomnitz, C.,** Earthquakes and reservoir impounding: state of the art, *Eng. Geol.,* 8, 191, 1974.

1334. **Bufe, C.G.,** The Anderson Reservoir seismic gap-induced aseismicity, *Eng. Geol.,* 10, 255, 1976.

1335. **Yaralioğlu, M.,** Seismic activity investigation at Keban Dam, Turkey, *Eng. Geol.,* 8, 53, 1974.

1336. **Brown, R.L.,** Seismic activity following impounding of Mangla Reservoir, *Eng. Geol.,* 8, 79, 1974.

1337. **Lane, R.G.T.,** Investigations of seismicity at dam/reservoir sites, *Eng. Geol.,* 8, 95, 1974.

1338. **Blum, R. and Fuchs, K.,** Observation of low-magnitude seismicity at a reservoir in the eastern Alps, *Eng. Geol.,* 8, 99, 1974.

1339. Tri-Cities Citizens Advisory Committee on Seismic Safety, Disaster preparedness findings, in The Seismic Safety Study for the General Plan, California Council on Intergovernmental Relations, Sacramento, 1973, 49.

1340. **Kates, R.W., Haas, J.E., Amaral, D.J., Olson, R.A., Ramos, R., and Olson, R.,** Human impact of the Managua earthquake, *Science,* 182, 981, 1973.

1341. **Wyllie, L.A., Jr., Wright, R.N., Sozen, M.A., Degenkolb, H.J., Steinbrugge, K.V., and Kramer, S.,** Effects on structures of the Managua earthquake of December 23, 1972, *Bull. Seismol. Soc. Am.,* 64, 1069, 1974.

1342. **Olson, R.A.,** Individual and organizational dimensions of the San Fernando earthquake, in San Fernando, California, Earthquake of February 9, 1971, Vol. 2, Benfer, N.A., Coffman, J.L., and Dees, L.T., Eds., U.S. Government Printing Office, Washington, D.C., 1973, 259.

1343. Hospital Council of Southern California, Earthquake, *Hosp. Forum,* 13, 3, 1971.

1344. **Steinbrugge, K.V., Schader, E.E., Bigglestone, H. C., and Weers, C. A.,** *San Fernando Earthquake, February 9, 1971,* Pacific Fire Rating Bureau, San Francisco, 1971.

1345. **Burton, I., Kates, R.W., and White, G.F.,** *The Environment as Hazard,* Oxford University Press, New York, 1978.

1346. **Anon.,** A planning tool for earthquakes: Tokyo, *Nat. Hazards Observer,* 1, 7, 1976.

1347. **T. Nakano,** personal communication, 1978.

1348. California Office of Emergency Services, Governor's Office of Emergency Services, State of California, Office of Emergency Services, Sacramento, no date.

1349. California Office of Emergency Services, Earthquake Response Plan, Office of Emergency Services, Sacramento, 1977.

1350. **Manfred, C.,** (Statement of) Charles Manfred, Director Office of Emergency Services, *Calif. Geol.,* 30, 158, 1977.

1351. California Earthquake Evaluation Council, Earthquake prediction evaluation guidelines, *Calif. Geol.,* 30, 158, 1977.

1352. **Manfred, C.,** Statement of Charles Manfred, Director, California State Office of Emergency Services, in EARTHQUAKE—Hearings before the Subcommittee on Science, Research, and Technology of the Committee on Science and Technology, U.S. House of Representatives, Ninety-Fourth Congress, U.S. Government Printing Office, Washington, D.C., 1976, 265.

1353. U.S. Senate and House of Representatives, Public Law 93-288, in United States Statutes at Large, Vol. 88, Part I, U.S. Government Printing Office, Washington, D.C., 1976, 143.

1354. **Dunne, T.P.,** Statement of Thomas Dunne, Director, Federal Disaster Assistance Administration, Department of Housing and Urban Development, in EARTHQUAKE—Hearings before the Subcommittee on Science, Research, and Technology of the Committee on Science and Technology, U.S. House of Representatives, Ninety-Fourth Congress, U.S. Government Printing Office, Washington, D.C., 1976, 109.

1355. U.S. Department of Housing and Urban Development, Programs of HUD, U.S. Department of Housing and Urban Deveopment, Washington, D.C., 1977.

1356. **Algermissen, S.T., Rinehart, W.A., and Dewey, J.,** A Study of Earthquake Losses in the San Francisco Bay Area, U.S. Government Printing Office, Washington, D.C., 1972.

1357. **Algermissen, S.T., Hopper, M., Campbell, K., Rinehart, W.A., and Perkins, D.,** A Study of Earthquake Losses in the Los Angeles, California Area, U.S. Government Printing Office, Washington, D.C., 1973.

1358. U.S. Geological Survey, A Study of Earthquake Losses in the Puget Sound, Washington Area, U.S. Geological Survey Open-File Report 75-375, 1975.

1359. U.S. Geological Survey, A Study of Earthquake Losses in the Salt Lake City, Utah Area, U.S. Geological Survey Open File Report 76—89, 1976.

1360. Department of the Interior, Geological Survey, Warning and preparedness for geologic-related hazards, proposed procedures, *Fed. Reg.,* 42, 19292, April 12, 1977.

1361. **Anon.,** Earthquake Prediction Council established, *Earthquake Inf. Bull.,* 9, 38, 1977.

1362. **Skeet, M.,** *Manual for Disaster Relief Work,* Churchill Livingstone, Edinburgh, 1977.

1363. Subcommittee to Investigate Problems Connected with Refugees and Escapees of the Committee on the Judiciary, United States Senate, Humanitarian Assistance to Earthquake Victims in Italy, U.S. Government Printing Office, Washington, D.C., 1976.

1364. **Duchek, W.J.**, Planning for seismic risks at the urban scale, in *Architects and Earthquakes: Research Needs,* AIA Research Corporation, Washington, D.C., 1976, 95.

1365. San Francisco Department of City Planning, A Proposal for Citizen Review, Community Safety, the Comprehensive Plan of San Francisco, San Francisco Dept. of City Planning, 1974.

1366. **Haas, J.E., Kates, R.W., and Bowden, M.J., Eds.,** *Reconstruction Following Disaster,* MIT Press, Cambridge, Mass., 1977.

1367. **Popkin, R.,** Summary, in *Reconstruction Following Disaster,* Haas, J.E., Kates, R.W., and Bowden, M.J., Eds., MIT Press, Cambridge, Mass., 1977, xxv.

1368. **Kates, R.W.,** Major insights: a summary and recommendations, in *Reconstruction Following Disaster,* Haas, J.E., Kates, R.W., and Bowden, M.J., Eds., MIT Press, Cambridge, Mass., 1977, 261.

1369. Panel on Geography, The Great Alaska Earthquake of 1964 — Human Ecology, National Academy of Sciences, Washington, D.C., 1970.

1370. **Smith, E.R.,** Book review, The Great Alaska Earthquake of 1964 — Human Ecology, *Bull. Seismol. Soc. Am.,* 63, 743, 1973.

1371. Office of Emergency Preparedness, Disaster insurance, in Disaster Preparedness, Vol, 1, U.S. Government Printing Office, Washington, D.C., 1972, 135.

1372. **Auerbach, A.,** Quake talk not expected to affect L.A. real estate, *Los Angeles Times,* April 22, 1976.

1373. **Kunreuther, H.,** Limited Knowledge and Insurance Protection, Working Paper No. 75-10-02, Department of Decision Sciences, University of Pennsylvania, Philadelphia, 1975.

1374. **Kunreuther, H.,** Limited knowledge and insurance protection, *Public Policy,* 24, 227, 1976.

1375. **Kunreuther, H., Ginsberg, R., Miller, L., Sagi, P., Slovic, P., Borkan, B., and Katz, N.,** *Limited Knowledge and Insurance Protection, Implications for Natural Hazard Policy, Executive Summary,* University of Pennsylvania, Philadelphia, 1977.

1376. **O'Riordan, T.,** The New Zealand natural hazard insurance scheme: application to North America, in *Natural Hazards,* White, G.F., Ed., Oxford University Press, New York, 1974, 217.

1377. **Jackson, E.L.,** Response to Earthquake Hazard: Factors Related to the Adoption of Adjustments by Residents of Three Earthquake Areas of the West Coast of North America, Ph.D. thesis, Department of Geography, University of Toronto, Canada, 1974.

1378. **Jackson, E.L.,** Public response to earthquake hazard, *Calif. Geol.,* 30, 278, 1977.

1379. **Slovic, P., Kunreuther, H., and White, G.F.,** Decision processes, rationality, and adjustment to natural hazards, in *Natural Hazards,* White, G. F., Ed., Oxford University Press, New York, 1974, 187.

1380. **Kates, R.W.,** The perception of storm hazard on the shores of megalopolis, in *Environmental Perception and Behavior,* Lowenthal, D., Ed., Department of Geography, University of Chicago, 1967, 60.

1381. **Jackson, E.L. and Mukerjee, T.,** Human adjustment to the earthquake hazard of San Francisco, California, in *Natural Hazards,* White, G.F., Ed., Oxford University Press, New York, 1974, 160.

1382. **Sullivan, R., Mustart, D.A., and Galehouse, J.S.,** Living in earthquake country, a summary of residents living along the San Andreas fault, *Calif. Geol.,* 30, 3, 1977.

1383. **Anon.,** The public wants to be informed, *Nat. Haz. Observer,* 2, 2, 1978.

1384. **Slosson, J.E.,** Legislation related to earthquakes, *Calif. Geol.,* 28, 37, 1975.

1385. **Haas, J.E. and Mileti, D.S.,** *Socioeconomic Impact of Earthquake Prediction on Government, Business, and Community,* Institute of Behavioral Science, University of Colorado, Boulder, 1976.

1386. **Haas, J.E. and Mileti, D.S.,** Socioeconomic impact of earthquake prediction on government, business, and community, *Calif. Geol.,* 30, 147, 1977.

1387. **Fritz, C.E.,** Disaster, in *Contemporary Social Problems,* Merton, R.K. and Nisbet, R.A., Eds., Harcourt, Brace & World, New York, 1961, 651.

1388. **Lachman, R., Tatsuoka, M., and Bonk, W.J.,** Human behavior during the tsunami of May 1960, *Science,* 133, 1405, 1961.

1389. **Moore, H.E., Bates, F.L., Layman, M.V., and Parenton, V.J.,** Before the Wind: A Study of the Response to Hurricane Carla, National Academy of Sciences, Washington, D.C., 1963.

1390. **Wilkinson, K.P. and Ross, P.J.,** Citizens' Responses to Warnings of Hurricane Camille, Social Science Research Center Report No. 35, Mississippi State University, State College, 1970.

1391. **Anderson, W.A.,** Tsunami warning in Crescent City, California, and Hilo, Hawaii, in The Alaska Earthquake of 1964 — Human Ecology, National Academy of Sciences, Washington, D.C., 1970, 116.

1392. **Quarantelli, E.L. and Dynes, R.R.,** When disaster strikes (It isn't like what you've heard and read about), *Psychol. Today,* 5, 66, 1972.

1393. **Smith, P.J.,** Geocomments. I. Camille and human nature, *Comments Earth Sci. Geophys.,* 3, 94, 1973.

1394. **Baker, E.J., Brigham, J.C., Paredes, J.A., and Smith, D.D.,** The Social Impact of Hurricane Eloise on Panama City, Florida, Technical Paper No. 1, Florida State University, Tallahassee, 1976.

1395. **Moore, H.E.,** . . . *and the Winds Blew,* University of Texas Press, Austin, 1964.

1396. **Dyness, R.R.,** *Organized Behavior in Disaster,* D.C. Heath & Co., Lexington, Mass., 1970.

1397. **McLuckie, B.J.,** The Warning System: A Social Science Perspective, U.S. Government Printing Office, Washington, D.C., 1973.

1398. **Mileti, D. and Krane, S.,** Countdown to the Unlikely, paper presented at the Annu. Meet. Am. Sociological Assoc., New York, 1973.

1399. **Gillette, R.,** Oil and gas resources: academy calls USGS math misleading, *Science,* 187, 723, 1975.

1400. **Anon.,** Earthquake prediction: is it better not to know?, *Natl. Sci. Found. Mosaic,* 8, 8, 1977.

1401. **Turner, R.,** Prediction of the Haicheng earthquake, mobilizing the masses, *EOS Trans. Am. Geophys. Union,* 58, 260, 1977.

1402. **Weigel, E.P.,** Killer from the bottom of the sea, *NOAA,* 4, 30, 1974.

1403. **Hawaiian Telephone Company,** *Oahu Telephone Directory,* Hawaiian Telephone Company, Honolulu, 1975.

1404. **Weller, J.M.,** Human response to tsunami warnings, in The Great Alaska Earthquake of 1964 — Oceanography and Coastal Engineering, National Academy of Sciences, Washington, D.C., 1972, 222.

1405. **Spaeth, M.G. and Berkman, S.C.,** The tsunamis as recorded at tide stations and the Seismic Wave Warning System, in The Great Alaska Earthquake of 1964 — Oceanography and Coastal Engineering, National Academy of Sciences, Washington, D.C., 1972, 38.

1406. **Haas, J.E.,** Lessons for coping with disaster, in The Great Alaska Earthquake of 1964 — Human Ecology, National Academy of Sciences, Washington, D.C., 1970, 39.

1407. Working Group Earthquake Hazards Reduction, Earthquake Hazards Prediction: Issues for an Implementation Plan, Office of Science and Technology Policy, Executive Office of the President, Washington, D.C., 1978.

1408. **Marshall, W.,** Biological reactions to earthquakes, *J. Abnorm. Psychol. Soc. Psychol.,* 30, 462, 1936.

1409. **Komine, and Maki,** Psychiatric observations of Tokyo earthquake, *Jpn. J. Neurol. Psychiatry,* 1, 1924.

1410. **Brussilowsky, L.,** Beeinfussung der neuropsychischen sphare durch erdbeben in der Krim 1926, *Z. Gesamte Neurol. Psychiatr.,* 116, 441, 1928.

1411. **Franz, S.I. and Norris, A.,** Human reactions in the Long Beach earthquake, *Bull. Seismol. Soc. Am.,* 24, 109, 1934.

1412. **Popovic, M. and Petrovic, D.,** After the earthquake, *Lancet,* 2, 1169, 1964.

1413. **Moric-Petrovic, S., Jojie-Milenkovic, M., and Marinkov, M.,** Mental hygene of children evacuated after catastrophic earthquake in Skopje, *Anal. Zavode Mentalno Zdravlje,* 4, 53, 1972.

1414. **Lantis, M.,** Impact of the earthquake on health and mortality, in The Great Alaska Earthquake of 1964 — Human Ecology, National Academy of Sciences, Washington, D.C., 1970, 77.

1415. **Langdon, J.R. and Parker, A.H.,** Psychiatric aspects of March 27, 1964, earthquake, *Alaska Med.,* 6, 33, 1964.

1416. **Deisher, J.B.,** Comments on Seward earthquake and tidal wave, *Alaska Med.,* 6, 49, 1964.

1417. **Scott, E.,** School nurses—Anchorage, *Alaska Nurse,* 13, 3, 1964.

1418. **Enches, H.G.,** The quake at Alaska Psychiatric Institute, *Alaska Nurse,* 13, 7, 1964.

1419. **Ray, H.,** The first hours after the Alaska earthquake, *Ment. Hosp.,* 15, 408, 1964.

1420. **Davis, N.Y.,** The role of the Russian Orthodox Church in five Pacific Eskimo villages as revealed by the earthquake, in The Great Alaska Earthquake of 1964 — Human Ecology, National Academy of Sciences, Washington, D.C., 1970, 125.

1421. **Levine, J.,** Response to emotional problems of the San Fernando earthquake, in Earthquake Risk, Joint Committee on Seismic Safety, California Legislature, Sacramento, 1971, 91.

1422. **Blaufarb, H. and Levine, J.,** Crisis intervention in an earthquake, *Soc. Work,* 17, 16, 1972.

1423. **Howard, S.J. and Gordon, N.S.,** Mental Health Intervention in a Major Disaster, Final Progress Report, San Fernando Valley Child Guidance Clinic, Van Nuys, Calif., no date.

1424. **Howard, S.J. and Gordon, N.S.,** Coping with Children's Reaction to Earthquakes and Other Disasters, San Fernando Valley Child Guidance Clinic, Van Nuys, Calif., no date.

1425. **Nichols, T.C., Jr.,** Global summary of human response to natural hazards: earthquakes, in *Natural Hazards,* White, G.F., Ed., Oxford University Press, New York, 1974, 274.

1426. **Anon.,** The earthquakes in Ancona, *Earthquake Inf. Bull.,* 5, 18, 1973.

1427. U.S. Geological Survey and Office of Emergency Preparedness, Safety and Survival in an Earthquake, U.S. Government Printing Office, Washington, D.C., 1969.

1428. **Hamilton, R.M.,** Earthquake Hazards Reduction Program — Fiscal Year 1978 Studies Supported by the U.S. Geological Survey, U.S. Geological Survey Circular 780, 1978.

1429. **Anon.,** The risk of earthquakes, *Earthquake Inf. Bull.,* 3, 3, 1971.

1430. National Science Foundation and U.S. Geological Survey, Earthquake Prediction and Hazard Mitigation—Options for USGS and NSF Programs, U.S. Government Printing Office, Washington, D.C., 1976.

1431. **Algermissen, S.T. and Perkins, D.M.,** A Probablistic Estimate of Maximum Acceleration in Rock in the Contiguous United States, U.S. Geological Survey Open-File Report 76-416, 1976.

1432. **Algermissen, S.T. and Perkins, D.M.,** Earthquake-hazard map of United States, *Earthquake Inf. Bull.,* 9, 20, 1977.

1433. **Kerr, R.A.,** U.S. earthquake hazards: real but uncertain in the east, *Science,* 201, 1001, 1978.

1434. **Steinbrugge, K.V.,** Seismic public policy and the design professional, in *Summer Seismic Institute for Architectural Faculty,* AIA Research Corp., Washington, D.C., 1977, 257.

1435. **Anon.,** California Seismic Safety Commission, *Calif. Geol.,* 29, 159, 1976.

1436. Ad Hoc Panel on Earthquake Prediction, Earthquake Prediction, A Proposal for a Ten-Year Program of Research, U.S. Office of Science and Technology, Washington, D.C., 1965.

1437. Ad Hoc Interagency Working Group for Earthquake Research, Proposal for a Ten-Year National Earthquake Hazards Program, U.S. Federal Council for Science and Technology, Washington, D.C., 1968.

1438. Committee on Seismology-National Research Council, Responsibilities and Requirements of a Growing Science. Part I, Summary and Recommendations; Part II, Problems and Prospects, National Academy of Sciences, Washington, D.C., 1969.

1439. Committee on Earthquake Engineering-National Research Council, Earthquake Engineering Research, National Academy of Sciences, Washington, D.C., 1969.

1440. U.S. Task Force on Earthquake Hazard Reduction-Office of Science and Technology, Report***Program Priorities, U.S. Government Printing Office, Washington, D.C., 1970.

1441. U.S. Task Force on Earthquake Hazard Reduction-Office of Science and Technology, In the Interest of Earthquake Safety, Institute of Governmental Studies, University of California, Berkeley, 1971.

1442. Joint Panel on the San Fernando Earthquake, The San Fernando Earthquake of February 9, 1971; Lessons from a Moderate Earthquake on the Fringe of a Densely Populated Region, National Academy of Sciences and National Academy of Engineering, Washington, D.C., 1971.

1443. Joint Panel on Problems Concerning Seismology and Rock Mechanics, Earthquakes Related to Reservoir Filling, National Academy of Sciences and National Academy of Engineering, Washington, D.C., 1972.

1444. Panel on Strong-Motion Seismology-National Research Council, Strong-Motion Engineering Seismology; The Key to Understanding and Reducing the Damaging Effects of Earthquakes, National Academy of Sciences, Washington, D.C., 1973.

1445. Joint Committee on Seismic Safety, Meeting the Earthquake Challenge, California Legislature, Sacramento, 1974.

1446. **Wallace, R.E.,** Goals, Strategy, and Tasks of the Earthquake Hazard Reduction Program, U.S. Geological Survey Circular 701, 1974.

1447. **White, G.F. and Haas, J.E.,** *Assessment of Research on Natural Hazards,* MIT Press, Cambridge, 1975.

1448. **Ayre, R.S., Mileti, D.S., and Trainer, P.B.,** Earthquake and Tsunami Hazards in the United States; A Research Assessment, Institute of Behavioral Science, University of Colorado, Boulder, 1975.

1449. U.S. Geological Survey, Earthquake Prediction—Opportunity to Avert Disaster, U.S. Geological Survey Circular 729, 1976.

1450. Committee on Seismology-National Research Council, Trends and Opportunities in Seismology, National Academy of Sciences, Washington, D.C., 1977.

1451. Panel on Seismograph Networks-National Research Council, Global Earthquake Monitoring; Its Uses, Potentials, and Support Requirements, National Academy of Sciences, Washington, D.C., 1977.

1452. U.S. 95th Congress, Earthquake Hazards Reduction Act of 1977, in Earthquake Hazards Reduction: Issues for an Implementation Plan, Office of Science and Technology Policy, Executive Office of the President, Washington, D.C., 1978, 85.

1453. U.S. 95th Congress, Earthquake Hazards Reduction Act of 1977, in Earthquake Hazards Reduction Program—Fiscal Year 1978 Studies Supported by the U.S. Geological Survey, U.S. Geological Survey Circular 780, 1978, 31.

1454. **Anon.,** Implementation plan for earthquake hazard reduction, *Natl. Hazards Observer,* 3, 10, 1978.

Appendixes

Appendices

Appendix A

SELECTED U.S. EARTHQUAKES, 1811—1971[a]

Name of earthquake	Date and (local) time	Epicenter location[b]	Maximum Modified Mercalli Intensity[c]	Richter magnitude[b]	Approximate length of surface faulting (km)	Lives lost[d]	Dollar loss[e]	Remarks
New Madrid, Mo.	Dec. 16, 1811 (about 2:15 a.m.); Jan. 23, 1812 (about 8:50 a.m.); Feb. 7, 1812 (about 10:10 a.m.)	36 N, 90 W	XII (for each shock)	Over 8	See remarks	1 death	—	Richter assigned a magnitude of greater than 8 based on observed effects; surface faulting possibly occurred
Charleston, S. C.	Aug. 31, 1886 (9:51 p.m.)	32.9 N, 80.0 W	X	—	None	27 killed outright, plus 83 or more from related causes	$5,000,000 to $6,000,000	—
San Francisco, Calif.	Apr. 18, 1906 (5:12 a.m., PST)	38 N, 123 W	XI	8.3	306 minimum, 435 possible	700 to 800 deaths	$400,000,000 incl. fire; $80,000,000 earthquake only	Portions of the San Andreas fault are under the Pacific Ocean
Santa Barbara; Calif.	June 29, 1925 (6:42 a.m.)	34.3 N, 119.8 W	VIII–IX	6.3	None	12 to 14 deaths	$ 6,500,000	The dollar loss is for Santa Barbara; losses elsewhere were slight

Appendix A (continued)

SELECTED U.S. EARTHQUAKES, 1811—1971[a]

Name of earthquake	Date and (local) time	Epicenter location[b]	Maximum Modified Mercalli Intensity[c]	Richter magnitude[b]	Approximate length of surface faulting (km)	Lives lost[d]	Dollar loss[e]	Remarks
Long Beach, Calif.	Mar. 10, 1933 (5:54 p.m., PST)	33.6 N, 118.0 W	IX	6.3	—	Coroner's reprt: 86. 102 killed is more probable	$40,000,000 to $50,000,000	Epicenter in ocean; associated with Inglewood fault
Helena, Mont.	Oct. 12, 1935 (0:51 a.m., MST)	46.6 N, 112.0 W	VII	—	None	—	$50,000	First of three destructive shocks: Oct. 12, 18, and 31
Helena, Mont.	Oct. 18, 1935 (9:48 p.m., MST)	46.6 N. 112.0 W	VIII	6.25	None	2 killed, "score" injured	$3,000,000 to over $4,000,000	—
Helena, Mont.	Oct. 31, 1935 (11:38 a.m., MST)	46.6 N, 112.0 W	VIII	6.0	None	2 killed, "score" injured	$3,000,000 to over $4,000,000	—
Imperial Valley, Calif.	May 18, 1940 (8:37 p.m., PST)	32.7 N, 115.5 W	X	7.1	64 minimum	8 killed outright, 1 died later of injuries	$5,000,000 to $6,000,000	M.M. IX for building damage and M.M. X for faulting
Santa Barbara, Calif.	June 30, 1941 (11:51 p.m., PST)	34.4 N, 119.6 W	VIII	5.9	—	None killed, 1 hospitalized	$250,000	Epicenter in ocean

Appendix A (continued)

SELECTED U.S. EARTHQUAKES, 1811—1971[a]

Name of earthquake	Date and (local) time	Epicenter location[b]	Maximum Modified Mercalli Intensity[c]	Richter magnitude[b]	Approximate length of surface faulting (km)	Lives lost[d]	Dollar loss[e]	Remarks
Olympia, Wash.	Apr. 13, 1949 (11:56 p.m., PST)	47.1 N, 122.7 W	VIII	7.1	None	8 deaths	$15,000,000 to $25,000,000	—
Kern County, Calif.	July 21, 1952 (4:52 a.m., PDT)	35.0 N, 119.0 W	XI	7.7	22	10 of 12 deaths in Tehachapi	$37,650,000 to buildings $48,650,000 total (includes Aug. 22 aftershock)	M.M. XI assigned to tunnel damage from faulting: vibration intensity to structures generally VIII, rarely IX; faulting probably longer, but covered by deep alluvium
Bakersfield, Calif.	Aug. 22, 1952 (3:41 p.m., PDT)	35.3 N, 118.9 W	VIII	5.8	None	2 killed and 35 injured in Bakersfield	See above	Aftershock of July 21, 1952 earthquake
Fallon—Stillwater, Nev.	July 6, 1954 (4:13 a.m., PDT)	39.4 N, 118.5 W	IX	6.6	18	No deaths, several injured	$500,000 to $700,000 includes $300,000 to irrigation system	M.M. IX assigned along fault trace; vibration intensity VIII; first of two shocks on same fault
Fallon—Stillwater, Nev.	Aug. 23, 1954 (10:52 p.m., PDT)	39.6 N, 118.5 W	IX	6.8	31	No deaths	$500,000 to $700,000, includes $300,000 to irrigation system	M.M. IX assigned along fault trace; vibration intensity VIII; second of two shocks on same fault

Appendix A (continued)

SELECTED U.S. EARTHQUAKES, 1811—1971[a]

Name of earthquake	Date and (local) time	Epicenter location[b]	Maximum Modified Mercalli Intensity[c]	Richter magnitude[b]	Approximate length of surface faulting (km)	Lives lost[d]	Dollar loss[e]	Remarks
Fairview Peak, Nev.	Dec. 16, 1954 (3:07 a.m., PST)	39.3 N, 118.1 W	X	7.1	56	No deaths	—	M. M. X assigned along fault trace; vibration intensity VII; two shocks considered as a single event from the engineering standpoint
Dixie Valley, Nev.	Dec. 16, 1954 (3:11 a.m., PST)	39.8 N, 118.1 W	X	6.8	48	No deaths	—	M.M. X assigned along fault trace; vibration intensity VII; two shocks considered as a single event from the engineering standpoint
Eureka, Calif.	Dec. 21, 1954 (11:56 a.m. PST)	40.8 N, 124.1 W	VII	6.6	None	1 killed	$1,000,000	—
Port Hueneme, Calif.	Mar. 18, 1957 (10:56 a.m., PST)	34.1 N, 119.2 W	VI	4.7	None	No deaths	—	Epicenter in ocean
San Francisco, Calif.	Mar. 22, 1957 (11:44 a.m., PST)	37.7 N, 122.5 W	VII	5.3	None	No deaths, about 40 minor injuries	$1,000,000	—

Appendix A (continued)

SELECTED U.S. EARTHQUAKES, 1811—1971[a]

Name of earthquake	Date and (local) time	Epicenter location[b]	Maximum Modified Mercalli Intensity[c]	Richter magnitude[b]	Approximate length of surface faulting (km)	Lives lost[d]	Dollar loss[e]	Remarks
Hebgen Lake, Mont.	Aug. 17, 1959 (11:37 p.m., MST)	44.8 N, 111.1 W	X	7.1	22	19 presumed buried by landslide, plus probably 9 others killed, mostly by landslide	$2,334,000 (roads and bridges); $150,000 (Hebgen Dam); $1,715,000 (landslide correction)	M.M. X assigned along fault trace; vibrational intensity was VIII maximum; faulting complex and regional warping occurred; dollar loss to buildings relatively small
Prince William Sound, Alaska	Mar. 27, 1964 (5:36 p.m., AST)	61.1 N, 147.5 W	—	8.4	645—806	110 killed by tsunami; 15 killed from all other causes	$311,192,000 (includes tsunami)	Also known as the "Good Friday Earthquake"; fault length derived from seismic data
Puget Sound, Wash.	Apr. 29, 1965 (8:29 a.m., PDT)	47.4 N, 122.3 W	VIII	6.5	None	3 killed outright, 3 died from heart attacks	$12,500,000	M.M. VII general, M.M. VIII rare

Appendix A (continued)

SELECTED U.S. EARTHQUAKES, 1811—1971[a]

Name of earthquake	Date and (local) time	Epicenter location[b]	Maximum Modified Mercalli Intensity[c]	Richter magnitude[b]	Approximate length of surface faulting (km)	Lives lost[d]	Dollar loss[e]	Remarks
Parkfield, Calif.	June 27, 1966 (9:26 p.m., PDT)	35.54 N, 120.54 W	VII	5.5	38 and 9	No deaths	Less than $50,000	Damaging earthquakes in same area in 1901, 1922, and 1934; the 1966 shock had peak acceleration of 0.5 × g
Santa Rosa, Calif.	Oct. 1, 1969 (9:57 p.m., PDT)	38.47 N, 122.69 W	VII—VIII	5.6	None	No deaths, 15 injuries, 1 heart attack	$6,000,000 to buildings $1,250,000 to contents	Two shocks considered as a single event from the engineering standpoint
Santa Rosa, Calif.	Oct. 1, 1969 (11:20 p.m., PDT)	38.45 N, 122.69 W	VII — VIII	5.7	None	No deaths, 15 injuries, 1 heart attack	$ 6,000,000 to buildings $1,250,000 to contents	Two shocks considered as a single enent from the engineering standpoint
San Fernando, Calif.	Feb. 9, 1971 (6:01 a.m., PST)	34.40 N, 118.40 W	VIII—IX	6.4	19	58 deaths, 5,000 reported injuries	$478,519,635	Many reported injuries were minor, but public or charitable services requested

[a]   Abbreviations: M.M. = Modified Mercalli Intensity, PST = Pacific Standard Time, PDT = Pacific Daylight Time (Subtract 1 hr for Pacific Standard Time), MST = Mountain Standard Time, AST = Alaska Standard Time.

[b]   Slight variations will be found in various publications.

[c]   Modified Mercalli Intensities are those assigned by the U.S. Coast and Geodetic Survey (now USGS) when available. Modified Mercalli Intensity Scale of 1931-Abridged: I. Not felt except by a very few under specially favorable circumstances. II. Felt only by a few persons at rest, especially on upper

Appendix A (continued)

SELECTED U.S. EARTHQUAKES, 1811—1971[a]

floors of buildings. Delicately suspended objects may swing. III. Felt quite noticeably indoors, especially on upper floors of buildings, but many people do not recognize it as an earthquake. Standing motorcars may rock slightly. Vibration like passing of truck. Duration estimated. IV. During the day felt indoors by many, outdoors by few. At night some awakened. Dishes, windows, doors disturbed; walls making creaking sound. Sensation like heavy truck striking building. Standing motorcars rocked noticeably. V. Felt by nearly everyone, many awakened, some dishes, windows, etc., broken; a few instances of cracked plaster; unstable objects overturned. Disturbances of trees, poles, and other tall objects sometimes noticed. Pendulum clocks may stop. VI. Felt by all, many frightened and run outdoors. Some heavy furniture moved; a few instances of fallen plaster or damaged chimneys. Damage slight. VII. Everybody runs outdoors. Damage negligible in buildings of good design and construction; slight to moderate in well-built ordinary structures; considerable in poorly built or badly designed structures; some chimneys broken. Noticed by persons driving motorcars. VIII. Damage slight in specially designed structures; considerable in ordinary substantial buildings with partial collapse; great in poorly built structures. Panel walls thrown out of frame structures. Fall of chimneys, factory stacks, columns, monuments, walls. Heavy furniture overturned. Sand and mud ejected in small amounts. Changes in well water. Persons driving motorcars disturbed. IX. Damage considerable in specially designed structures; well-designed frame structures thrown out of plumb; great in substantial buildings, with partial collapse. Buildings shifted off foundations. Ground cracked conspicuously. X. Some well-built wooden structures destroyed; most masonry and frame structures destroyed with foundations; ground badly cracked. Rails bent. Landslides considerable from riverbanks and steep slopes. Shifted sand and mud. Water splashed (slopped) over banks. XI. Few, if any, (masonry) structures remain standing. Bridges destroyed. Broad fissures in ground. Underground pipelines completely out of service. Earth slumps and land slips in soft ground. Rails bent greatly. XII. Damage total. Waves seen on ground surfaces. Lines of sight and level distorted. Objects thrown upward into air.

[a] Original sources do not always clearly indicate if deaths include those attributable to exposure, unattended injury, heart attacks, and other nonimmediate deaths.

[b] Value of dollar at time of earthquake. Use of these figures requires a critical examination of reference materials, since the basis for the estimates vary.

From Algermissen, S.T., Rinehart, W.A., and Dewey, J., A Study of Earthquake Losses in the San Francisco Bay Area, National Oceanic and Atmospheric Administration, Washington, D.C., 1972, 10.

Appendix B

FELT EARTHQUAKES BY STATE, 1969—1977

State	1969	1970	1971	1972	1973	1974	1975	1976	1977
Alabama	—	—	1	—	—	1	2	1	—
Alaska	51	68	90	113	90	52	86	45	57
Arizona	1	1	1	1	—	2	4	8	4
Arkansas	1	1	1	2	2	3	3	4	1
California	97	130	118	117	52	41	92	81	103
Colorado	7	3	7	2	5	1	2	1	3
Connecticut	—	—	—	—	—	—	—	1	—
Delaware	—	—	—	—	—	—	—	—	1
Florida	—	—	—	—	1	—	1	—	—
Georgia	—	—	—	—	—	1	—	2	1
Hawaii	43	1	5	4	10	14	26	187	158
Idaho	4	—	1	1	1	—	8	4	1
Illinois	—	—	1	1	—	3	1	—	1
Indiana	—	—	—	—	—	—	—	1	1
Iowa	—	—	—	—	—	—	1	—	—
Kentucky	—	—	—	—	—	—	2	3	—
Maine	—	—	—	—	1	—	2	1	1
Massachusetts	—	—	1	—	—	—	2	2	2
Michigan	—	—	—	—	—	—	—	1	—
Mississippi	—	—	—	—	1	—	3	1	1
Missouri	1	—	—	2	2	4	3	6	1
Montana	83	22	93	19	—	2	13	4	18
Nebraska	—	—	—	1	—	—	1	—	—
Nevada	1	4	9	—	9	—	2	6	4
New Hampshire	—	—	—	—	—	—	1	—	1
New Jersey	1	—	—	—	1	—	—	2	—
New Mexico	1	3	16	—	3	—	1	6	2
New York	1	—	4	1	—	1	5	—	2
North Carolina	1	1	—	—	—	—	2	2	1
North Dakota	—	—	—	—	—	—	1	—	—
Ohio	—	—	—	1	—	2	1	—	1
Oklahoma	1	—	—	—	—	2	1	8	1
Oregon	1	1	1	—	1	1	1	3	2
Pennsylvania	—	—	—	1	—	—	—	—	—
Rhode Island	—	—	—	—	—	—	—	1	—
South Carolina	—	—	3	3	1	1	4	3	4
South Dakota	—	—	1	—	—	—	2	—	—
Tennessee	1	1	2	—	4	1	3	5	1
Texas	4	—	—	1	1	—	—	4	2
Utah	2	1	2	7	—	6	8	3	6
Vermont	—	—	—	—	—	—	1	—	—
Virginia	—	—	1	1	—	1	2	2	1
Washington	5	4	3	1	3	3	7	7	4
West Virginia	1	1	—	1	—	1	—	1	—
Wyoming	—	—	—	1	8	9	4	8	18
Totals	308	242	361	281	196	152	299	414	404

From National Earthquake Information Service, U.S. Geological Survey, Denver, Colorado.

Appendix C

MAJOR STRATIGRAPHIC AND TIME DIVISIONS

Subdivisions in use by the U.S. Geological Survey			Age estimates commonly used for boundaries (in million years)[a]	
Era or Erathem	System or Period	Series or Epoch	(A)	(B)
Cenozoic	Quaternary	Holocene		
		Pleistocene		
		— 1.5—2 —	— 1.8 —	
		Pliocene		
		— ca. 7 —	— 5.0 —	
	Tertiary	Miocene		
		— 26 —	— 22.5 —	
		Oligocene		
		— 37—38 —	— 37.5 —	
		Eocene		
		— 53—54 —	— 53.5 —	
		Paleocene		
		— 65 —	— 65 —	
Mesozoic	Cretaceous[b]	Upper (Late)		
		Lower (Early)		
		— 136 —		
	Jurassic	Upper (Late)		
		Middle (Middle)		
		Lower (Early)		
		— 190—195 —		
	Triassic	Upper (Late)		
		Middle (Middle)		
		Lower (Early)		
		— 225 —		
	Permian[b]	Upper (Late)		
		Lower (Early)		
		— 280 —		
	Pennsylvanian[b]	Upper (Late)		
		Middle (Middle)		
		Lower (Early)		
		— 320[c] —		
	Mississippian[b]	Upper (Late)		
		Lower (Early)		
		— 345 —		
...ozoic	Devonian	Upper (Late)		
		Middle (Middle)		
		Lower (Early)		
		— 395 —		
	Silurian[b]	Upper (Late)		
		Middle (Middle)		
		Lower (Early)		
		— 430—440 —		
	Ordovician[b]	Upper (Late)		
		Middle (Middle)		
		Lower (Early)		
		— ca. 500 —		
	Cambrian[b]	Upper (Late)		
		Middle (Middle)		
		Lower (Early)		
		— 570 —		
Precambrian	Time subdivisions of the Precambrian: Precambrian Z — base of Cambrian to 800 m.y.[d] Precambrian Y — 800 to 1600 m.y. Precambrian X — 1600 to 2500 m.y. Precambrian W — older than 2500 m.y.			

Note: Terms designating time are in parentheses. Informal time terms — early, middle, and late — may be used for the eras, for periods where there is no formal subdivision into Early, Middle, and Late,

Appendix C (continued)

MAJOR STRATIGRAPHIC AND TIME DIVISIONS

and for epochs. Informal rock terms — lower, middle, and upper — may be used where there is no formal subdivision of an era, system, or series.

[a]   Estimates for ages of time boundaries are under continuous study and subject to refinement and controversy. Two scales are given for comparison:

> (A) Geological Society of London, The Phanerozoic time-scale; a symposium, *Geol. Soc. London Q. J.*, 120 (Suppl.), 260—262, 1964.
>
> (B) Berggren, W.A., A Cenozoic time-scale — some implications for regional geology and paleobiogeography, *Lethaia*, 5 (2), 195—215, 1972.

In addition to these, a useful time scale for North American mammalian stages is given by:

> Evernden, J.F., Savage, D.E., Curtis, G.H., and James, G.T., Potassium-argon dates and the Cenozoic mammalian chronology of North America, *Am. J. Sci.*, 262, 145—198, 1964.

[b]   Includes provincial series accepted for use in U.S. Geological Survey reports.

[c]   From Table 1: Correlation chart for the Carboniferous of north-west Europe, Russia, and North America, *Geol. Soc. London Q.J.*, 120 (Suppl.), 222, 1964.

[d]   Million years.

From Geologic Names Committee, Stratigraphic Nomenclature in Reports of the U.S. Geological Survey, U.S. Geological Survey, Reston, Va., 1974, 18.

Appendix D

EARTHQUAKE REGULATIONS — UNIFORM BUILDING CODE*

Earthquake Regulations
Sec. 2312.
(a) General

Every building or structure and every portion thereof shall be designed and constructed to resist stresses produced by lateral forces as provided in this Section. Stresses shall be calculated as the effect of a force applied horizontally at each floor or roof level above the base. The force shall be assumed to come from any horizontal direction.

Structural concepts other than set forth in this Section may be approved by the Building Official when evidence is submitted showing that equivalent ductility and energy absorption are provided.

Where prescribed wind loads produce higher stresses, such loads shall be used in lieu of the loads resulting from earthquake forces.

(b) Definitions

The following definitions apply only to the provisions of this Section:

BASE is the level at which the earthquake motions are considered to be imparted to the structure or the level at which the structure as a dynamic vibrator is supported.

BOX SYSTEM is a structural system without a complete vertical load-carrying space frame. In this system the required lateral forces are resisted by shear walls or braced frames as hereinafter defined.

BRACED FRAME is a truss system or its equivalent which is provided to resist lateral forces in the frame system and in which the members are subjected primarily to axial stresses.

DUCTILE MOMENT RESISTING SPACE FRAME is a moment resisting space frame complying with the requirements for a ductile moment resisting space frame as given in Section 2312(j).

ESSENTIAL FACILITIES — See Section 2312 (k).

LATERAL FORCE RESISTING SYSTEM is that part of the structural system assigned to resist the lateral forces prescribed in Section 2312 (d) 1.

MOMENT RESISTING SPACE FRAME is a vertical load carrying space frame in which the members and joints are capable of resisting forces primarily by flexure.

SHEAR WALL is a wall designed to resist lateral forces parallel to the wall.

SPACE FRAME is a three-dimensional structural system without bearing walls, composed of interconnected members laterally supported so as to function as a complete self-contained unit with or without the aid of horizontal diaphragms or floor bracing systems.

VERTICAL LOAD-CARRYING SPACE FRAME is a space frame designed to carry all vertical loads.

(c) Symbols and Notations

The following symbols and notations apply only to the provisions of this Section:

$C =$ Numerical coefficient as specified in Section 2312 (d) 1.

$C_p =$ Numerical coefficient as specified in Section 2312 (g) and as set forth in Table No. 23-J.

$D =$ The dimension of the structure, in feet, in a direction parallel to the applied forces.

$\delta_i \delta_n =$ Deflections at levels i and n respectively, relative to the base, due to applied lateral forces or as determined in Section 2312 (h).

$F_i F_n F_x =$ Lateral force applied to level i, n, or x, respectively.

$F_p =$ Lateral forces on a part of the structure and in the direction under consideration.

$F_t =$ That portion of V considered concentrated at the top of the structure in addition to F_n.

$g =$ Acceleration due to gravity.

$h_i h_n h_x =$ Height in feet above the base to level i, n, or x respectively.

$I =$ Occupancy Importance Factor as specified in Table No. 23-K.

$K =$ Numerical coefficient as set forth in Table No. 23-I.

Level i

$l =$ Level of the structure referred to by the subscript i.

$i =$ l designates the first level above the base.

Level n

$=$ That level which is uppermost in the main portion of the structure.

Level x

$=$ That level which is under design consideration.

$x =$ l designates the first level above the base.

$N =$ The total number of stories above the base to level n.

$S =$ Numerical coefficient for site-structure resonance.

$T =$ Fundamental elastic period of vibration of the building or structure in seconds in the direction under consideration.

$T_s =$ Characteristic site period.

$V =$ The total lateral force or shear at the base.

$W =$ The total dead load as defined in Section 2302 including the partition loading specified in Section 2304 (d) where applicable. **EXCEPTION:** " W " shall be equal to the total dead load plus 25 percent of the floor live load in storage and warehouse occupancies. Where the design snow load is 30 psf or less, no part need be included in the value of " W." Where the snow load is greater than 30 psf, the snow load shall be included; however, where the snow load duration warrants, the Building Official may allow the snow load to be reduced up to 75 percent.

$w_i w_x =$ That portion of W which is located at or is assigned to level i or x respectively.

$W_p =$ The weight of a portion of a structure.

$Z =$ Numerical coefficient dependent upon the zone as determined by Figures No. 1, No. 2 and No. 3 in this Chapter. For locations in Zone No. 1, $Z = {}^3/_{16}$. For locations in Zone No. 2, $Z = {}^3/_8$. For locations in Zone No. 3, $Z = \frac{3}{4}$. For locations in Zone No. 4, $Z = 1$.

(d) Minimum Earthquake Forces for Structures

Except as provided in Section 2312 (g) and (i), every structure shall be designed and constructed to resist minimum total lateral seismic forces assumed to act nonconcurrently in the direction of each of the main axes of the structure in accordance with the following formula:

$$V = ZIKCSW \qquad (12\text{-}1)$$

The value of K shall be not less than that set forth in Table No. 23-I. The value of C and S are as indicated hereafter except that the product of CS need not exceed 0.14. The value of C shall be determined in accordance with the following formula:

$$C = \frac{1}{15\sqrt{T}} \qquad (12\text{-}2)$$

The value of C need not exceed 0.12.

The period T shall be established using the structural properties and deformational characteristics of the resisting elements in a properly substantiated analysis such as the following formula:

$$T = 2\pi \sqrt{\left(\sum_{i=1}^{n} \omega_i \delta_i^2\right) \div g \left[\sum_{i=1}^{n-1} F_i \delta_i + (F_t + F_n)\delta_n\right]} \qquad (12\text{-}3)$$

where the values of F_i, F_t, δ_i and δ_n shall be determined from the base shear V, distributed approximately in accordance with the principles of Formulas (12-5), (12-6), and (12-7) or any arbitrary base shear with a rational distribution.

In the absence of a determination as indicated above, the value of T for buildings may be determined by the following formula:

$$T = \frac{0.05 h_n}{\sqrt{D}} \qquad (12\text{-}3\text{A})$$

Or in buildings in which the lateral force resisting system consists of ductile moment-resisting space frames capable of resisting 100 percent of the required lateral forces and such system is not enclosed by or adjoined by more rigid elements tending to prevent the frame from resisting lateral forces:

$$T = 0.10N \qquad (12\text{-}3\text{B})$$

The value of S shall be determined by the following formulas, but shall be not less than 1.0:

$$\text{For } T/T_s = 1.0 \text{ or less} \quad S = 1.0 + \frac{T}{T_s} - 0.5 \left[\frac{T}{T_s}\right]^2 \qquad (12\text{-}4)$$

For T/T_s greater than 1.0 $\quad S =$

$$1.2 + 0.6 \frac{T}{T_s} - 0.3 \left[\frac{T}{T_s}\right]^2 \qquad (12\text{-}4\text{A})$$

where: T in Formulas (12-4) and (12-4A) shall be established by a properly substantiated analysis but T shall be not less than 0.3 second.

The range of values of T_s may be established from properly substantiated geotechnical data, in accordance with U.B.C. Standard No. 23-1, except that T_s shall not be taken as less than 0.5 second nor more than 2.5 seconds. T_s shall be that value within the range of site periods, as determined above, that is nearest to T.

When T_s is not properly established, the value of S shall be 1.5. **EXCEPTION:** Where T has been established by a properly substantiated analysis and exceeds 2.5 seconds, the value of S may be determined by assuming a value of 2.5 seconds for T_s.

(e) Distribution of Lateral Forces

1. Structures Having Regular Shapes or Framing Systems. The total lateral force V shall be distributed over the height of the structure in accordance with Formulas (12-5), (12-6) and (12-7).

$$V = F_t + \sum_{i=1}^{n} F_i \qquad (12\text{-}5)$$

The concentrated force at the top shall be determined according to the following formula:

$$F_t = 0.07TV \qquad (12\text{-}6)$$

F_t need not exceed $0.25\,V$ and may be considered as 0 where T is 0.7 second or less. The remaining portion of the total base shear V shall be distributed over the height of the structure including level n according to the following formula:

$$F_x = \frac{(V - F_t)\,\omega_x h_x}{\sum\limits_{i=1}^{n} \omega_i h_i} \qquad (12\text{-}7)$$

At each level designated as x, the force F_x shall be applied over the area of the building in accordance with the mass distribution on that level.

2. Setbacks. Buildings having setbacks wherein the plan dimension of the tower in each direction is at least 75 percent of the corresponding plan dimension of the lower part may be considered as uniform buildings without setbacks providing other irregularities as defined in this Section do not exist.

3. Structures having irregular shapes or framing systems. The distribution of the lateral forces in structures which have highly irregular shapes, large differences in lateral resistance or stiffness between adjacent stories or other unusual structural features shall be determined considering the dynamic characteristics of the structure.

4. Distribution of horizontal shear. The total shear in any horizontal plane shall be distributed to the various elements of the lateral force resisting system in proportion to their rigidities considering the rigidity of the horizontal bracing system or diaphragm.

Rigid elements that are assumed not to be part of the lateral force resisting system may be incorporated into buildings provided that their effect on the action of the system is considered and provided for in the design.

5. Horizontal torsional moments. Provisions shall be made for the increase in shear resulting from the horizontal torion due to an eccentricity between the center of mass and the center of rigidity. Negative torsional shears shall be neglected. Where the vertical resisting elements depend on diaphragm action for shear distribution at any level, the shear-resisting elements shall be capable of resisting a torsional moment assumed to be equivalent to the story shear acting with an eccentricity of not less than 5 percent of the maximum building dimension at that level.

(f) Overturning.

Every building or structure shall be designed to resist the overturning effects caused by the wind forces and related requirements specified in Section 2311, or the earthquake forces specified in this Section, which ever governs.

At any level the incremental changes of the design overturning moment, in the story under consideration, shall be distributed to the various resisting elements in the same proportion as the distribution of the shears in the resisting system. Where other vertical members are provided which are capable of partially resisting the overturning moments, a redistribution may be made to these members if framing members of sufficient strength and stiffness to transmit the required loads are provided.

Where a vertical resisting element is discontinuous, the overturning moment carried by the lowest story of that element shall be carried down as loads to the foundation.

(g) Lateral Force on Elements of Structures

Parts or portions of structures and their anchorage shall be designed for lateral forces in accordance with the following formula:

$$F_p = ZIC_pSW_p \qquad (12\text{-}8)$$

EXCEPTION: Where C_p in Table No. 23-J is 1.0 or more the value of I and S need not exceed 1.0.

The distribution of these forces shall be according to the gravity loads pertaining thereto.

(h) Drift and Building Separations

Lateral deflections or drift of a story relative to its adjacent stories shall not exceed 0.005 times the story height unless it can be demonstrated that greater drift can be tolerated. The displacement calculated from the application of the required lateral forces shall be multiplied by $(1.0/K)$ to obtain the drift. The ratio $(1.0/K)$ shall be not less than 1.0.

All portions of structures shall be designed and constructed to act as an integral unit in resisting horizontal forces unless separated structurally by a distance sufficient to avoid contact under deflection from seismic action or wind forces.

(i) Alternate Determination and Distribution of Seismic Forces

Nothing in Section 2312 shall be deemed to prohibit the submission of properly substantiated technical data for establishing the lateral forces and distribution by dynamic analyses, in such analyses the dynamic characteristics of the structure must be considered.

(j) Structural Systems
1. Ductility requirements.

A. All buildings designed with a horizontal force factor $K = 0.67$ or 0.80 shall have ductile moment resisting space frames.
B. Buildings more than 160 feet in height (48.8 m) shall have ductile moment resisting space frames capable of resisting not less than 25 percent of the required seismic forces for the structure as a whole. **EXCEPTION:** Buildings more than 160 feet (48.8 m) in height in Seismic Zone No. 1 may have concrete shear walls designed in conformance with Section 2627 of this Code in lieu of a ductile moment resisting space frame, provided a K value of 1.00 or 1.33 is utilized in the design.

C. In Seismic Zones No. 2, No. 3 and No. 4 all concrete space frames required by design to be part of the lateral force resisting system and all concrete frames located in the perimeter line of vertical support shall be ductile moment resisting space frames. **EXCEPTION:** Frames in the perimeter line of the vertical support of buildings designed with shear walls taking 100 percent of the design lateral forces need only conform with Section 2312(j) 1D.

D. In Seismic Zones No. 2, No. 3 and No. 4 all framing elements not required by design to be part of the lateral force resisting system shall be investigated and shown to be adequate for vertical load-carrying capacity and induced moment due to $3/K$ times the distortions resulting from the Code required lateral forces. The rigidity of other elements shall be considered in accordance with Section 2312 (e)4.

E. Moment resisting space frames and ductile moment resisting space frames may be enclosed by or adjoined by more rigid elements which would tend to prevent the space frame from resisting lateral forces where it can be shown that the action or failure of the more rigid elements will not impair the vertical and lateral load resisting ability of the space frame.

F. The necessary ductility for a ductile moment resisting space frame shall be provided by a frame of structural steel with moment resisting connections (complying with Section 2722 for buildings in Seismic Zones No. 2, No. 3 and No. 4 or Section 2723 for buildings in Seismic Zone No. 1) or by a reinforced concrete frame (complying with Section 2626 for buildings in Seismic Zones No. 2, No. 3 and No. 4 or Section 2625 for buildings in Seismic Zone No. 1).

G. In Seismic Zones No. 2, No. 3 and No. 4 all members in braced frames shall be designed for 1.25 times the force determined in accordance with Section 2312 (d). Connections shall be designed to develop the full capacity of the members or shall be based on the above forces without the one-third increase usually permitted for stresses resulting from earthquake forces.

   Braced frames in buildings shall be composed of axially loaded bracing members of A36, A440, A441, A501, A572 (except Grades 60 and 65) or A588 structural steel; or reinforced concrete members conforming to the requirements of Section 2627.

H. Reinforced concrete shear walls for all buildings shall conform to the requirements of Section 2627.

I. In structures where $K = 0.67$ and $K = 0.80$, the special ductility requirements of structural steel (complying with Section 2722 for buildings in Seismic Zones No. 2, No. 3 and No. 4 or Section 2723 for buildings in Seismic Zone No. 1) or by reinforced concrete (complying with Section 2626 for buildings in Seismic Zone No. 1), as appropriate, shall apply to all structural elements below the base which are required to transmit to the foundation the forces resulting from lateral loads.

2. Design Requirements

A. **Minor alterations.** Minor structural alterations may be made in existing buildings and other structures, but the resistance to lateral forces shall be not less than that before such alterations were made, unless the building as altered meets the requirements of this Section.

B. **Reinforced masonry or concrete.** All elements within structures located in Seismic Zones No. 2, No. 3 and No. 4 which are of masonry or concrete shall be reinforced so as to qualify as reinforced masonry or concrete under the provisions of Chapters 24 and 26. Principal reinforcement in masonry shall be spaced 2 feet (0.6 m) maximum on center in buildings using a moment resisting space frame.

C. **Combined vertical and horizontal forces.** In computing the effect of seismic force in combination with vertical loads, gravity load stresses induced in members by dead load plus design live load, except roof live load, shall be considered. Consideration should also be given to minimum gravity loads acting in combination with lateral forces.

D. **Diaphragms.** Floor and roof diaphragms shall be designed to resist the forces set forth in Table No. 23-J. Diaphragms supporting concrete or masonry walls shall have continuous ties between diaphragm chords to distribute, into the diaphragm, the anchorage forces specified in this Chapter. Added chords may be used to form sub-diaphragms to transmit the anchorage forces to the main cross ties. Diaphragm deformations shall be considered in the design of the supported walls. See Section 2312 (j) 3 A for special anchorage requirements of wood diaphragms.

3. Special Requirements

A. **Wood diaphragms providing lateral support for concrete or masonry walls.** Where wood diaphragms are used to laterally support concrete or masonry walls the anchorage shall conform to Section 2310. In Zones No. 2, No. 3 and No. 4 anchorage shall not be accomplished by use of toe nails, or nails subjected to withdrawal; nor shall wood framing be used in cross grain bending or cross grain tension.

B. **Pile caps and caissons.** Individual pile caps and caissons of every building or structure shall be interconnected by ties, each of which can carry by tension and compression a minimum horizontal force equal to 10 percent of the larger pile cap or caisson loading, unless it can be demonstrated that equivalent restraint can be provided by other approved methods.

C. **Exterior elements.** Precast, nonbearing, nonshear wall panels or similar elements which are attached to or enclose the exterior, shall accommodate movements of the structure resulting from lateral forces or temperature changes. The concrete panels or other elements shall be supported by means of cast-in-place concrete or by mechanical fasteners in accordance with the following provisions.

Connections and panel joints shall allow for a relative movement between stories of not less than two times story drift caused by wind or $(3.0/K)$ times story drift caused by required seismic forces; or ¼ inch (0.6 cm), whichever is greater.

Connections shall have sufficient ductility and rotation capacity so as to preclude fracture of the concrete or brittle failures at or near welds. Inserts in concrete shall be attached to, or hooked around reinforcing steel, or otherwise terminated so as to effectively transfer forces to the reinforcing steel.

Connections to permit movement in the plane of the panel for story drift shall be properly designed sliding connections using slotted or oversize holes or may be connections which permit movement by bending of steel or other connections providing equivalent sliding and ductility capacity.

(k) Essential Facilities

Essential facilities are those structures or buildings which must be safe and usable for emergency purposes after an earthquake in order to preserve the health and safety of the general public. Such facilities shall include but not be limited to:

1. Hospitals and other medical facilities having surgery or emergency treatment areas.
2. Fire and police stations.
3. Municipal government disaster operation and communication centers deemed to be vital in emergencies.

The design and detailing of equipment which must remain in place and be functional following a major earthquake shall be based upon the requirements of Section 2312(g) and Table No. 23-J. In addition, their design and detailing shall consider effects induced by structure drifts of not less than $(2.0/K)$ times the story drift caused by required seismic forces nor less than the story drift caused by wind. Special consideration shall also be given to relative movements at separation joints.

(1) Earthquake Recording Instrumentations

For earthquake recording instrumentations see Appendix, Section 2312 (1).

TABLE 23-I

Horizontal Force Factor "K" for Buildings or Other Structures[a]

Type or arrangement of resisting elements	Value[b] of K
1. All building framing systems except as hereinafter classified	1.00
2. Buildings with a box system as specified in Section 2312(b)	1.33
3. Buildings with a dual bracing system consisting of a ductile moment resisting space frame and shear walls or braced frames using the following design criteria:	0.80
a. The frames and shear walls shall resist the total lateral force in accordance with their relative rigidities considering the interaction of the shear walls and frames	
b. The shear walls acting independently of the ductile moment resisting portions of the space frame shall resist the total required lateral forces	
c. The ductile moment resisting space frame shall have the capacity to resist not less than 25 percent of the required lateral force	
4. Buildings with a ductile moment resisting space frame designed in accordance with the following criteria: The ductile moment resisting space frame shall have the capacity to resist the total required lateral force	0.67
5. Elevated tanks plus fill contents, on four or more cross-braced legs and not supported by a building	2.5[c]
6. Structures other than buildings and other than those set forth in Table No. 23-J	2.00

Note: The tower shall be designed for an accidental torsion of five percent as specified in Section 2312 (e)
     5. Elevated tanks which are supported by buildings or do not conform to type or arrangement of supporting elements as described above shall be designed in accordance with Section 2312 (g) using "C_p" = 0.2.

[a] Where wind load as specified in Section 2311 would produce higher stresses, this load shall be used in lieu of the loads resulting from earthquake forces.
[b] See Figure Nos. 1, 2 and 3 this chapter and definition of "Z" as specified in Section 2312(c).
[c] The minimum value of "KC" shall be 0.12 and the maximum value of "KC" need not exceed 0.25.

TABLE 23-J

Horizontal Force Factor "C_p" for Elements of Structures

Part or portion of buildings	Direction of force	Value of C_p
1. Exterior bearing and nonbearing walls, interior bearing walls and partitions, interior nonbearing walls and partitions. Masonry or concrete fences	Normal to flat surface	0.20[a]
2. Cantilever parapet	Normal to flat surface	1.00
3. Exterior and interior ornamentations and appendages	Any direction	1.00

TABLE 23-J (continued)

Horizontal Force Factor " C_p " for Elements of Structures

4. When connected to, part of, or housed within a building:

a. Towers, tanks, towers and tanks plus contents, chimneys, smokestacks and penthouse	Any direction	0.20^b
b. Storage racks with the upper storage level at more than 8 feet (2.4 m) in height plus contents	Any direction	$0.20^{b,c}$
c. Equipment or machinery not required for life safety systems or for continued operations of essential facilities	Any direction	$0.20^{b,d}$
d. Equipment or machinery required for life safety systems or for continued operation of essential facilities	Any direction	$0.50^{d,e}$

5. When resting on the ground, tank plus effective mass of its contents.   Any direction   0.12

6. Suspended ceiling framing systems (Applies to Seismic Zones Nos. 2, 3 and 4 only)   Any direction 0.20^f

7. Floors and roofs acting as diaphragms   Any direction 0.12^g

8. Connections for exterior panels or for elements complying with Section 2312 (j) 3C   Any direction   2.00

9. Connections for prefabricated structural elements other than walls, with force applied at center of gravity of assembly   Any direction 0.30^h

[a] See also Section 2309 (b) for minimum load on deflection criteria for interior partitions.

[b] When located in the upper portion of any building where the h_n/D ratio is five-to-one or greater the value shall be increased by 50 percent.

[c] W_p for storage racks shall be the weight of the racks plus contents. The value of C_p for racks over two storage support levels in height shall be 0.16 for the levels below the top two levels. In lieu of the tabulated values steel storage racks may be designed in accordance with U.B.C. Standard No. 27-11.

Where a number of storage rack units are interconnected so that there are a minimum of four vertical elements in each direction on each column line designed to resist horizontal forces, the design coefficients may be as for a building with K values from Table No. 23-I, $CS = 0.20$ for use in the formula $V = ZIKCSW$ and W equal to the total dead load plus 50 percent of the rack rated capacity. Where the design and rack configurations are in accordance with this paragraph the design provisions in U.B.C. Standard No. 27-11 do not apply.

[d] For flexible and flexibly mounted equipment and machinery, the appropriate values of C_p shall be determined with consideration given to both the dynamic properties of the equipment and machinery and to the building or structure in which it is placed but shall not be less than the listed values. The design of the equipment and machinery and their anchorage is an integral part of the design and specification of such equipment and machinery.

[e] For Essential Facilities and life safety systems, the design and detailing of equipment which must remain in place and be functional following a major earthquake shall consider drifts in accordance with Section 2312 (k). The product of IS need not exceed 1.5.

[f] Ceiling weight shall include all light fixtures and other equipment which are laterally supported by the ceiling. For purposes of determining the lateral force, a ceiling weight of not less than 4 pounds per square foot shall be used.

[g] Floors and roofs acting as diaphragms shall be designed for a minimum force resulting from a C_p of 0.12 applied to w_x unless a greater force results from the distribution of lateral forces in accordance with Section 2312 (e).

[h] The W_p shall include 25 percent of the floor live load in storage and warehouse occupancies.

TABLE 23-K

Values for Occupancy Importance Factor "I"

Type of occupancy	I factor
Essential Facilities[a]	1.5
Any building where the primary occupancy is for assembly use for more than 300 persons (in one room)	1.25
All others	1.0

[a]   See Section 2312 (k) for definition and additional requirements for essential facilities.

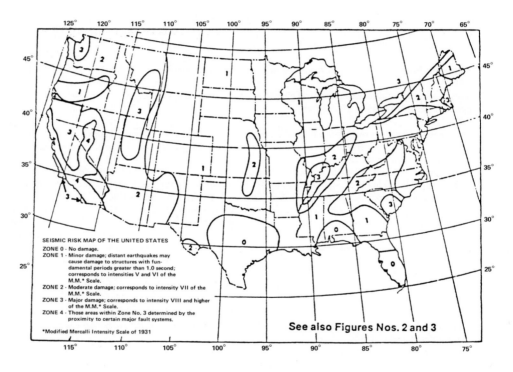

SEISMIC RISK MAP OF THE UNITED STATES
ZONE 0 - No damage.
ZONE 1 - Minor damage; distant earthquakes may cause damage to structures with fundamental periods greater than 1.0 second; corresponds to intensities V and VI of the M.M.* Scale.
ZONE 2 - Moderate damage; corresponds to intensity VII of the M.M.* Scale.
ZONE 3 - Major damage; corresponds to intensity VIII and higher of the M.M.* Scale.
ZONE 4 - Those areas within Zone No. 3 determined by the proximity to certain major fault systems.

*Modified Mercalli Intensity Scale of 1931

See also Figures Nos. 2 and 3

FIGURE 1.   Seismic risk map of the conterminous U.S.

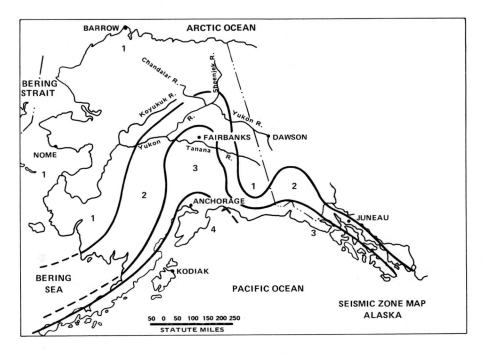

FIGURE 2.   Seismic risk map of Alaska.

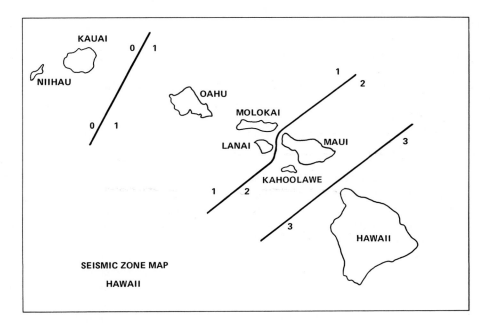

FIGURE 3.   Seismic risk map of Hawaii.

Appendix E

UNIFORM BUILDING CODE STANDARD NO. 23-1*

DETERMINATION OF THE CHARACTERISTIC SITE PERIOD, T$_s$

Based on Recommended Standard of the International Conference of Building Officials

See Section 2312, Uniform Building Code

Scope

Sec. 23.101. This Standard defines procedures for determining the characteristic site periods which are to be used in establishing the seismic site-structure resonance coefficient, S, of Section 2312 of the Uniform Building Code.

Definitions

Sec. 23.102. The following definitions apply to the provisions of this Standard.

BEDROCK is any underlying material that has a shear wave velocity at low strains (0.0001 percent) equal to or in excess of 2500 feet/second (762 m/sec) and is not underlaid by materials having significantly lower shear wave velocities.

CHARACTERISTIC SITE PERIOD is the natural period of the geotechnical profile at a site, determined in accordance with the methods specified herein.

GEOTECHNICAL PROFILE is a profile which lists the various layers of soil and rock at a site in order of increasing depth from the ground surface together with the location of the water table. It includes a brief description of each layer and its physical properties used in determining the characteristic site period.

SHEAR WAVE VELOCITY is the speed at which a shear (distortional) wave propagates through a material.

ZONE is a Seismic Zone as used in Section 2312, Uniform Building Code.

Symbols and Notations

Sec. 23.103. The following symbols and notations apply only to the provisions of this Standard.

C_{si} = Shear wave velocity, in feet/second, at low (0.0001 percent) strain levels of soil layer *i* which is part of the geotechnical profile.

C_s = Strain corrected shear wave velocity which represents an equivalent uniform layer having the same characteristic period and total depth to bedrock as the soils overlying bedrock in the geotechnical profile.

g = Acceleration due to gravity.

H_i = Thickness of a soil layer *i*, in feet, which is part of the geotechnical profile.

H = The total depth to bedrock in feet of the geotechnical profile at the site. For the purposes of this Standard, where the actual depth to bedrock exceeds 500 feet, H shall be taken as 500 feet (152.4 m) and the materials deeper than that depth considered bedrock for analytical purposes.

i = Layer within the geologic profile referred to by this subscript.

n = Final level of the profile referred to by this subscript.

R = A factor used to convert the average value of C_{si} to a velocity consistent with the strain levels of the code seismic zone. See Table No. 23-1-A.

T_s = Characteristic site period, seconds.

$T_s{}'$ = Combined characteristic site period of two layers in a multi-layered system.

Reporting

Sec.23.104. The characteristic site period shall be determined as required by Sections 23.105 through 23.108 and this Section of the Standard. The characteristic site period shall be reported as a reasonable range of values together with appropriate substantiating data. Reported values of T_s need not be less than 0.5 second nor greater than 2.5 seconds.

Geotechnical Profiles

Sec. 23.105. When required to determine the characteristic site period, a geotechnical profile representative of the site shall be prepared taking into account the variability of the subsurface conditions at the site and the reliability with which those conditions can be established. In areas where subsurface conditions are generally known or may be reasonably determined by analysis of geologic structure in the surrounding areas it is not necessary to explore the site to large depths.

Method of Calculating T_s

Sec. 23.106. Values of characteristic site period T_s shall be determined by any of the following methods, as appropriate:

1. **Site Response Analysis Method.** This method requires that a dynamic response analysis be made of a representative soil profile for the site. The analytical model may consist of a one-dimensional layered soil deposit on bedrock or a two-dimensional system where it is considered necessary to include the effects of laterally varying soil properties or depth to bedrock in the analysis. The soil profile may be assumed to undergo shearing deformations only. The analytical model should conform to the principles of dynamics but it is only necessary to include frequencies in the range of 15 to 0.25 hertz in the analysis. The model should consider the effect of non-uniform material properties in the geotechnical profile and the effects of variations in shear wave velocities and damping with strain levels developed during the earthquake motions.

   For the purposes of this analysis, to determine the strain levels, the base rock excitation used in the anaysis shall be representative of horizontal rock motions for earthquakes having the following characteristics:

   Zone No. 1. A magnitude 6 earthquake producing a peak acceleration of 0.1 g.
   Zone No. 2. A magnitude 6 earthquake producing a peak acceleration of 0.2 g.
   Zone No. 3. A magnitude 7 earthquake producing a peak acceleration of 0.3 g.
   Zone No. 4. A magnitude 7 earthquake producing a peak acceleration of 0.4 g.

   Based on the dynamic response analysis, the characteristic site period should be determined in any of the following ways:

   A. If modal analysis is used, the site period is the period of the first mode computed from a solution of the characteristic value (eigen value) problem for the soil system.

   B. If a wave propagation analysis procedure is used, the site period is the period at which the peak occurs in the amplification spectrum (transfer function) for the soil system.

C. If ground surface motions are calculated directly, acceleration response spectra shall be determined for both the bedrock input motion and the motion of a point near the ground surface or at a significant foundation level. These two spectra shall be calculated for the same oscillator damping ratio, and the ratio between them using the bedrock motion spectral values as divisors, shall be determined for every period of calculation. The period at which the maximum spectral ratio is obtained shall be taken as the characteristic site period. Sufficient periods of calculation shall be employed, to determine the maximum spectral ratio.

2. **Equivalent Single Layer Method.** This method assumes that a weighted average strain corrected shear wave velocity, C_s, may be used in the formula for a homogeneous layer as follows:

$$T_s = \frac{4H}{\bar{c}_s}$$

where:

$$\bar{c}_s = \frac{R \sum_{i=1}^{n} c_{si} H_i}{H}$$

and:

$$H = \sum_{i=1}^{n} H_i$$

This method shall be used only for geotechnical profiles in which the values of C_{si} do not decrease markedly with depth at any point in the profile, and for which no layer below a depth of 15 feet (4.6 m) has a value of C_{si} less than 300 feet/second (91.4 m/sec). The values of R which are applicable to the various seismic zones are shown in Table No. 23-1-A.

3. **Multi-layer Methods.** In this method, the soil layers within the geotechnical profile shall be assumed to undergo shearing deformations only.

The fundamental period of vibration of the profile shall be obtained by methods consistent with the principles of dynamics. The location of each soil layer within the profile as well as its thickness and shearing modulus shall be considered. Consideration of the effects of strain dependence of material properties shall also be included.

Alternatively, the following procedure may be used. The natural period of each layer in the soil profile is calculated according to the following formula:

$$T_i = \frac{4H_i}{C_{si}}$$

The fundamental period of the entire profile is calculated by taking the top two layers of the profile, 1 and 2, and calculating their combined period, T'_s, using the following expression:

$$\frac{H_2}{H_1} \times \frac{T_2}{T_s} = \tan\left(\frac{\pi}{2} \times \frac{T_1}{T_s'}\right) \times \tan\left(\frac{\pi}{2} \times \frac{T_2}{T_s'}\right)$$

or its graphical equivalent shown in Figure 23-1-1.

The uppermost two layers are then considered as a new single layer of period $T_{1-2} = T'_s$ and depth equal to $H_1 + H_2$ and the process repeated until the final period of the profile is found. Note that this particular procedure assumes a constant soil density throughout the profile. The resultant period of the profile shall be corrected for the effects of strain level by the following expression:

$$T_s = \frac{T_s'}{R}$$

4. **Firm Sites.** Firm sites, as defined below, shall be considered to have a T_s equal to 0.5 second. Firm sites, for the purpose of this Standard, are defined as:
    A. Any site where bedrock exists at a depth of 10 feet (3 m) or less.
    B. Any site where only dense to very dense granular soils overlie bedrock, and the depth to bedrock is 30 feet (9.1 m) or less.
    C. Any site where only very dense cemented granular soils overlie bedrock and where the depth to bedrock is 70 feet (21.3 m) or less.
5. **Other Methods.** Procedures and methods other than those of this Standard may be used to determine T_s. Such procedures and methods shall be consistent with the principles of this Standard.

Qualification

Sec. 23.107. The Building Official may require that any or all parts of the determination of the characteristic site period, including but not limited to the preparation of the geotechnical profile, determination of soil classifications or properties, determination of bedrock and analysis of the soil profile for T_s be made and reported by an engineer, engineering geologist and/or geologist licensed by the state to practice as such for each portion of the work applicable to his discipline.

Micro-regionalization

Sec. 23.108. Nothing in this Standard shall be interpreted as prohibiting the preparation of maps which delineate areas of constant characteristic site periods so long as those periods have been determined in accordance with this Standard. The establishment of such maps shall not act to prevent the submission of evidence supporting characteristic site periods for specific sites which are at variance with those of the maps.

TABLE 23-1-A

Seismic zone[a]	R
1	0.9
2	0.8
3	0.67
4	0.67

[a]See Section 2312 of the Uniform Building Code.

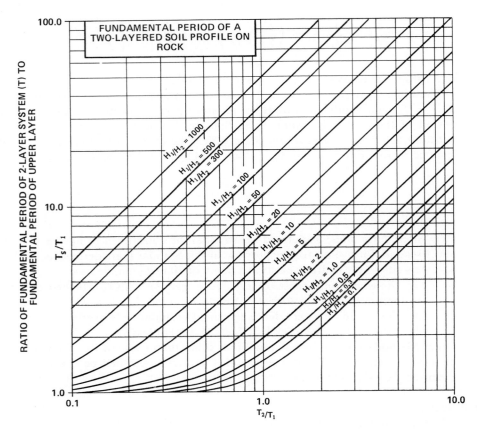

RATIO OF FUNDAMENTAL PERIOD OF LOWER LAYER (T₂) TO
FUNDAMENTAL PERIOD OF UPPER LAYER (T₁)

FIGURE 23-1-1.   (Reproduced from the 1976 edition of the Uniform Building Code Standards, copyright © 1976. With permission of the International Conference of Building Officials, Whittier, Calif.)

Appendix F

SEISMIC AND GEOLOGIC SITING CRITERIA FOR NUCLEAR POWER PLANTS*

I. PURPOSE

General Design Criterion 2 of Appendix A to Part 50 of this chapter requires that nuclear power plant structures, systems, and components important to safety be designed to withstand the effects of natural phenomena such as earthquakes, tornadoes, hurricanes, floods, tsunami, and seiches without loss of capability to perform their safety functions. It is the purpose of these criteria to set forth the principal seismic and geologic considerations which guide the commission in its evaluation of the suitability of proposed sites for nuclear power plants and the suitability of the plant design bases established in consideration of the seismic and geologic characteristics of the proposed sites.

These criteria are based on the limited geophysical and geological information available to date concerning faults and earthquake occurrence and effect. They will be revised as necessary when more complete information becomes available.

II. SCOPE

These criteria, which apply to nuclear power plants, describe the nature of the investigations required to obtain the geologic and seismic data necessary to determine site suitability and provide reasonable assurance that a nuclear power plant can be constructed and operated at a proposed site without undue risk to the health and safety of the public. They describe procedures for determining the quantitative vibratory ground motion design basis at a site due to earthquakes and describe information needed to determine whether and to what extent a nuclear power plant need be designed to withstand the effects of surface faulting. Other geologic and seismic factors required to be taken into account in the siting and design of nuclear power plants are identified.

The investigations described in this appendix are within the scope of investigations permitted by § 50.10 (c) (1) of this chapter.

Each applicant for a construction permit shall investigate all seismic and geologic factors that may affect the design and operation of the proposed nuclear power plant irrespective of whether such factors are explicitly included in these criteria. Additional investigations and/or more conservative determinations than those included in these criteria may be required for sites located in areas having complex geology or in areas of high seismicity. If an applicant believes that the particular seismology and geology of a site indicate that some of these criteria, or portions thereof, need not be satisfied, the specific sections of these criteria should be identified in the license application, and supporting data to justify clearly such departures should be presented.

These criteria do not address investigations of volcanic phenomena required for sites located in areas of volcanic activity. Investigations of the volcanic aspects of such sites will be determined on a case-by-case basis.

---

* From Office of the Federal Register, Appendix A — Seismic and geologic siting criteria for nuclear power plants, in Code of Federal Regulations 10, Energy, Parts 0 to 199, U.S. Government Printing Office, Washington, D.C., 1977, 412.

III. DEFINITIONS

As used in these criteria:

(a) The "magnitude" of an earthquake is a measure of the size of an earthquake and is related to the energy released in the form of seismic waves. "Magnitude" means the numerical value on a Richter scale.

(b) The "intensity" of an earthquake is a measure of its effects on man, on man-built structures, and on the earth's surface at a particular location. "Intensity" means the numerical value on the Modified Mercalli scale.

(c) The "Safe Shutdown Earthquake"[1] is that earthquake which is based upon an evaluation of the maximum earthquake potential considering the regional and local geology and seismology and specific characteristics of local subsurface material. It is that earthquake, which produces the maximum vibratory ground motion for which certain structures, systems, and components are designed to remain functional. These structures, systems, and components are those necessary to assure:

(1)   The integrity of the reactor coolant pressure boundary,
(2)   The capability to shut down the reactor and maintain it in a safe shutdown condition,
(3)   The capability to prevent or mitigate the consequences of accidents which could result in potential off-site exposures comparable to the guideline exposures of this part.

(d) The "Operating Basis Earthquake" is that earthquake which, considering the regional and local geology and seismology and specific characteristics of local subsurface material, could reasonably be expected to affect the plant site during the operating life of the plant; it is that earthquake which produces the vibratory growth motion for which those features of the nuclear power plant necessary for continued operation without undue risk to the health and safety of the public are designed to remain functional.

(e) A "fault" is a tectonic structure along which differential slippage of the adjacent earth materials has occurred parallel to the fracture plane. It is distinct from other types of ground disruptions such as landslides, fissures, and craters. A fault may have gouge or breccia between its two walls and includes any associated monoclinal flexure or other similar geologic structural feature.

(f) "Surface faulting" is differential ground displacement at or near the surface caused directly by fault movement and is distinct from nontectonic types of ground disruptions, such as landslides, fissures, and craters.

(g) A "capable fault" is a fault which has exhibited one or more of the following characteristics:

(1)   Movement at or near the ground surface at least once within the past 35,000 years or movement of a recurring nature within the past 500,000 years.
(2)   Macro-seismicity instrumentally determined with records of sufficient precision to demonstrate a direct relationship with the fault.
(3)   A structural relationship to a capable fault according to characteristics (1) or (2) of this paragraph such that movement on one could be reasonably expected to be accompanied by movement on the other.

---

[1]   The "Safe Shutdown Earthquake" defines that earthquake which has commonly been referred to as the "Design Basis Earthquake."

In some cases, the geologic evidence of past activity at or near the ground surface along a particular fault may be obscured at a particular site. This might occur, for example, at a site having a deep overburden. For these cases, evidence may exist elsewhere along the fault from which an evaluation of its characteristics in the vicinity of the site can be reasonably based. Such evidence shall be used in determining whether the fault is a capable fault within this definition.

Notwithstanding the foregoing paragraphs III (g) (1), (2) and (3), structural association of a fault with geologic structural features which are geologically old (at least pre-Quaternary) such as many of those found in the Eastern region of the United States shall, in the absence of conflicting evidence, demonstrate that the fault is not a capable fault within this definition.

(h) A "tectonic province" is a region of the North American continent characterized by a relative consistency of the geologic structural features contained therein.

(i) A "tectonic structure" is a large scale dislocation or distortion within the earth's crust. Its extent is measured in miles.

(j) A "zone requiring detailed faulting investigation" is a zone within which a nuclear power reactor may not be located unless a detailed investigation of the regional and local geologic and seismic characteristics of the site demonstrates that the need to design for surface faulting has been properly determined.

(k) The "control width" of a fault is the maximum width of the zone containing mapped fault traces, including all faults which can be reasonably inferred to have experienced differential movement during Quaternary times and which join or can reasonably be inferred to join the main fault trace, measured within 10 miles (16.1 km) along the fault's trend in both directions from the point of nearest approach to the site. (See Figure 1 of this appendix.)

(l) A "response spectrum" is a plot of the maximum responses (acceleration, velocity or displacement) of a family of idealized single-degree-of-freedom damped oscillators against natural frequencies (or periods) of the oscillators to a specified vibratory motion input at their supports.

IV. REQUIRED INVESTIGATIONS

The geologic, seismic and engineering characteristics of a site and its environs shall be investigated in sufficient scope and detail to provide reasonable assurance that they are sufficiently well understood to permit an adequate evaluation of the proposed site, and to provide sufficient information to support the determinations required by these criteria and to permit adequate engineering solutions to actual or potential geologic and seismic effects at the proposed site. The size of the region to be investigated and the type of data pertinent to the investigations shall be determined by the nature of the region surrounding the proposed site. The investigations shall be carried out by a review of the pertinent literature and field investigations and shall include the steps outlined in paragraphs (a) through (c) of this section.

(a) Required Investigation for Vibratory Ground Motion

The purpose of the investigations required by this paragraph is to obtain information needed to describe the vibratory ground motion produced by the Safe Shutdown Earthquake. All of the steps in paragraphs (a) (5) through (a) (8) of this section need not be carried out if the Safe Shutdown Earthquake can be clearly established by investigations and determinations of a lesser scope. The investigations required by this paragraph provide an adequate basis for selection of an Operating Basis Earthquake. The investigations shall include the following:

(1) Determination of the lithologic, stratigraphic, hydrologic, and structural geologic conditions of the site and the region surrounding the site, including its geologic history;

(2) Identification and evaluation of tectonic structures underlying the site and the region surrounding the site, whether buried or expressed at the surface. The evaluation should consider the possible effects caused by man's activities such as withdrawal of fluid from or addition of fluid to the subsurface, extraction of minerals, or the loading effects of dams or reservoirs;

(3) Evaluation of physical evidence concerning the behavior during prior earthquakes of the surficial geologic materials and the substrata underlying the site from the lithologic, stratigraphic, and structural geologic studies;

(4) Determination of the static and dynamic engineering properties of the materials underlying the site. Included should be properties needed to determine the behavior of the underlying material during earthquakes and the characteristics of the underlying material in transmitting earthquake-induced motions to the foundations of the plant, such as seismic wave velocities, density, water content, porosity, and strength;

(5) Listing of all historically reported earthquakes which have affected or which could reasonably be expected to have affected the site, including the date of occurrence and the following measured or estimated data: magnitude or highest intensity, and a plot of the epicenter or location of highest intensity. Where historically reported earthquakes could have caused a maximum ground acceleration of at least one-tenth the acceleration of gravity (0.1 g) at the foundations of the proposed nuclear power plant structures, the acceleration or intensity and duration of ground shaking at these foundations shall also be estimated. Since earthquakes have been reported in terms of various parameters such as magnitude, intensity at a given location, and effect on ground, structures, and people at a specific location, some of these data may have to be estimated by use of appropriate empirical relationships. The comparative characteristics of the material underlying the epicentral location or region of highest intensity and of the material underlying the site in transmitting earthquake vibratory motion shall be considered;

(6) Correlation of epicenters or locations of highest intensity of historically reported earthquakes, where possible, with tectonic structures any part of which is located within 200 miles (322 km) of the site. Epicenters or locations of highest intensity which cannot be reasonably correlated with tectonic structures shall be identified with tectonic provinces any part of which is located within 200 miles of the site;

(7) For faults, any part of which is within 200 miles[2] of the site and which may be of significance in establishing the Safe Shutdown Earthquake, determination of whether these faults are to be considered as capable faults.[3,4] This determination is required in order to permit appropriate consideration of the geologic history of such faults in establishing the Safe Shutdown Earthquake. For guidance in

[2] If the Safe Shutdown Earthquake can be associated with a fault closer than 200 miles (322 km) to the site, the procedures of paragraphs (a) (7) and (a) (8) of this section need not be carried out for successively more remote faults.

[3] In the absence of absolute dating, evidence of recency of movement may be obtained by applying relative dating technique to ruptured, offset, warped or otherwise structurally disturbed surface or near surface materials or geomorphic features.

[4] The applicant shall evaluate whether or not a fault is a capable fault with respect to the characteristics outlined in paragraphs III(g) (1), (2), and (3) by conducting a reasonable investigation using suitable geologic and geophysical techniques.

determining which faults may be of significance in determining the Safe Shutdown Earthquake, Table 1 of this appendix presents the minimum length of fault to be considered versus distance from site. Capable faults of lesser length than those indicated in Table 1 and faults which are not capable faults need not be considered in determining the Safe Shutdown Earthquake, except where unusual circumstances indicate such consideration is appropriate;

TABLE 1

Distance from the site (miles)	Minimum length[a]
0 to 20 (0 to 32.2 km)	1 (1.6 km)
Greater than 20 to 50 (32.2 to 80.4 km)	5 (8.0 km)
Greater than 50 to 100 (80.4 to 160.9 km)	10 (16.1 km)
Greater than 100 to 150 (160.9 to 241.4 km)	20 (32.2 km)
Greater than 150 to 200 (241.4 to 321.9 km)	40 (64.4 km)

[a]   Minimum length of fault (miles) which shall be considered in establishing Safe Shutdown Earthquake.

(8)   For capable faults, any part of which is within 200 miles[2] (322 km) of the site and which may be of significance in establishing the Safe Shutdown Earthquake, determination of:
   (i)   The length of the fault;
   (ii)   The relationship of the fault to regional tectonic structures; and
   (iii)   The nature, amount, and geologic history of displacements along the fault, including particularly the estimated amount of the maximum Quaternary displacement related to any one earthquake along the fault.

(b) Required Investigation for Surface Faulting

The purpose of the investigations required by this paragraph is to obtain information to determine whether and to what extent the nuclear power plant need be designed for surface faulting. If the design basis for surface faulting can be clearly established by investigations of a lesser scope, not all of the steps in paragraphs (b) (4) through (b) (7) of this section need be carried out. The investigations shall include the following:

(1)   Determination of the lithologic, stratigraphic, hydrologic, and structural geologic conditions of the site and the region surrounding the site, including its geologic history;

(2)   Evaluation of tectonic structures underlying the site, whether buried or expressed at the surface, with regard to their potential for causing surface displacement at or near the site. The evaluation shall consider the possible effects caused by man's activities such as withdrawal of fluid from or addition of fluid to the subsurface, extraction of minerals, or the loading effects of dams or reservoirs;

(3)   Determination of geologic evidence of fault offset at or near the ground surface at or near the site;

(4)   For faults greater than 1000 feet (304.8 m) long, any part of which is within 5 miles[5] (8 km) of the site, determination of whether these faults are to be considered as capable faults;[6,7]

[7]   The applicant shall evaluate whether or not a fault is a capable fault with respect to the characteristics outlined in paragraphs III(g) (1), (2), and (3) by conducting a reasonable investigation using suitable geologic and geophysical techniques.

[5]   If the design basis for surface faulting can be determined from a fault closer than 5 miles (8 km) to the site, the procedures of paragraphs (b) (4) through (b) (7) of this section need not be carried out for successively more remote faults.

[6]   In the absence of absolute dating, evidence of recency of movement may be obtained by applying relative dating techniques to ruptured, offset, warped or otherwise structurally disturbed surface of near-surface materials or geomorphic features.

(5)   Listing of all historically reported earthquakes which can reasonably be associated with capable faults greater than 1000 feet (304.8 m) long, any part of which is within 5 miles[5] (8 km) of the site, including the date of occurrence and the following measured or estimated data: magnitude or highest intensity, and a plot of the epicenter or region of highest intensity;

(6)   Correlation of epicenters or locations of highest intensity of historically reported earthquakes with capable faults greater than 1000 feet (304.8 km) long, any part of which is located within 5 miles[5] (8 km) of the site;

(7)   For capable faults greater than 1000 feet (304.8 km) long, any part of which is within 5 miles[5] (8 km) of the site, determination of:
   (i)   The length of the fault;
   (ii) The relationship of the fault to regional tectonic structures ;
   (iii) The nature, amount, and geologic history of displacements along the fault, including particularly the estimated amount of the maximum Quaternary displacement related to any one earthquake along the fault; and
   (iv) The outer limits of the fault established by mapping Quaternary fault traces for 10 miles (16.1 km) along its trend in both directions from the point of its nearest approach to the site.

(c) Required Investigation for Seismically Induced Floods and Water Waves

(1)   For coastal sites, the investigations shall include the determination of:

(i)   Information regarding distantly and locally generated waves or tsunami which have affected or could have affected the site. Available evidence regarding the runup and drawdown associated with historic tsunami in the same coastal region as the site shall also be included;

(ii)   Local features of coastal topography which might tend to modify tsunami runup or drawdown. Appropriate available evidence regarding historic local modifications in tsunami runup or drawdown at coastal locations having topography similar to that of the site shall also be obtained; and

(iii) Appropriate geologic and seismic evidence to provide information for establishing the design basis for seismically induced floods or water waves from a local offshore earthquake, from local offshore effects of an onshore earthquake, or from coastal subsidence. This evidence shall be determined, to the extent practical, by a procedure similar to that required in paragraphs (a) and (b) of this section. The probable slip characteristics of offshore faults shall also be considered as well as the potential for offshore slides in submarine material.

(2)   For sites located near lakes and rivers, investigations similar to those required in paragraph (c) (1) of this section shall be carried out, as appropriate, to determine the potential for the nuclear power plant to be exposed to seismically induced floods and water waves as, for example, from the failure during an earthquake of an upstream dam or from slides of earth or debris into a nearby lake.

V. SEISMIC AND GEOLOGIC DESIGN BASES

(a) Determination of Design Basis for Vibratory Ground Motion

The design of each nuclear power plant shall take into account the potential effects of vibratory ground motion caused by earthquakes. The design basis for the maximum vibratory ground motion and the expected vibratory ground motion should be determined through evaluation of the seismology, geology, and the seismic and geologic history of the site and the surrounding region. The most severe earthquakes associated with tectonic structures or tectonic provinces in the region surrounding the site should be identified, considering those historically reported earthquakes that can be associated with these structures or provinces and other relevant factors. If faults in the region surrounding the site are capable faults, the most severe earthquakes associated with these faults should be determined by also considering their geologic history. The vibratory ground motion at the site should be then determined by assuming that the epicenters or locations of highest intensity of the earthquakes are situated at the point on the tectonic structures or tectonic provinces nearest to the site. The earthquake which could cause the maximum vibratory ground motion at the site should be designated the Safe Shutdown Earthquake. The specific procedures for determining the design basis for vibratory ground motion are given in the following paragraphs.

(1) Determination of the Safe Shutdown Earthquake

The Safe Shutdown Earthquake shall be identified through evaluation of seismic and geologic information developed pursuant to the requirements of paragraph IV(a), as follows:

(i)   The historic earthquakes of greatest magnitude or intensity which have been correlated with tectonic structures pursuant to the requirements of paragraph (a) (6) of Section IV shall be determined. In addition, for capable faults, the information required by paragraph (a) (8) of Section IV shall also be taken into account in determining the earthquakes of greatest magnitude related to the faults. The magnitude or intensity of earthquakes based on geologic evidence may be larger than that of the maximum earthquakes historically recorded. The accelerations at the site shall be determined assuming that the epicenters of the earthquakes of greatest magnitude or the locations of highest intensity related to the tectonic structures are situated at the point on the structures closest to the site;

(ii)   Where epicenters or locations of highest intensity of historically reported earthquakes cannot be reasonably related to tectonic structures but are identified pursuant to the requirements of paragraph (a) (6) of Section IV with tectonic provinces in which the site is located, the accelerations at the site shall be determined assuming that these earthquakes occur at the site.

(iii)   Where epicenters or locations of the highest intensity of historically reported earthquakes cannot be reasonably related to tectonic structures but are identified pursuant to the requirements of paragraph (a) (6) of Section IV with tectonic provinces in which the site is not located, the accelerations at the site shall be determined assuming that the epicenters or locations of highest intensity of these earthquakes are at the closest point to the site on the boundary of the tectonic province;

(iv)   The earthquake producing the maximum vibratory acceleration at the site, as determined from paragraph (a) (1) (i) through (iii) of this section shall be designated the Safe Shutdown Earthquake for vibratory ground motion, except as noted in paragraph (a) (1) (v) of this section. The characteristics of the Safe Shutdown Earthquake shall be derived from more than one earthquake determined from paragraph (a) (1) (i) through (iii) of this section, where necessary to assure that the maximum vibratory acceleration at the site throughout the frequency range of interest is included. In the case where a causative fault is near the site, the effect of proximity of an earthquake on the spectral characteristics of the

Safe Shutdown Earthquake shall be taken into account. In order to compensate for the limited data, the procedures in paragraphs (a) (1) (i) through (iii) of this section shall be applied in a conservative manner. The maximum vibratory accelerations of the Safe Shutdown Earthquake at each of the various foundation locations of the nuclear power plant structures at a given site shall be determined taking into account the characteristics of the underlying soil material in transmitting the earthquake-induced motions, obtained pursuant to paragraphs (a) (1), (3), and (4) of section IV. The Safe Shutdown Earthquake shall be defined by response spectra corresponding to the maximum vibratory accelerations as outlined in paragraph (a) of section VI; and

(v)   Where the maximum vibratory accelerations of the Safe Shutdown Earthquake at the foundations of the nuclear power plant structures are determined to be less than one-tenth the acceleration of gravity (0.1 g) as a result of the steps required in paragraphs (a) (1) (i) through (iv) of this section, it shall be assumed that the maximum vibratory accelerations of the Safe Shutdown Earthquake at these foundations are at least 0.1 g.

(2) Determination of Operating Basis Earthquake

The Operating Basis Earthquake shall be specified by the applicant after considering the seismology and geology of the region surrounding the site. If vibratory ground motion exceeding that of the Operating Basis Earthquake occurs, shutdown of the nuclear power plant will be required. Prior to resuming operations, the licensee will be required to demonstrate to the Commission that no functional damage has occurred to those features necessary for continued operation without undue risk to the health and safety of the public. The maximum vibratory ground acceleration of the Operating Basis Earthquake shall be at least one-half the maximum vibratory ground acceleration of the Safe Shutdown Earthquake.

(b) Determination of Need to Design for Surface Faulting

In order to determine whether a nuclear power plant is required to be designed to withstand the effects of surface faulting, the location of the nuclear power plant with respect to capable faults shall be considered. The area over which each of these faults has caused surface faulting in the past is identified by mapping its fault traces in the vicinity of the site. The fault traces are mapped along the trend of the fault for 10 miles (16.1 km) in both directions from the point of its nearest approach to the nuclear power plant because, for example, traces may be obscured along portions of the fault. The maximum width of the mapped fault traces, called the control width, is then determined from this map. Because surface faulting has sometimes occurred beyond the limit of mapped fault traces or where fault traces have not been previously recognized, the control width of the fault is increased by a factor which is dependent upon the largest potential earthquake related to the fault. This larger width delineates a zone, called the zone requiring detailed faulting investigation, in which the possibility of surface faulting is to be determined. The following paragraphs outline the specific procedures for determining the zone requiring detailed faulting investigation for a capable fault.

(1) Determination of Zone Requiring Detailed Faulting Investigation

The zone requiring detailed faulting investigation for a capable fault which was investigated pursuant to the requirement of paragraph (b) (7) of Section IV shall be determined through use of the following table:

TABLE 2

Determination of Zone Requiring Detailed Faulting Investigation

Magnitude of earthquake	Width of zone requiring detailed faulting investigation (see Fig. 1)
Less than 5.5	1 × control width
5.5—6.4	2 × control width
6.5—7.5	3 × control width
Greater than 7.5	4 × control width

The largest magnitude earthquake related to the fault shall be used in Table 2. This earthquake shall be determined from the information developed pursuant to the requirements of paragraph (b) of Section IV for the fault, taking into account the information required by paragraph (b) (7) of Section IV. The control width used in Table 2 is determined by mapping the outer limits of the fault traces from information developed pursuant to paragraph (b) (7) (iv) of section IV. The control width shall be used in Table 2 unless the characteristics of the fault are obscured for a significant portion of the 10 miles (16.1 km) on either side of the point of nearest approach to the nuclear power plant. In this event, the use in Table 2 of the width of mapped fault traces more than 10 miles (16.1 km) from the point of nearest approach to the nuclear power plant may be appropriate.

The zone requiring detailed faulting investigation, as determined from Table 2, shall be used for the fault except where:

(i) The zone requiring detailed faulting investigation from Table 2 is less than one-half mile (0.8 km) in width. In this case the zone shall be at least one-half mile (0.8 km) in width; or

(ii) Definitive evidence concerning the regional and local characteristics of the fault justifies use of a different value. For example, thrust or bedding-plane faults may require an increase in width of the zone to account for the projected dip of the fault plane; or

(iii) More detailed three-dimensional information, such as that obtained from precise investigative techniques, may justify the use of a narrower zone. Possible examples of such techniques are the use of accurate records from closely spaced drill holes or from closely spaced, high-resolution offshore geophysical surveys.

In delineating the zone requiring detailed faulting investigation for a fault, the center of the zone shall coincide with the center of the fault at the point of nearest approach of the fault to the nuclear power plant as illustrated in Figure 1.

(c) Determination of Design Bases for Seismically Induced Floods and Water Waves

The size of seismically induced floods and water waves which could affect a site from either locally or distantly generated seismic activity shall be determined, taking into consideration the results of the investigation required by paragraph (c) of section IV. Local topographic characteristics which might tend to modify the possible runup and drawdown at the site shall be considered. Adverse tide conditions shall also be taken into account in determining the effect of the floods and waves on the site. The characteristics of the earthquake to be used in evaluating the offshore effects of local earthquakes shall be determined by a procedure similar to that used to determine the characteristics of the Safe Shutdown Earthquake in paragraph V (a).

(d) Determination of Other Design Conditions
(1) Soil Stability

Vibratory ground motion associated with the Safe Shutdown Earthquake can cause soil instability due to ground disruption such as fissuring, differential consolidation, liquefaction, and cratering which is not directly related to surface faulting. The following geologic features which could affect the foundations of the proposed nuclear power plant structures shall be evaluated, taking into account the information concerning the physical properties of materials underlying the site developed pursuant to paragraphs (a) (1), (3), and (4) of Section IV and the effects of the Safe Shutdown Earthquake:

(i)   Areas of actual or potential surface or subsurface subsidence, uplift, or collapse resulting from:
    (a)   Natural features such as tectonic depressions and cavernous or karst terrains, particularly those underlain by calcareous or other soluble deposits;
    (b)   Man's activities such as withdrawal of fluid from or addition of fluid to the subsurface, extraction of minerals, or the loading effects of dams or reservoirs; and
    (c)   Regional deformation.
(ii)   Deformational zones such as shears, joints, fractures, folds, or combinations of these features.
(iii) Zones of alteration or irregular weathering profiles and zones of structural weakness composed of crushed or disturbed materials.
(iv)   Unrelieved residual stresses in bedrock.
(v)   Rocks or soils that might be unstable because of their mineralogy, lack of consolidation, water content, or potentially undesirable response to seismic or other events. Seismic response characteristics to be considered shall include liquefaction, thixotropy, differential consolidation, cratering, and fissuring.

(2) Slope Stability

Stability of all slopes, both natural and artificial, the failure of which could adversely affect the nuclear power plant, shall be considered. An assessment shall be made of the potential effects of erosion or deposition and of combinations of erosion or deposition with seismic activity, taking into account information concerning the physical property of the materials underlying the site developed pursuant to paragraph (a) (1), (3), and (4) of Section IV and the effects of the Safe Shutdown Earthquake.

(3) Cooling Water Supply

Assurance of adequate cooling water supply for emergency and long-term shutdown decay heat removal shall be considered in the design of the nuclear power plant, taking into account information concerning the physical properties of the materials underlying the site developed pursuant to paragraphs (a) (1), (3), and (4) of section IV and the effects of the Safe Shutdown Earthquake and the design basis for surface faulting. Consideration of river blockage or diversion or other failures which may block the flow of cooling water, coastal uplift or subsidence, or tsunami runup and drawdown, and failure of dams and intake structures shall be included in the evaluation, where appropriate.

(4) Distant Structures

Those structures which are not located in the immediate vicinity of the site but which are safety related shall be designed to withstand the effect of the Safe Shutdown Earthquake and the design basis for surface faulting determined on a comparable basis to that of the nuclear power plant, taking into account the material underlying the structures and the different location with respect to that of the site.

VI. APPLICATION TO ENGINEERING DESIGN

(a) Vibratory Ground Motion

(1) Safe Shutdown Earthquake

The vibratory ground motion produced by the Safe Shutdown Earthquake shall be defined by response spectra corresponding to the maximum vibratory accelerations at the elevations of the foundations of the nuclear power plant structures determined pursuant to paragraph (a) (1) of Section V. The response spectra shall relate the response of the foundations of the nuclear power plant structures to the vibratory ground motion, considering such foundations to be single-degree-of-freedom damped oscillators and neglecting soil-structure interaction effects. In view of the limited data available on vibratory ground motions of strong earthquakes, it usually will be appropriate that the response spectra be smoothed design spectra developed from a series of response spectra related to the vibratory motions caused by more than one earthquake.

The nuclear power plant shall be designed so that, if the Safe Shutdown Earthquake occurs, certain structures, systems, and components will remain functional. These structures, systems, and components are those necessary to assure (i) the integrity of the reactor coolant pressure boundary, (ii) the capability to shut down the reactor and maintain it in a safe condition, or (iii) the capability to prevent or mitigate the consequences of accidents which could result in potential offsite exposures comparable to the guideline exposures of this part. In addition to seismic loads, including aftershocks, applicable concurrent functional and accident-induced loads shall be taken into account in the design of these safety-related structures, systems, and components. The design of the nuclear power plant shall also take into account the possible effects of the Safe Shutdown Earthquake on the facility foundations by ground disruption, such as fissuring, differential consolidation, cratering, liquefaction, and landsliding, as required in paragraph (d) of section V.

The engineering method used to ensure that the required safety functions are maintained during and after the vibratory ground motion associated with the Safe Shutdown Earthquake shall involve the use of either a suitable dynamic analysis or a suitable qualification test to demonstrate that structures, systems and components can withstand the seismic and other concurrent loads, except where it can be demonstrated that the use of an equivalent static load method provides adequate conservatism.

The analysis or test shall take into account soil-structure interaction effects and the expected duration of vibratory motion. It is permissible to design for strain limits in excess of yield strain in some of these safety-related structures, systems, and components during the Safe Shutdown Earthquake and under the postulated concurrent conditions, provided that the necessary safety functions are maintained.

(2) Operating Basis Earthquake

The Operating Basis Earthquake shall be defined by response spectra. All structures, systems, and components of the nuclear power plant necessary for continued operation without undue risk to the health and safety of the public shall be designed to remain functional and within applicable stress and deformation limits when subjected to the effects of the vibratory motion of the Operating Basis Earthquake in combination with normal operating loads. The engineering method used to ensure that these structures, systems, and components are capable of withstanding the effects of the Operating Basis Earthquake shall involve the use of either a suitable dynamic analysis or a suitable qualification test to demonstrate that the structures, systems and components can withstand the seismic and other concurrent loads, except where it can be demonstrated that the use of an equivalent static load method provides adequate conservatism. The analysis or test shall take into account soil-structure interaction effects and the expected duration of vibratory motion.

(3) Required Seismic Instrumentation

Suitable instrumentation shall be provided so that the seismic response of nuclear power plant features important to safety can be determined promptly to permit comparison of such response with that used as the design basis. Such a comparison is needed to decide whether the plant can continue to be operated safely and to permit such timely action as may be appropriate.

These criteria do not address the need for instrumentation that would automatically shut down a nuclear power plant when an earthquake occurs which exceeds a predetermined intensity. The need for such instrumentation is under consideration.

(b) Surface Faulting

(1) If the nuclear power plant is to be located within the zone requiring detailed faulting investigation, a detailed investigation of the regional and local geologic and seismic characteristics of the site shall be carried out to determine the need to take into account surface faulting in the design of the nuclear power plant. Where it is determined that surface faulting need not be taken into account, sufficient data to clearly justify the determination shall be presented in the license application.

(2) Where it is determined that surface faulting must be taken into account, the applicant shall, in establishing the design basis for surface faulting on a site take into account evidence concerning the regional and local geologic and seismic characteristics of the site and from any other relevant data.

(3) The design basis for surface faulting shall be taken into account in the design of the nuclear power plant by providing reasonable assurance that in the event of such displacement during faulting certain structures, systems, and components will remain functional. These structures, systems, and components are those necessary to assure (i) the integrity of the reactor coolant pressure boundary, (ii) the capability to shut down the reactor and maintain it in a safe shutdown condition, or (iii) the capability to prevent or mitigate the consequences of accidents which could result in potential offsite exposures comparable to the guideline exposures of this part. In addition to seismic loads, including aftershocks, applicable concurrent functional and accident-induced loads shall be taken into account in the design of such safety features. The design provisions shall be based on an assumption that the design basis for surface faulting can occur in any direction and azimuth and under any part of the nuclear power plant unless evidence indicates this assumption is not appropriate, and shall take into account the estimated rate at which the surface faulting may occur.

(c) Seismically Induced Floods and Water Waves and Other Design Conditions

The design basis for seismically induced floods and water waves from either locally or distantly generated seismic activity and other design conditions determined pursuant to paragraphs (c) and (d) of Section V, shall be taken into account in the design of the nuclear power plant so as to prevent undue risk to the health and safety of the public.

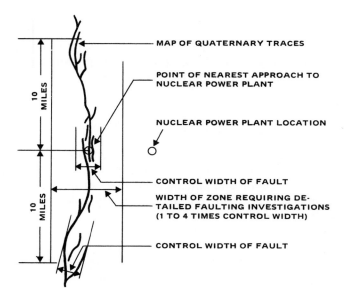

FIGURE 1. Diagrammatic illustration of delineation of width of zone requiring detailed faulting investigations for specific nuclear power plant location.

Appendix G

GEOLOGIC AND GEOPHYSICAL ELEMENTS FOR THE SAN FRANCISCO BAY REGION ENVIRONMENT AND RESOURCES PLANNING STUDY*

ACTIVE FAULTS

The San Francisco Bay region is traversed by three major earthquake-generating fault zones that have been active in historic time and by several smaller active or possibly active faults. The location and anticipated activity of these faults are two of the dominating environmental restraints on the location of buildings and structures and on land use in the Bay region.

The primary hazards of active faults are:

1. Slip along the active faults periodically displaces the earth's surface and disrupts overlying structures. This slip may occur as large sudden displacements or as small semicontinuous displacements called fault creep.
2. Fault movement is commonly, although not always and everywhere, a sudden event that generates earthquakes.

    The secondary hazards of active fault zones result from the patches of boggy ground that they typically contain, and their particular susceptibility to landslides.

The magnitude of the active fault problem in the Bay region may be gaged from the fact that active zones a few tens of feet to 1½ miles (2.4 km) wide traverse 285 miles (459.7 km) of Bay region terrain. In addition, destructive earthquakes (Modified Mercalli Intensity VII and above) have struck the region 36 times since the first earthquake was recorded in 1800.

Proposed Investigations

The location and the historical and recent geologic history of the active faults in the Bay region will be determined by surface and subsurface geologic techniques, distribution of microearthquakes, and measurement of crustal strain. It is hoped that from these data the location of most future disruptions of the earth's surface along active faults in the Bay region can be predicted, and an estimate of the frequency of future episodes of fault slip and destructive earthquakes can be made. The width of the active fault zones that should be considered hazardous because of branching faults or subordinate geologic structures and the location of sag ponds, boggy ground, and landslides in the active fault zones will also be determined. The last phase of the project will be the practical application of the data on active faults to land-use planning and decision-making in the Bay region. In this phase Geological Survey personnel, a structural engineer, and planners will collaborate.

The San Andreas fault and the Hayward fault will have the highest priority for the current investigations. Both have moved in historic time, and both have produced major earthquakes and major surface rupture in the San Francisco Bay area. The Calaveras, Rodgers Creek, Healdsburg, and Sargent faults exhibit convincing evidence of

---

* From Wallace, R. E. (Project Director), San Francisco Bay Region Environment and Resources Planning Study, U.S. Department of Interior and the U.S. Department of Housing and Urban Development, Washington, D.C., 1972, 28.

current or historic activity, but surface geologic evidence indicates a lower rate of movement than on the San Andreas and Hayward faults. Nevertheless, data are needed to evaluate these features. A number of faults, including the Pilarcitos, San Gregorio, and Franklin faults, are not known to be active but will be evaluated geologically, because they cut very young deposits or because they are potentially active.

Specific investigations designed to provide the essential data are:

1. The preparation of strip maps based on areal geologic and geomorphic investigations of suspected active faults and of sedimentary deposits and geologic structures in active fault zones that are potentially hazardous to buildings and structures. Data are to be based on geologic interpretation of aerial photographs, field investigations, and possibly geophysical studies of selected areas. Goals are to identify and map active faults, and to collect geologic and geomorphic evidence bearing on movement rates and level of risk.

2. Studies of microearthquakes based on a broadly distributed telemetered seismic net in the nine Bay counties to provide seismic data on known and suspected active faults. Goals are to determine the distribution of microearthquakes in space and time, to identify and map sections of faults that are currently moving, and to monitor sections that are known to be geologically active but appear to be presently aseismic.

3. Physical exploration of active fault zones. Data are to be derived from trenches, auger drill holes, and possibly rotary drill holes (shallow shot-hole rig) located so as to provide data for detailed stratigraphic and structural studies. Goals are to obtain data on fault zone widths, frequency of historical and geologically young fault movements, and, where possible, the age and amount of displacement of those fault movements.

4. Strain measurements and monitoring of known active fault zones. Data are obtained from such relatively low cost nonrecording devices as quadrilaterals and alinement arrays. Minimum new installations are needed in the Point Reyes segment of the San Anreas fault; the Point Arena — Fort Ross segment of the San Andreas fault; and the Healdsburg — Rodgers Creek fault. Goals are to compare local surface strain on known active faults with regional strain as determined from Coast and Geodetic triangulation nets, geodimeter lines, and other precise survey data; also to detect changes from established long-term patterns of strain.

5. Interpretation of marine geophysical surveys in order to locate active faults, and to obtain stratigraphic data pertaining to the age of the last movement. Highest priority is to determine the northern extent of the Hayward fault in San Pablo Bay. Marine surveys will also help locate and date movements on the offshore parts of the San Andreas and San Gregorio faults between Montara Mountain and Point Reyes. Goals are to identify and map submarine segments of active faults, and to collect marine geologic and geophysical evidence bearing on movement rates and level of risk.

Products
Basic Data Contributions
1. Strip maps showing active faults breaks:[1]
    a. The San Andreas fault from Paicines to San Francisco. Scale 1:24,000; open filed mid-1971, to be published later as Miscellaneous Geologic Investigation (MGI).

[1] Strip maps showing active breaks along the southern Hayward and southern Calaveras faults have already been prepared.

b.   The San Andreas fault from Bolinas Lagoon to Point Delgada. Scale 1:24,000; open filed mid-1970, to be published later as MGI.
c.   The Healdsburg-Rodgers Creek fault (to be prepared only if field and photo interpretive data identify these faults as potentially major sources of earthquake risk).
d.   The Hayward fault northwest of Pinole Point (contingent on availability of geophysical and geologic evidence needed to locate fault in San Pablo Bay and to the north).
e.   Other active faults that may be identified during the course of the investigation.
2.   Fault-plane geometry and history of geologically recent faulting in the Bay region as exposed in trenches and drill holes.
3.   Distribution and character of microearthquakes in the Bay region from 1968 to 1973.
4.   Crustal strain along the active faults of the Bay region from 1968 to 1973.
5.   Brief reports summarizing environmentally significant findings with time value, obtained from physical exploration program, field investigations, seismology program, or other parts of the active-fault program.
6.   Scientific and technical reports for journal or U.S. Geological Survey publication summarizing or describing important research findings. Where such research data are important in planning and decision-making they will be released first under Item 5.

Technical Reports

Tectonic synthesis and evaluation of the known and suspected active faults of San Francisco Bay region, accompanied by a discussion of the relative level of risk and an estimate of future activity, as judged from geologic and geophysical evidence and from historic records. To be accompanied by an active-fault map of the Bay region, scale 1: 125,000.

Interpretive Reports

In the various interpretive reports to be prepared (see section on Focus and Methods) the risk of the region's various active faults will be rated in terms of the probable recurrence of major earthquakes and the nature and amount of displacement to be expected. Guidelines for land use in and near fault zones will be provided; for example, set-back limits will be suggested, model ordinance language will be developed, and suggestions of regulatory procedures for public agencies will be prepared.

Slope Stability and Engineering Behavior of Bedrock Areas

The behavior of the ground from an engineering standpoint places varying constraints on man's use of the land, and the possibility of landslides, particularly, will become an increasingly dominant constraint as the need for housing sites extends into the hillside areas. Earth materials in the Bay region range from hard solid rock to swelling clays and various unconsolidated sediments. This range, together with varying geologic structures, topography, and other factors, leads to great differences in the behavior of the ground from place to place. If these differences are ignored, building foundations may collapse, and such public facilities as roads, sewer lines, water and gas pipelines, and bridges may be severely damaged.

Landslides constitute the greatest problem in the region. Large to small masses of earth move downhill at varying rates of speed, removing support for structures above them, impinging on land and structures below them, and multiplying sediment loads in adjacent streams. Fast-moving landslides can destroy life as well as property.

Bearing strength, swelling potential, ease of excavation, and other characteristics of earth materials are of lesser importance but cannot be ignored. These characteristics can cause excessive construction and maintenance costs and structural failures.

In order to make the best and most economical use of the land, unstable slopes must be identified, grading design must be accommodated to the local materials and slopes, foundation and building design must be fitted to local foundation conditions, and other physical aspects of the ground must be anticipated and planned for. One measure of the significance of these factors is the cost to the state highway system as a result of landslides. Landslides are common in the Bay region and in some places constitute as much as one-quarter of the land area. The cost of emergency cleanup alone of landslides on the 1,115 miles (1,798.4 km) of state highways throughout California averaged almost $3,000,000 per year between 1946 and 1967, and the total direct costs of landslides in State Highway District 5 in the Coast Ranges amounted to about $18,000,000, or 10 percent of annual construction costs, in fiscal year 1966—1967.

Proposed Investigations

The distribution and general physical characteristics of the various earth materials in the region will be established through literature study, interviews, and field and laboratory observations and testing. From this information, the most critical engineering factors, such as strength, ease of excavation, and erodibility, will be predicted for the region.

The distribution and character of existing landslides in the region will be determined by study of aerial photographs and by limited field investigations. The occurrence of the landslides will be related to the natural and artificial processes which cause them, including lithology, geologic structure, amount and height of slope, rainfall, and land-use practices. The results will be used to predict slope stability over the whole region.

The critical geologic and hydrologic processes and land-use practices that produce landslides and other slope failures are not fully known. Accordingly, selection of the specific physical characteristics to be studied and the procedural details of the slope-stability study will depend on the early results of the investigation.

The last phase of the project will be to apply the information gathered concerning the physical characteristics of earth materials and the slope stability to land-use planning and decision-making in the Bay region. This phase will involve collaboration between the planners and the geologists, hydrologists, and engineers of the Geological Survey. It is hoped that the results of the project will provide a basis for planning and land-use decisions on a regional and subregional scale, and will provide useful guidance to more detailed studies. However, the results are not intended for site planning and design, or for other local planning decisions.

Specific investigations designed to provide data and geologic interpretation are:

1. Existing landslides will be identified and mapped through interpretation of aerial photographs, supplemented by field investigations in selected areas. The factors responsible for the landslides will be determined, and the occurrence of small landslides in the control areas will be established.
2. Maps will be prepared to show the kinds of earth materials present in the Bay region, and their spatial relations, structural character, and areal distribution. The maps will be based on compilation of published and unpublished studies, supplemented by new field investigations. In addition to providing the basic information needed in the landslide and engineering behavior studies, the geologic mapping will provide the essential framework for the active fault, resource, seismic, and hydrologic studies.
3. The physical properties of the earth materials in the Bay region that have critical

engineering and land-use significance will be determined and tabulated. Sources of information include the technical literature, selected unpublished files, and field and laboratory testing. The engineering behavior of the earth materials will be predicted, based on the physical properties and methods determined during the investigation.

4.   Analysis of the relative stability of hillsides in the Bay region will be based on the occurrence and character of existing landslides and on the geologic, hydrologic, and topographic features and physical processes that cause them. These features and processes include the character, properties, and geologic structure of the earth materials, the height, steepness, and exposure of slopes, groundwater conditions and rainfall, and seismic shaking.

Products
Basic Data Contributions

1.   Landslide inventory.
      a.   Preliminary maps at a scale of 1:62,500 will show identifiable landslides in selected parts of the Bay region larger than about 300 feet (91.5 m) across. These maps will be released to open file.
      b.   A final map at a scale of 1:125,000 will show landslide occurrence and type, and distribution of those physical factors that appear to control landslides in the Bay region.
2.   Geologic maps.
      a.   Preliminary geologic maps of the Bay region at a scale of 1:62,500 will be compiled from the literature and a minimum amount of field investigation. These maps will be released to open file according to the same tentative schedule as the preliminary landslide maps.
      b.   A final geologic map at a scale of 1:125,000 will show the distribution and character of earth materials, faults, and the general orientation of bedding.
3.   Engineering geologic map. A map at a scale of 1:125,000 will interpret map units according to their physical properties and engineering behavior and will also show faults and generalized structural characteristics. A tabular text listing the physical properties and engineering behavior of the map units will accompany the map.

Technical Reports

1.   Landslides and other slope instabilities in the San Francisco Bay region will be described together with their occurrence, formation, and relation to physical properties and geologic processes.
2.   A map, at a scale of 1:125,000, will give a prediction of the future stability of slopes in the San Francisco Bay region, and will be accompanied by a discussion of the map and the data and methods used in its preparation.

Interpretive Reports

In the various interpretive reports to be prepared (see section on Focus and Methods), guidelines will be formulated for developments in areas having slope-stability problems. Maps will be prepared on a regional scale to show the relative stability of various terrains. Factors such as density of development, alternative land uses, and other methods of minimizing the slope-stability problem will be discussed.

PHYSICAL PROPERTIES OF UNCONSOLIDATED DEPOSITS

Most of the population in the San Francisco Bay region is concentrated in lowland areas underlain by unconsolidated sedimentary deposits. As population increases and greater demands are made on these areas for residential, agricultural, industrial, and recreational use, it is imperative that various geological conditions be considered. Pertinent geological data should be incorporated into regional and local plans to avoid or minimize the adverse affects of natural processes, to prevent the artificial intensification of any natural problems, and to avoid creating new problems. Several of the most important geologic conditions affecting land use in lowland areas are summarized below.

Ground motion resulting from earthquakes may be amplified as a result of the low shear strength of unconsolidated materials. Consequently, earthquakes pose a more serious problem in areas underlain by sediments than in areas underlain by consolidated bedrock. Because the San Francisco Bay region is traversed by three major earthquake-generating fault zones that have been active in historic time, this problem is of particular significance. The 1906 San Francisco earthquake clearly illustrates the actual hazards, and the 1964 Alaska earthquake, in which property damage was much more severe on unconsolidated ground than on bedrock, illustrates the potential hazard.

Poor foundation conditions such as liquefaction and compressibility (low bearing strength) may be associated with certain sedimentary deposits such as the Bay mud. Expansive soil is a common problem in the lowland areas, though it is not restricted to this particular geologic or geographic setting. When ignored these properties may cause safety hazards, excessive maintenance costs, or even condemnation as a result of structural failures.

The unconsolidated deposits in lowland areas form the natural reservoirs supplying most of the locally derived water for domestic, agricultural, and industrial uses. The quantity and quality of this water is both directly and indirectly related to the physical characteristics (such as porosity, permeability, bedding, and lateral extent) and the structural features (such as faults and folds) of the sedimentary deposits. To protect this valuable water resource, land use and operational practices should be carefully evaluated. Pumping rates, if too high, could deplete reservoirs and cause regional subsidence as in San Jose (as much as 8 feet [2.4 m] in a 30-year period) or allow encroachment of saline water into fresh-water aquifers. Solid- and liquid-waste disposal, if not properly effected, could contaminate ground water or clog natural aquifers. Inadequately designed or improperly situated percolation ponds would inhibit recharge of ground water reservoirs.

Areas susceptible to the various problems outlined above must be identified in order to make the best and most economical use of the lowland regions. The thickness and the gross physical characteristics of the sedimentary prism are among the most important parameters required for a seismic-intensity (ground-motion amplification) map of the Bay region. Contour maps of the bedrock surface and isopach maps of the sedimentary deposits will provide essential data for identifying areas of high seismic intensity. Areal geologic maps and cross sections of the sedimentary deposits will provide data for identifying areas with potential foundation and ground-water problems. With these important data readily available, land use and building design can be better adapted to local geological conditions.

Proposed Investigations

The gross physical aspects of the sedimentary deposits — lateral extent, thickness, bedding characteristics, structural features, compaction, grain size, etc. — will be determined as completely as possible by surface and subsurface geologic methods. Compilations of published and unpublished maps and well logs will be checked and augmented by limited field mapping, geophysical investigations, drilling, and laboratory tests, such as age dating and fossil identification. These basic data will be synthesized to provide an insight into the late Cenozoic geologic history of the Bay region. A knowledge of the geology as recorded in the sedimentary deposits is essential not only to the problems previously outlined but also to many other aspects of the program.

The initial phase of the project will involve collecting and synthesizing available geologic data. The second phase will focus on selected topics or strategic areas with particular problems or inadequate information. Therefore, the detailed execution of the second phase will depend on the first. The final phase will involve interpretation of the basic data with the focus on applying the results to land-use planning. This phase will involve collaboration between planners and engineers from local and regional governments, and earth scientists of the Geological Survey.

It is hoped that the proposed maps and reports will provide adequate information for regional and subregional planning. They probably will not provide adequate information for local planning or specific site appraisal, but should provide useful guidance for detailed studies.

Specific investigations designed to provide the basic data and geologic interpretations are:

1. Data will be derived from well logs, geophysical surveys, and field mapping, the goals being to define the depth and configuration of the upper bedrock surface primarily as an aid to the prediction of variations of ground-motion intensity during earthquakes.

2. A study of the thickness and lithology of unconsolidated sedimentary units will be based on surface and subsurface mapping and geophysical surveys. These data are needed to predict the seismic and engineering behavior of the unconsolidated deposits. They are basic to understanding the past movements and predicting the future activity of earthquake faults. They are also essential for understanding present ground-water problems (supply as well as resulting subsidence) and predicting future problems. They also provide information for construction-material inventories.

3. The characteristics of the unconsolidated deposits that determine their engineering behavior will be studied and tabulated. Data will be derived from field and laboratory engineering tests and geologic maps compiled from the literature, augmented by field mapping. From a knowledge of the lateral distribution at or near the ground surface, potential engineering problem areas will be delineated.

Products
Basic Data Contributions

1. Contour map of bedrock surface; scale 1:62,500 or larger.
2. Isopach map of unconsolidated deposits; scale 1:62,500 or larger.
3. Map of historic limits of San Francisco Bay marshlands; scale 1:62,500 or larger.
4. Geologic maps and cross sections by counties; scale 1:62,500.
5. Engineering geologic map; scale probably 1:62,500 or larger; with tabular data listing physical properties and engineering behavior.
6. Miscellaneous maps; scale probably 1:62,500 or larger. Maps show distribution of specific characteristics or problems (such as land subsidence). Topics to be determined as project progresses.

Interpretive Reports

The results of this program element will be incorporated into general interpretive reports or manuals (see section on Focus and Methods), and will serve as background for a more meaningful appraisal of such factors as land subsidence, seismic damage potential, availability of sand and gravel, foundation conditions for construction, and tunneling excavation and drainage problems. These data will be fundamental to the land-use guidelines which will be developed in the interpretive reports.

SEISMICITY AND GROUND MOTION

The rapidly expanding urban region surrounding San Francisco Bay lies on one of the earth's most active tectonic features — the San Andreas fault system. Along this fault system the crustal plate that floors the Pacific Ocean is sliding northwestward past the crustal plate supporting the North American continent at a rate of several centimeters per year. The opposing plates slide past each other quite smoothly in some regions where their relative motion is accommodated by aseismic creep and frequent small-to-moderate earthquakes. In other regions, including the Bay area, they are more firmly locked together and slip suddenly, after long but irregular intervals of time, when stresses across the fault induced by slowly accumulating elastic strains along the plate edges build up to levels that exceed the "strength" of the fault. The great earthquakes that can result from this process, such as the 1906 San Francisco earthquake, constitute a major hazard to life and property. The historical record suggests that the Bay area should expect at least one great earthquake ($M \geq 8$), several major ones ($M \geq 7$), and many destructive ones ($M \geq 5.5$) per century.

The San Andreas fault system is quite complex in the Bay area. It splits southeast of San Francisco Bay and major branches run along both sides of the Bay. The principal "San Andreas" branch, which produced the 1906 quake, runs up the peninsula west of the Bay and continues northwestward to Cape Mendocino. The subsidiary Calaveras and Hayward branches can be traced up the east side of the Bay to Carquinez Strait but their relationship to major faults in the Coast Ranges north of the strait is not clear. Although a major earthquake occurred on the Hayward fault in 1868, both the Hayward and Calaveras faults appear to be creeping sporadically at the present time.

Most of the destruction caused by earthquakes results from damage to, or collapse of, structures built by man — destruction caused either directly by ground shaking produced by seismic waves spreading outward from the generating fault, or indirectly by failures in the underlying materials (foundation failures, landslides, etc.) induced by ground shaking. Measures to lessen or avoid earthquake damage usually are directed toward preventing damage to structures. The success of these measures is dependent upon the accuracy of the estimates of the character and intensity of ground shaking in specific regions (or even at specific sites) that will be produced by earthquakes originating on the major faults.

Traditional methods of arriving at such estimates are mostly empirical. They involve examination and analysis of damage produced by specific earthquakes on the terrain and on the structures that were by chance subjected to them. Progress by these methods has been slow, and the results that they have produced have been semiquantitative at best. Even a great historical earthquake, such as the 1906 earthquake in the Bay area, fails to provide an adequate basis for sufficiently detailed, accurate predictions of the effects of a future great earthquake in the same region. Urbanized areas have expanded enormously, and much of the growth has been on ground of uncertain geologic stability. Moreover, far more sophisticated, reliable predictions of ground shaking are required to insure a safe design for a large structure than for the simpler frame structures against which the 1906 quake was measured.

In the search for a better understanding of how earthquakes are generated and for more reliable and direct methods of predicting their effects, the Geological Survey has undertaken a broad range of studies on the San Andreas fault system in central California. The long-range goal of this program is the prediction of the time, place, and magnitude of earthquakes in the region as well as the character, intensity, and duration of strong ground motion experienced at sites throughout the shaken area.

Proposed Investigations

Different instrument systems and methods of study are employed to attack various aspects of the problem, as outlined below.

1. The mechanics of earthquake generation are being investigated by:
    a. Detailed study of microearthquakes and aftershocks by the permanent telemetered seismic network and by special portable seismic networks to map faults in three dimensions and to determine the nature and time history of movement along the fault surface.
    b. Detailed measurements of strain along the San Andreas fault system. The measurements are designed to detect the elastic strain accumulating across a broad region containing the fault zone as well as to detect the concentrated, nonstored strain (or slip) associated with creep and microearthquakes.
    c. Analysis of the space-time history of microearthquakes and strain to define the outlines of local crustal "plates" and to develop methods for short-term earthquake prediction.
    d. Laboratory studies of the mechanical properties and behavior of crustal rocks under the conditions of temperature and stress encountered in the crust.
    e. Model studies of the mechanics of "faulting" in a heterogeneous strained elastic medium.
2. Seismic wave propagation in the crust is being investigated by:
    a. Explosion-refraction studies of gross crustal structures and of geometrical spreading and dissipative attenuation along various paths.
    b. Same as above, but with precisely located microearthquake sources at depth within the crust instead of surface explosions.
    c. "Mapping" of seismically sensitive sedimentary deposits in basins by seismic refraction and reflection techniques. Determination of in situ elastic constants by the same methods.
3. Influence of local geologic conditions on the character, intensity, and duration of strong ground shaking are being investigated through:
    a. Studies of the relative amplification and spectral ratios ("site" to "bedrock") of seismic waves from distant sources as a function of the geology of the recording site. These are "small-motion" measurements.
    b. Analysis of telemeter and portable seismograph station magnitude residuals to determine variations in average amplitude response for local earthquakes as a function of recording-site geology.
    c. Multilevel down-hole seismograph installations to record waves from moderate local earthquakes to provide data for detailed studies of the amplification of seismic waves by weak near-surface materials. On occasion, the down-hole installation will be augmented by a surface tripartite array of three-component portable stations to establish the identity of the wave types associated with large amplitudes and particle velocities at the surface of a sedimentary section.
    d. Theoretical studies of the interaction of seismic waves emerging from the

"basement" with localized basins of low-velocity sediments.

e.   Strong-motion versus weak-motion amplification ratio studies in the vicinity of nuclear tests to establish the range of applicability of "small-motion" studies.

Most of these studies are being made along the San Andreas fault between San Francisco and Cholame, where the opportunity to study the widest variety of conditions with a minimum of effort presented itself. For the purposes of the San Francisco Bay Region Study, the telemetered seismic net should be strengthened north of the Bay and strain-monitoring networks should be established there to extend the work now carried on in the southern counties to the rest of the nine-county pilot program area.

Seismic studies to be added to answer specific questions pertinent to the San Francisco Bay Region Study include:

1.   Ground-motion studies; primarily install and operate down-hole seismograph at edge of Bay.
2.   Extension of microearthquake and strain networks into the area north of Carquinez Strait.
3.   Special seismic reflection/refraction studies of sedimentary basins.
4.   Preparation of seismic intensity estimate maps.

Products

Although the primary USGS earthquake study program is open ended and largely financed under programs other than the cooperative program with HUD, substantial results of great importance to the San Francisco Bay Region Study are anticipated during the next 3 years.

Basic Data Contributions

1.   Precise delineation of microearthquake-generating active faults in the Bay area and detailed monitoring of quiet sections of known major active faults for signs of reactivation. Both the seismic and strain nets will contribute to this study. Seismic and strain observations will be analyzed and evaluated in the light of geologic evidence on active faulting developed under the Active Faults Project, and the results of the coordinated study will be presented in the reports described under that project. Supporting lists of earthquakes, giving locations, occurrence times, and magnitudes, will be provided in addition to the epicenter maps.
2.   Development and field evaluation of active seismic techniques for mapping the distribution and thickness of low-velocity sediments in basins and of estimating the elastic constants of the sediments. This work will be closely coordinated with the Unconsolidated Deposits Project, and many of the results will be published as refinements and extensions to maps presenting the work of that project. In addition, it is anticipated that cross-section profiles will be published for the major basins in the nine Bay area counties.
3.   The relative ground amplification map of the margins of the Bay (originally published in a paper by Borcherdt) will be refined and extended as the availability of Nevada Test Site shots of suitable size permits.

Technical Reports

Technical papers will be published in scientific journals to report the major theoretical and experimental results of the earthquake study program.

Interpretive Reports

The seismic risk factor will be incorporated into various interpretive reports. A special interpretive report will include maps showing the extent of expected damage from earthquakes of various magnitudes originating on the major active faults of the region. The expected damage will be expressed in terms of intensity measured by the Modified Mercalli scale — an arbitrary scale that extends from I (shock generally not felt by anyone) to XII (total destruction of manmade structures). These intensities will be estimated from their relation to: (1) observed historic damage as a function of distance from the earthquake epicenter and (2) variations in damage intensity with variations in rock and soil types. The expected ground-motion data will be generated by computer and can be updated as new information is accumulated. In collaboration with structural engineers, an attempt will be made to develop model code requirements in different earthquake risk zones.

Appendix H

EARTHQUAKE PREDICTION EVALUATION GUIDELINES, CALIFORNIA EARTHQUAKE PREDICTION EVALUATION COUNCIL, FEBRUARY 22, 1977*

I. PRECEPTS

1. The California Emergency Services Act and Senate Bill 1950 (adding Section 955.1 to the Government Code) authorize and require certain actions with respect to earthquake events by the California Office of Emergency Services. These actions include pre-earthquake preparation and warning, coordination of post-earthquake operations, and recommending designation of areas to be included in a "State of Emergency" proclamation by the Governor.

2. Two recent reports, *Earthquake Prediction and Public Policy*[1] and *Predicting Earthquakes: A Scientific and Technical Evaluation with Implications for Society*[2] provide a useful overview of the expected development of earthquake prediction technology and attendant social implications in the United States.

3. As used here, "earthquake prediction" means the specification of the time, place, magnitude, and probability of occurrence of an anticipated earthquake, with sufficient precision that actions to minimize loss of life and reduce damage to property are possible.

4. An earthquake prediction will be significant only if large numbers of people take it seriously.[3] Some people take earthquake predictions from psychics or amateurs seriously but, for the most part, the impact of such pronouncements is relatively short-lived. If, however, reputable scientists known to be experts on earthquake prediction agree as to the reliability of the data, methods of analysis, and interpretation of the results, the prediction will be taken quite seriously by large numbers of people.

5. Earthquake prediction, as a developing branch of the earth sciences, is presently subject to many uncertainties as to its basic hypotheses, methodology, and data interpretation. Therefore, earth scientists generally believe it may be several years before scientifically-based earthquake predictions having a high degree of certainty will be issued in California and the United States.

6. Earthquake prediction statements are likely to be issued at any time by individuals or agencies whether qualified or not. Before government can act responsibly on any earthquake prediction, its scientific validity must first be determined. Increased public concern and the immense potential for life and property loss require that California State Government establish a mechanism for official response to earthquake predictions now, despite the uncertainties involved.

* From California Earthquake Prediction Evaluation Council, Earthquake prediction evaluation guidelines, *Calif. Geol.*, 30, 158, 1977.
[1] Panel on the Public Policy Implications of Earthquake Prediction, Earthquake Prediction and Public Policy, National Academy of Sciences, Washington, D.C., 1975.
[2] Panel on Earthquake Prediction, Predicting Earthquakes: A Scientific and Technical Evaluation with Implications for Society, National Academy of Sciences, Washington, D.C., 1976, 7.
[3] Haas, J. Eugene and Mileti, Dennis S., *Socioeconomic Impact of Earthquake Prediction on Government, Business, and Community*, Institute of Behavior Science, University of Colorado, Boulder, 1976, 5.

II. CALIFORNIA EARTHQUAKE PREDICTION EVALUATION COUNCIL

A. Objectives

1.   The California Earthquake Prediction Evaluation Council is appointed under existing administrative authority to advise the Director of the Office of Emergency Services (OES) on the validity of predictions of earthquakes capable of causing damage in California.
2.   The Council will evaluate and provide the Director with professional opinion as to the reliability of the data and the scientific validity of the technique used to arrive at a specific prediction.

B. Membership

1.   Council membership is composed of nine earth scientists as follows:
     a.   Eight scientists in the fields of geology, seismology and geophysics, appointed by the Director, OES, whose tenure is determined by the Director.
     b.   The State Geologist, whose tenure is concurrent with his term of office.
2.   The State Geologist serves as Chairman of the Council, ex officio.
3.   The Chairman, with the concurrence of the Director, OES, may appoint a Vice-Chairman from among the members.
4.   If the Chairman or Vice-Chairman is not available, an interim chairman may be designated by the Director, OES, from among the members.
5.   A meeting quorum consists of half or more of current Council members. Discussion may be held with less than a quorum present, but no official evaluation may be issued by the Council.
6.   Staff, appropriate to support Council activities, is designated by the Director from among OES employees.

C. Meetings

1.   Council meetings generally are called by the Chairman, but may be called at the request of any other member, or the Director, OES.
2.   Robert's *Rules of Order* will be in effect.
3.   Council meetings during which the results of the deliberations are announced will be open to the public.
4.   An official record will be kept of all Council meeting proceedings. Copies will be furnished to members and the Director, OES.
5.   Generally, discussion will be limited to Council members. Participation by others will be by invitation or permission of the Chairman.

D. Results

The principal product of Council meetings will be a statement, agreed to by a majority of the Council, evaluating a prediction, for the use of the Director, OES. Minority opinions will be included.

III. PROCEDURES

A. Screening of Predictions

1.   Predictions or similar information predicting specific earthquake events will be

screened as described in 2. and 3. below, prior to being accepted by the Council for official evaluation. (This applies to all prediction information, whether received directly by the Council or indirectly, e.g., by appearance in the public media.)

2. All prediction statements for Council consideration should be brought immediately to attention of the Council Chairman.
3. The Chairman, in consultation with such members of the Council as he deems appropriate, and with the Director, OES, will determine whether to present a prediction statement to the full Council for formal evaluation, or to declare it to be without sufficient merit to warrant Council deliberation.
4. The results of the screening process will be reported by the Council Chairman to the Director, OES, and Council Members.

B. Council Procedures

When a prediction statement has been accepted for Council evaluation, copies of all appropriate material will be provided by the Chairman to Council Members. Normally, a Council meeting will be scheduled by the Chairman as soon as feasible — ordinarily as soon as a quorum can be assembled, but after allowing sufficient time for members to make individual assessments of the materials furnished.

C. Evaluation

1. Approach
    a. The Council's test of scientific validity is primarily to evaluate the accuracy and completeness of the predictor's data, the logic and applicability of the scientific method used, and the predictor's accuracy in applying them in arriving at the announced results. In effect, it is an assessment based on the Council's evaluation, of the probability that the predicted event will indeed occur.
    b. A confidence level should be included in a prediction statement — an expressed level of probability or certainty that the event will occur as predicted.
    c. Any prediction statement may be considered, but only those with a scientific basis will be evaluated.
    d. The Council will evaluate only predictions for potentially damaging earthquakes; normally those of Richter magnitude 5.5 or greater.
    e. The methods of analysis and sources of data should be given and made available for Council use.
    f. The predictor should be willing to meet reasonably soon with the Council and fully disclose and discuss all aspects of the work (methods, data, etc.) leading to a prediction.

2. Criteria
    a. A prediction statement should specify the expected magnitude range, the geographical area, and the time interval with sufficient precision so that the accuracy of the prediction can readily be judged.[2]
    b. A confidence level should be included in a prediction statement — an expressed level of probability or certainty that the event will occur as predicted.
    c. Any prediction statement may be considered, but only those with a scientific basis will be evaluated.
    d. The Council will evaluate only predictions for potentially damaging earthquakes; normally those of Richter magnitude 5.5 or greater.

e. The methods of analysis and sources of data should be given and made available for Council use.

f. The predictor should be willing to meet reasonably soon with the Council and fully disclose and discuss all aspects of the work (methods, data, etc.) leading to a prediction.

3. Other Considerations

Public concern, or other circumstances outside the prediction statement itself, may make it advisable for the Council to consider a specific prediction, despite the statement's failure to meet established criteria.

EARTHQUAKE PREDICTIONS: INFORMATION FLOW

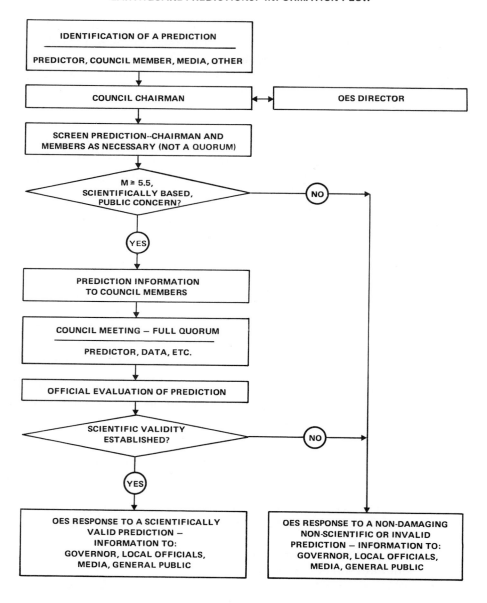

Appendix I

DISASTER RELIEF ACT OF 1974: FEDERAL DISASTER ASSISTANCE AND ECONOMIC RECOVERY PROGRAMS FOR DISASTER AREAS*

TITLE IV — FEDERAL DISASTER ASSISTANCE PROGRAMS

Federal Facilities

SEC. 401. (a) The President may authorize any Federal agency to repair, reconstruct, restore, or replace any facility owned by the United States and under the jurisdiction of such agency which is damaged or destroyed by any major disaster if he determines that such repair, reconstruction, restoration, or replacement is of such importance and urgency that it cannot reasonably be deferred pending specific authorizing legislation or the making of an appropriation for such purposes, or the obtaining of congressional committee approval.

(b) In order to carry out the provisions of this section, such repair, reconstruction, restoration, or replacement may be begun notwithstanding a lack or an insufficiency of funds appropriated for such purpose, where such lack or insufficiency can be remedied by the transfer, in accordance with law, of funds appropriated to that agency for another purpose.

(c) In implementing this section, Federal agencies shall evaluate the natural hazards to which these facilities are exposed and shall take appropriate action to mitigate such hazards, including safe land-use and construction practices, in accordance with standards prescribed by the President.

Repair and Restoration of Damaged Facilities

SEC. 402. (a) The President is authorized to make contributions to State or local governments to help repair, restore, reconstruct, or replace public facilities belonging to such State or local governments which were damaged or destroyed by a major disaster.

(b) The President is also authorized to make grants to help repair, restore, reconstruct, or replace private nonprofit educational, utility, emergency, medical, and custodial care facilities, including those for the aged or disabled, and facilities on Indian reservations as defined by the President, which were damaged or destroyed by a major disaster.

(c) For those facilities eligible under this section which were in the process of construction when damaged or destroyed by a major disaster, the grant shall be based on the net costs of restoring such facilities substantially to their predisaster condition.

(d) For the purposes of this section, "public facility" includes any publicly owned flood control, navigation, irrigation, reclamation, public power, sewage treatment and collection, water supply and distribution, watershed development, or airport facility, any non-Federal-aid street, road, or highway, any other public building, structure, or system including those used for educational or recreational purposes, and any park.

(e) The Federal contribution for grants made under this section shall not exceed 100 per centum of the net cost of repairing, restoring, reconstructing, or replacing any such facility on the basis of the design of such facility as it existed immediately prior to such disaster and in conformity with current applicable codes, specifications, and standards.

---

* From U.S. Senate and House of Representatives, Public Law 93-288, in United States Statutes at Large, Vol. 88 (Part 1), U.S. Government Printing Office, Washington, D.C., 1976, 153.

(f) In those cases where a State or local government determines that public welfare would not be best served by repairing, restoring, reconstructing, or replacing particular public facilities owned or controlled by that State or that local government which have been damaged or destroyed in a major disaster, it may elect to receive, in lieu of the contribution described in subsection (e) of this section, a contribution based on 90 per centum of the Federal estimate of the total cost of repairing, restoring, reconstructing, or replacing all damaged facilities owned by it within its jurisdiction. The cost of repairing, restoring, reconstructing, or replacing damaged or destroyed public facilities shall be estimated on the basis of the design of each such facility as it existed immediately prior to such disaster and in conformity with current applicable codes, specifications and standards. Funds contributed under this subsection may be expended either to repair or restore certain selected damaged public facilities or to construct new public facilities which the State or local government determines to be necessary to meet its needs for governmental services and functions in the disaster-affected area.

Debris Removal

SEC. 403. (a) The President, whenever he determines it to be in the public interest, is authorized —

(1)   through the use of Federal departments, agencies, and instrumentalities, to clear debris and wreckage resulting from a major disaster from publicly and privately owned lands and waters; and

(2)   to make grants to any State or local government for the purpose of removing debris or wreckage resulting from a major disaster from publicly or privately owned lands and waters.

(b) No authority under this section shall be exercised unless the affected State or local government shall first arrange an unconditional authorization for removal of such debris or wreckage from public and private property, and, in the case of removal of debris or wreckage from private property, shall first agree to indemnify the Federal Government against any claim arising from such removal.

Temporary Housing Assistance

SEC. 404. (a) The President is authorized to provide, either by purchase or lease, temporary housing, including, but not limited to, unoccupied habitable dwellings, suitable rental housing, mobile homes or other readily fabricated dwellings for those who, as a result of a major disaster, require temporary housing. During the first twelve months of occupancy no rentals shall be established for any such accommodations, and thereafter rentals shall be established, based upon fair market value of the accommodations being furnished, adjusted to take into consideration the financial ability of the occupant. Any mobile home or readily fabricated dwelling shall be placed on a site complete with utilities provided either by the State or local government, or by the owner or occupant of the site who was displaced by the major disaster, without charge to the United States. The President may authorize installation of essential utilities at Federal expense and he may elect to provide other more economical or accessible sites when he determines such action to be in the public interest.

(b) The President is authorized to provide assistance on a temporary basis in the form of mortage or rental payments to or on behalf of individuals and families who, as a result of financial hardship caused by a major disaster, have received written notice of dispossession or eviction from a residence by reason of foreclosure of any mortgage or lien, cancellation of any contract of sale or termination of any lease, entered into prior to such disaster. Such assistance shall be provided for a period of not to exceed one year or for the duration of the period of financial hardship, whichever is the lesser.

(c) In lieu of providing other types of temporary housing after a major disaster, the President is authorized to make expenditures for the purpose of repairing or restoring to a habitable condition owner-occupied private residential structures made uninhabitable by a major disaster which are capable of being restored quickly to a habitable condition with minimal repairs. No assistance provided under this section may be used for major reconstruction or rehabilitation of damaged property.

(d) (1) Notwithstanding any other provision of law, any temporary housing acquired by purchase may be sold directly to individuals and families who are occupants of temporary housing at prices that are fair and equitable, as determined by the President.

(2) The President may sell or otherwise make available temporary housing units directly to States, other governmental entities, and voluntary organizations. The President shall impose as a condition of transfer under this paragraph a covenant to comply with the provisions of section 311 of this Act requiring nondiscrimination in occupancy of such temporary housing units. Such disposition shall be limited to units purchased under the provisions of subsection (a) of this section and to the purposes of providing temporary housing for disaster victims in emergencies or in major disasters.

Protection of Environment

SEC. 405. No action taken or assistance provided pursuant to sections 305, 306, or 403 of this Act, or any assistance provided pursuant to section 402 or 419 of this Act that has the effect of restoring facilities substantially as they existed prior to the disaster, shall be deemed a major Federal action significantly affecting the quality of the human environment within the meaning of the National Environmental Policy Act of 1969 (83 Stat. 852). Nothing in this section shall alter or effect the applicability of the National Environmental Policy Act of 1969 (83 Stat. 852) to other Federal actions taken under this Act or under any other provision of law.

Minimum Standards for Public and Private Structures

SEC. 406. As a condition of any disaster loan or grant made under the provisions of this Act, the recipient shall agree that any repair or construction to be financed therewith shall be in accordance with applicable standards of safety, decency, and sanitation and in conformity with applicable codes, specifications, and standards, and shall furnish such evidence of compliance with this section as may be required by regulation. As a further condition of any loan or grant made under the provisions of this Act, the State or local government shall agree that the natural hazards in the areas in which the proceeds of the grants or loans are to be used shall be evaluated and appropriate action shall be taken to mitigate such hazards, including safe land-use and reconstruction practices, in accordance with standards prescribed or approved by the President after adequate consultation with the appropriate elected officials of general purpose local governments, and the State shall furnish such evidence of compliance with this section as may be required by regulation.

Unemployment Assistance

SEC. 407. (a) The President is authorized to provide to any individual unemployed as a result of a major disaster such benefit assistance as he deems appropriate while such individual is unemployed. Such assistance as the President shall provide shall be available to an individual as long as the individual's unemployment caused by the major disaster continues or until the individual is reemployed in a suitable position, but no longer than one year after the major disaster is declared. Such assistance for a week of unemployment shall not exceed the maximum weekly amount authorized under the unemployment compensation law of the State in which the disaster occurred, and the amount of assistance under this section to any such individual for a week of unemployment shall be reduced by any amount of unemployment compensation or of private income protection insurance compensation available to such individual for such week of unemployment.The President is directed to provide such assistance through agreements with States which, in his judgment, have an adequate system for administering such assistance through existing State agencies.

(b) The President is further authorized for the purposes of this Act to provide reemployment assistance services under other laws to individuals who are unemployed as a result of a major disaster.

Individual and Family Grant Programs

SEC. 408. (a) The President is authorized to make a grant to a State for the purpose of such State making grants to meet disaster-related necessary expenses or serious needs of individuals or families adversely affected by a major disaster in those cases where such individuals or families are unable to meet such expenses or needs through assistance under other provisions of this Act, or from other means. The Governor of a State shall administer the grant program authorized by this section.

(b) The Federal share of a grant to an individual or a family under this section shall be equal to 75 per centum of the actual cost of meeting such an expense or need and shall be made only on condition that the remaining 25 per centum of such cost is paid to such individual or family from funds made available by a State. Where a State is unable immediately to pay its share, the President is authorized to advance to such State such 25 per centum share, and any such advance is to be repaid to the United States when such State is able to do so. No individual and no family shall receive any grant or grants under this section aggregating more than $5,000 with respect to any one major disaster.

(c) The President shall promulgate regulations to carry out this section and such regulations shall include national criteria, standards, and procedures for the determination of eligibility for grants and the administration of grants made under this section.

(d) A State may expend not to exceed 3 per centum of any grant made by the President to it under subsection (a) of this section for expenses of administering grants to individuals and families under this section.

(e) This section shall take effect as of April 20, 1973.

Food Coupons and Distribution

SEC. 409. (a) Whenever the President determines that, as a result of a major disaster, low-income households are unable to purchase adequate amounts of nutritious food, he is authorized, under such terms and conditions as he may prescribe, to distribute through the Secretary of Agriculture or other appropriate agencies coupon allotments to such households pursuant to the provisions of the Food Stamp Act of 1964 (P.L. 91 —671; 84 Stat. 2048) and to make surplus commodities available pursuant to the provisions of this Act.

(b) The President, through the Secretary of Agriculture or other appropriate agencies, is authorized to continue to make such coupon allotments and surplus commodities available to such households for so long as he determines necessary, taking into consideration such factors as he deems appropriate, including the consequences of the major disaster on the earning power of the households, to which assistance is made available under this section.

(c) Nothing in this section shall be construed as amending or otherwise changing the provisions of the Food Stamp Act of 1964 except as they relate to the availability of food stamps in an area affected by a major disaster.

Food Commodities

SEC. 410. (a) The President is authorized and directed to assure that adequate stocks of food will be ready and conveniently available for emergency mass feeding or distribution in any area of the United States which suffers a major disaster or emergency.

(b) The Secretary of Agriculture shall utilize funds appropriated under section 32 of the Act of August 24, 1935 (7 U.S.C. 612c), to purchase food commodities necessary to provide adequate supplies for use in any area of the United States in the event of a major disaster or emergency in such area.

Relocation Assistance

SEC. 411. Notwithstanding any other provision of law, no person otherwise eligible for any kind of replacement housing payment under the Uniform Relocation Assistance and Real Property Acquisition Policies Act of 1970 (P.L. 91—646) shall be denied such eligibility as a result of his being unable, because of a major disaster as determined by the President, to meet the occupancy requirements set by such Act.

Legal Services

SEC. 412. Whenever the President determines that low-income individuals are unable to secure legal services adequate to meet their needs as a consequence of a major disaster, consistent with the goals of the programs authorized by this Act, the President shall assure that such programs are conducted with the advice and assistance of appropriate Federal agencies and State and local bar associations.

Crisis Counseling Assistance and Training

SEC. 413. The President is authorized (through the National Institute of Mental Health) to provide professional counseling services, including financial assistance to State or local agencies or private mental health organizations to provide such services or training of disaster workers, to victims of major disasters in order to relieve mental health problems caused or aggravated by such major disaster or its aftermath.

Community Disaster Loans

SEC. 414. (a) The President is authorized to make loans to any local government which may suffer a substantial loss of tax and other revenues as a result of a major disaster, and has demonstrated a need for financial assistance in order to perform its governmental functions. The amount of any such loan shall be based on need, and shall not exceed 25 per centum of the annual operating budget of that local government for the fiscal year in which the major disaster occurs. Repayment of all or any part of such loan to the extent that revenues of the local government during the three full fiscal year period following the major disaster are insufficient to meet the operating budget of the local government, including additional disaster-related expenses of a municipal operation character shall be cancelled.

(b) Any loans made under this section shall not reduce or otherwise affect any grants or other assistance under this Act.

(c) (1) Subtitle C of title I of the State and Local Fiscal Assistance Act of 1972 (P.L. 92-512; 86 Stat. 919) is amended by adding at the end thereof the following new section:

"SEC. 145. Entitlement Factors Affected by Major Disasters"

"In the administration of this title the Secretary shall disregard any change in data used in determining the entitlement of a State government or a unit of local government for a period of 60 months if that change —

"(1) results from a major disaster determined by the President under section 301 of the Disaster Relief Act of 1974, and

"(2) reduces the amount of the entitlement of that State government or unit of local government."

(2) The amendment made by this section takes effect on April 1, 1974.

Emergency Communications

SEC. 415. The President is authorized during, or in anticipation of, an emergency or major disaster to establish temporary communications systems and to make such communications available to State and local government officials and other persons as he deems appropriate.

Emergency Public Transportation

SEC. 416. The President is authorized to provide temporary public transportation service in an area affected by a major disaster to meet emergency needs and to provide transportation to governmental offices, supply centers, stores, post offices, schools, major employment centers, and such other places as may be necessary in order to enable the community to resume its normal pattern of life as soon as possible.

Fire Suppression Grants

SEC. 417. The President is authorized to provide assistance, including grants, equipment, supplies, and personnel, to any State for the suppression of any fire on publicly or privately owned forest or grassland which threatens such destruction as would constitute a major disaster.

Timber Sale Contracts

SEC. 418. (a) Where an existing timber sale contract between the Secretary of Agriculture or the Secretary of the Interior and a timber purchaser does not provide relief from major physical change not due to negligence of the purchaser prior to approval of construction of any section of specified road or of any other specified development facility and, as a result of a major disaster, a major physical change results in additional construction work in connection with such road or facility by such purchaser with an estimated cost, as determined by the appropriate Secretary, (1) of more than $1,000 for sales under one million board feet, (2) of more than $1 per thousand board feet for sales of one to three million board feet or (3) of more than $3,000 for sales over three million board feet, such increased construction cost shall be borne by the United States.

(b) If the appropriate Secretary determines that damages are so great that restoration, reconstruction, or construction is not practical under the cost-sharing arrangement authorized by subsection (a) of this section, he may allow cancellation of a contract entered into by his Department notwithstanding contrary provisions therein.

(c) The Secretary of Agriculture is authorized to reduce to seven days the minimum period of advance public notice required by the first section of the Act of June 4, 1897 (16 U.S.C. 476), in connection with the sale of timber from national forests, whenever the Secretary determines that (1) the sale of such timber will assist in the construction of any area of a State damaged by a major disaster, (2) the sale of such timber will assist in sustaining the economy of such area, or (3) the sale of such timber is necessary to salvage the value of timber damaged in such major disaster or to protect undamaged timber.

(d) The President, when he determines it to be in the public interest is authorized to make grants to any State or local government for the purpose of removing from privately owned lands timber damaged as a result of a major disaster, and such State or local government is authorized upon application, to make payments out of such grants to any person for reimbursement of expenses actually incurred by such person in the removal of damaged timber, not to exceed the amount that such expenses exceed the salvage value of such timber.

In-Lieu Contribution

SEC. 419. In any case in which the Federal estimate of the total cost of (1) repairing, restoring, reconstructing, or replacing, under section 402, all damaged or destroyed public facilities owned by a State or local government within its jurisdiction, and (2) emergency assistance under section 306 and debris removed under section 403, is less than $25,000, then on application of a State or local government, the President is authorized to make a contribution to such State or local government under the provisions of this section in lieu of any contribution to such State or local government under section 306, 402, or 403. Such contribution shall be based on 100 per centum of such total estimated cost, which may be expended either to repair, restore, reconstruct, or replace all such damaged or destroyed public facilities, to repair, restore, reconstruct, or replace certain selected damaged or destroyed public facilities, to construct new public facilities which the State or local government determines to be necessary to meet its needs for governmental services and functions in the disaster-affected area, or to undertake disaster work as authorized in section 306 or 403. The cost of repairing, restoring, reconstructing, or replacing damaged or destroyed public facilities shall be estimated on the basis of the design of each such facility as it existed immediately prior to such disaster and in conformity with current applicable codes, specifications and standards.

TITLE V — ECONOMIC RECOVERY FOR DISASTER AREAS

Amendment to Public Works and Economic Development Act of 1965

SEC. 501. The Public Works and Economic Development Act of 1965, as amended, is amended by adding at the end thereof the following new title:

"TITLE VIII — ECONOMIC RECOVERY FOR DISASTER AREAS

"Purpose of Title

"SEC. 801. (a) It is the purpose of this title to provide assistance for the economic recovery, after the period of emergency aid and replacement of essential facilities and services, of any major disaster area which has suffered a dislocation of its economy of sufficient severity to require (1) assistance in planning for development to replace that lost in the major disaster; (2) continued coordination of assistance available under Federal-aid programs; and (3) continued assistance toward the restoration of the employment base.

"(b) As used in this title, the term 'major disaster' means a major disaster declared by the President in accordance with the Disaster Relief Act of 1974.

"Disaster Recovery Planning

"SEC. 802. (a) (1) In the case of any area affected by a major disaster the Governor may request the President for assistance under this title. The Governor, within thirty days after authorization of such assistance by the President, shall designate a Recovery Planning Council for such area or for each part thereof.

"(2) Such Recovery Planning Council shall be composed of not less than five members, a majority of whom shall be local elected officials of political subdivisions within the affected areas, at least one representative of the State, and a representative of the Federal Government appointed by the President in accordance with paragraph (3) of this subsection. During the major disaster, the Federal coordinating officer shall also serve on the Recovery Planning Council.

"(3) The Federal representative on such Recovery Planning Council may be the Chairman of the Federal Regional Council for the affected area, or a member of the Federal Regional Council designated by the Chairman of such Regional Council. The Federal representative on such Recovery Planning Council may be the Federal Cochairman of the Regional Commission established pursuant to title V of this Act, or the Appalachian Regional Development Act of 1965, or his designee, where all of the area affected by a major disaster is within the boundaries of such Commission.

"(4) The Governor may designate an existing multijurisdictional organization as the Recovery Planning Council where such organization complies with paragraph (2) of this subsection with the addition of State and Federal representatives except that if all or part of an area affected by a major disaster is within the jurisdiction of an existing multijurisdictional organization established under title IV of this Act or title III of the Appalachian Regional Development Act of 1965, such organization, with the addition of State and Federal representatives in accordance with paragraph (2) of this subsection, shall be designated by the Governor as the Recovery Planning Council. In any case in which such title III or IV organization is designated as the Recovery Planning Council under this paragraph, some local elected officials of political subdivisions within the affected areas must be appointed to serve on such Recovery Planning Council. Where possible, the organization designated as the Recovery Planning Council shall be or shall be subsequently designated as the appropriate agency required by section 204 of the Demonstration Cities and Metropolitan Development Act of 1966 (42 U.S.C. 3334) and by the Intergovernmental Cooperation Act of 1968 (P.L. 90—577; 82 Stat. 1098).

"(5) The Recovery Planning Council shall include private citizens as members to the extent feasible, and shall provide for and encourage public participation in its deliberations and decisions.

"(b) The Recovery Planning Council (1) shall review existing plans for the affected area; and (2) may recommend to the Governor and responsible local governments such revisions as it determines necessary for the economic recovery of the area, including the development of new plans and the preparation of a recovery investment plan for the 5-year period following the declaration of the major disaster. The Recovery Planning Council shall accept as one element of the recovery investment plans determinations made under section 402 (f) of the Disaster Relief Act of 1974.

"(c) (1) A recovery investment plan prepared by a Recovery Planning Council may recommend the revision, deletion, reprograming, or additional approval of Federal-aid projects and programs within the area —

"(A) for which application has been made but approval not yet granted;
"(B) for which funds have been obligated or approval granted but construction not yet begun;
"(C) for which funds have been or are scheduled to be apportioned within the five years after the declaration of the disaster;
"(D) which may otherwise be available to the area under any State schedule or revised State schedule of priorities; or
"(E) which may reasonably be anticipated as becoming available under existing programs.

"(2) Upon the recommendation of the Recovery Planning Council and the request of the Governor, any funds for projects or programs identified pursuant to paragraph (1) of this subsection may, to any extent consistent with appropriation Acts, be placed in reserve by the responsible Federal agency for use in accordance with such recommendations. Upon the request of the Governor and with the concurrence of affected local governments, such funds may be transferred to the Recovery Planning Council to be expended in the implementation of the recovery investment plan, except that no such transfer may be made unless such expenditure is for a project or program for which such funds originally were made available by an appropriation Act.

"Public Works and Development Facilities Grants and Loans

"SEC. 803. (a) The President is authorized to provide funds to any Recovery Planning Council for the implementation of a recovery investment plan by public bodies. Such funds may be used —

"(1) to make loans for the acquisition or development of land and improvements for public works, public service, or development facility usage, including the acquisition or development of parks or open spaces, and the acquisition, construction, rehabilitation, alteration, expansion, or improvement of such facilities, including related machinery and equipment, and

"(2) to make supplementary grants to increase the Federal share for projects for which funds are reserved pursuant to subsection (c) (2) of section 802 of this Act, or other Federal-aid projects in the affected area.

"(b) Grants and loans under this section may be made to any State, local government, or private or public nonprofit organization representing any area or part thereof affected by a major disaster.

"(c) No supplementary grant shall increase the Federal share of the cost of any project to greater than 90 per centum, except in the case of a grant for the benefit of Indians or Alaska Natives, or in the case of any State or local government which the President determines has exhausted its effective taxing and borrowing capacity.

"(d) Loans under this section shall bear interest at a rate determined by the Secretary of the Treasury taking into consideration the current average market yield on outstanding marketable obligations of the United States with remaining periods to maturity comparable to the average maturities of such loans, adjusted to the nearest one-eighth of 1 per centum, less 1 per centum per annum.

"(e) Financial assistance under this title shall not be extended to assist establishments relocating from one area to another or to assist subcontractors whose purpose is to divest, or whose economic success is dependent upon divesting, other contractors or subcontractors of contracts therefore customarily performed by them. Such limitations shall not be construed to prohibit assistance for the expansion of an existing business entity through the establishment of a new branch, affiliate, or subsidiary of such entity if the Secretary of Commerce finds that the establishment of such branch, affiliate, or subsidiary will not result in an increase in unemployment of the area of original location or in any other area where such entity conducts business operations, unless the Secretary has reason to believe that such branch, affiliate, or subsidiary is being established with the intention of closing down the operations of the existing business entity in the area of its original location or in any other area where it conducts such operations.

"Loan Guarantees

"SEC. 804. The President is authorized to provide funds to Recovery Planning Councils to guarantee loans made to private borrowers by private lending institutions

(1) to aid in financing any project within an area affected by a major disaster for the purchase or development of land and facilities (including machinery and equipment) for industrial or commercial usage including the construction of new buildings, and rehabilitation of abandoned or unoccupied buildings, and the alteration, conversion, or enlargement of existing buildings; and (2) for working capital in connection with projects in areas assisted under paragraph (1), upon application of such institution and upon such terms and conditions as the President may prescribe. No such guarantee shall at any time exceed 90 per centum of the amount of the outstanding unpaid balance of such loan.

"Technical Assistance

"SEC. 805. (a) In carrying out the purposes of this title the President is authorized to provide technical assistance which would be useful in facilitating economic recovery in areas affected by major disasters. Such assistance shall include project planning and feasibility studies, management and operational assistance, and studies evaluating the needs of, and developing potentialities for economic recovery of such areas. Such assistance may be provided by the President directly, through the payment of funds authorized for this title to other departments or agencies of the Federal Government, through the employment of private individuals, partnerships, firms, corporations, or suitable institutions, under contracts entered into for such purposes, or through grants-in-aid to appropriate public or private nonprofit State, area, district, or local organizations.

"(b) The President is authorized to make grants to defray not to exceed 75 per centum of the administrative expenses of Recovery Planning Councils designated pursuant to section 802 of this Act. In determining the amount of the non-Federal share of such costs or expenses, the President shall give due consideration to all contributions both in cash and in kind, fairly evaluated, including but not limited to space, equipment, and services. Where practicable, grants-in-aid authorized under this subsection shall be used in conjunction with other available planning grants, to assure adequate and effective planning and economical use of funds.

" Authorization of Appropriations.

"SEC. 806. There is authorized to be appropriated not to exceed $250,000,000 to carry out this title."

Appendix J

PROPOSED PROCEDURES OF THE U.S. GEOLOGICAL SURVEY FOR REPORTING HAZARDOUS CONDITIONS*

3. Provisional Procedures to Report Hazardous Conditions

When and where information is obtained that suggests the development of a hazardous condition, the U.S. Geological Survey will attempt to authenticate it, and communicate such information to appropriate State, local and Federal authorities and to the public. The U.S. Geological Survey recognizes that providing earth-science information, in accordance with its expertise, is only the first of the inputs needed by State and local governments and the public in mitigating the effects of geologic hazards. The actual adoption of the most effective mitigation measures by local authorities will result from a cooperative effort by agencies at all governmental levels and by non-governmental organizations and the public. Decisions for adoption of such mitigation measures should be based upon a broad range of earth-science, engineering, and socio-economic information.

a. *Hazard Identification.* — Information acquired by Geological Survey personnel that indicates a region, area, or locality may be susceptible to geologic or hydrologic conditions or processes that could pose a significant potential hazard to life or property will be conveyed promptly to the Director of the Geological Survey with all supporting evidence and documentation.

b. *Hazard Evaluation.*

(1) The Director will submit information pertaining to potentially hazardous conditions or events to carefully selected scientific evaluation panels for review of the scientific basis for the hazard identification. Such panels may be established formally, such as the Survey's Earthquake Prediction Council, which relies on scientific expertise pertaining to a specific type of hazard; or informally, with scientist members changing according to their expertise with different types or areas of potential hazards. Upon review of the evidence, the evaluation panel(s) will transmit the findings and recommendations to the Director. The panel may find that:

(a) A hazard to life or property is unlikely or insufficiently defined to justify a Notice of Potential Hazard without additional information;

(b) A potential hazard to life and/or property exists;

(c) The potential hazard exists and that monitoring by the Geological Survey could lead to a better definition of location or magnitude, extent, or timing of the hazard; or

(d) The hazard conditions are sufficiently well defined as to location, magnitude, and time to warrant the issuance of a Hazard Watch or a Hazard Warning.

(2) Similarly, the Director will also undertake to have reviewed and evaluated identifications or predictions of potentially hazardous events made by scientists outside the Geological Survey, as deemed appropriate or upon the request of the head of an appropriate State of Federal agency. The requestor will be notified promptly of the findings of the evaluation panel and, if appropriate, a Notice of Potential Hazard, a Hazard Watch, or a Hazard Warning will be issued.

* From the U.S. Department of the Interior, U.S. Geological Survey, Warning and preparedness for geologic-related hazards, proposed procedures, *Fed. Reg.,* 42, 19292, 1977.

c.   *Notice of Potential Hazard.*

(1) Where the Director has authenticated identification of an area as susceptible to a potentially hazardous condition, but available evidence is insufficient to suggest that a hazardous event is imminent or evidence has not been developed to determine the time of occurrence, the information will be prepared for normal publication.

(2) The Director or his designee will transmit such information, as soon as possible, as a Notice of Potential Hazard to appropriate Federal, State, and local officials responsible for the public safety and welfare and to the public by a press release. The reports and maps cited earlier that show the distribution of earthquake, volcanic, landslide, and subsidence hazards are examples of identifications of potentially hazardous conditions that will form the basis for notices of potential hazards.

(3) Notices of Potential Hazard will be accompanied by a description of the geologic and hydrologic conditions that exist, the factors that suggest that such conditions constitute a potential hazard, and the location or area they may affect. In most instances, it will not be possible to estimate the severity of the hazard or the time it might occur. Information such as possible earthquake recurrence intervals will be given, however, if justified by available information.

(4) Where available evidence suggests that a hazardous event could occur and that precursory phenomena exists that will better define the time, location, and magnitude of the event, the geologic conditions or processes likely to trigger a hazardous event will be monitored by the U.S. Geological Survey within the limits of available funds and manpower.

d.   *Hazard Watch.*

(1) If existing or new information indicates that a region, area, locality, or geologic condition is undergoing change that may be interpreted as a precursor to a potentially hazardous event within an unspecified period of time (possibly months or years), such information will be evaluated and, if authenticated, the Director will assure that such information is transmitted promptly to civil authorities and the public as a Hazards Watch.

(2) Federal, State and local officials responsible for public safety will be notified in advance of the intent to issue a Hazard Watch to enable them to invoke emergency preparedness plans for an orderly public response.

(3) Hazard Watches will be accompanied, to the extent possible, by a definition of the parameters of the expected event, including, in addition to the place, magnitude, and general time, the possible geologic or hydrologic effects and the uncertainties associated with each.

e.   *Hazard Warning.*

(1) When developing information from precursory phenomena, which have been monitored through an experimental or operational hazard assessment program, appears to signal a potentially hazardous event within a specific period of time (possibly days or hours), the information will be conveyed promptly to the Director for evaluation and consideration as a hazard prediction.

(2) The Director or his designee will determine whether or not the prediction has a sound scientific basis and is authenticated by a comprehensive evaluation. If a prediction is issued as a result of this review and authentication process, any uncertainties that may exist will be evaluated and stated.

(3) The Director, upon authentication of a prediction of an event of possible catastrophic proportions, will assure that such information is promptly transmitted as a Hazard Warning, first to Federal, State, and local officials

responsible for public safety, to enable them to invoke emergency prepared-ness plans for an orderly public response, and then to the news media.

(4) Hazard Warnings will be accompanied, to the extent possible, by a definition of the parameters of the expected event including, in addition to the time, place, and magnitude, the possible geologic or hydrologic effects, and the uncertainties associated with each.

f. *Communication of Notices of Potential Hazard, Hazard Watches, and Hazard Warnings.*

(1) Information leading to a Notice of Potential Hazard or a Hazard Watch will generally be obtained well in advance of an event and can be transmitted directly to concerned officials by letters and to the public by press releases to the news media.

(2) Where potentially hazardous conditions are monitored, local, State, and Federal authorities will be informed periodically of the results of such inves-tigations and technical assistance, to the extent possible, will be extended as requested by these officials to assit in developing possible mitigation meas-ures.

(3) At the present time, a capability to predict a geologic event of possible cata-strophic proportions within days or hours does not exist except in rare cases. In such cases, where the information becomes available that suggests a po-tentially disastrous event may be imminent, public officials will be notified by telephone and such information will be transmitted directly to the public as a Hazard Warning. Public and existing Federal communication facilities, such as the Department of Commerce's Weather Radio System and the De-partment of Defense's National Warning System will be utilized whenever possible and appropriate.

(4) The Geological Survey will also communicate to responsible Federal Agen-cies and State and local governments, as soon as practicable, all available new knowledge as to geologic conditions or processes that may affect or alter public response to Notices of Potential Hazard, monitoring programs, Haz-ard Watches, or Hazard Warnings. This may result in the cancellation of the notice, watch, or warning, or a change in the hazard classification to better reflect an increased degree of uncertainty as to the time of occurrence of the event or a lessened sense of urgency.

(5) Notices of Potential Hazard, Hazard Watches, and Hazard Warnings to gov-ernmental agencies will also include:

(a) A statement of the authority of the U.S. Geological Survey for issuing the notice, watch, or warning;

(b) Copies of scientific papers or authentication reports that form the basis of the notice, watch, or warning;

(c) An offer to consult with any reviewers that the Governor or Governors of affected States may wish to appoint;

(d) An offer to provide appropriate technical assistance within areas of exper-tise in the Geological Survey in evaluating possible geologic hazards, as they may affect people and property;

(e) A statement of what additional steps, if any, the U.S. Geological Survey proposes to take to better define the degree or area of hazard; and

(f) A list of all parties to whom the notice, watch, or warning is being trans-mitted.

g. *Technical assistance.*

(1) As used in this statement, technical assistance pertains to:

    (a) Advice of available Geological Survey personnel on subjects within their area of expertise — geology, hydrology, chemistry, and, to a limited extent, soil engineering, structural engineering, and land use planning; and

    (b) Deployment of available instruments to better define hazardous conditions, processes, or events.

(2) Technical assistance should not be interpreted to refer to:

    (a) Funding for public works or hazard mitigation projects for which funds have not been allocated to the Geological Survey:

    (b) Assignment of personnel or equipment to assess hazardous conditions outside geographical or topical areas of on-going research or mapping programs except for unusual or compelling reasons.

Appendix K

THE EARTHQUAKE PREDICTION IMPACT STATEMENT (EPIS) PROCESS*

There are three basic objectives in the EPIS process:

- To develop and promulgate a formal set of earthquake-prediction cases that might emanate from an earthquake-prediction system in a given region.
- To analyze and evaluate the outcome and impacts of responding to this range of earthquake-prediction cases with appropriate tactics for reducing property damage and saving lives.
- To adopt rules for governmental response to the range of possible earthquake-prediction cases based on the analyses and evaluation.

Example of Earthquake Prediction Case

Lead time (months)	6
Time window (weeks)	+3
Epicenter or region of fault ruptures	San Juan Bautista to Los Gatos along the San Andreas Fault
Magnitude (Richter)	7.0 to 7.2
Confidence that event will occur (percent)	85
Contingent effects	Possible 8.3 Richter magnitude along entire "locked" San Francisco Bay section of the San Andreas Fault (no confidence judgment possible)

When the scientific program in a region generates a hypothesis concerning a future earthquake, the scientific data would first be validated and then translated into a set of prediction facts by an independent committee having technical expertise and public representation. These facts would then be identified with one of the preestablished earthquake-prediction cases. The governmental response would accordingly be governed by the rules that have been adopted for that specific case.

The discussion that follows describes how the process would be applied in a region. The developing technology of earthquake prediction is the background and basis of the entire process on which (1) through consideration of the geology and seismology of the region to be instrumented for earthquake prediction a set of possible prediction cases is developed and (2) a scientific program for developing earthquake-prediction capabilities in the region is initiated.

Under the authority of the USGS, an Earthquake Prediction Case Development and Evaluation Committee consisting of representatives of the scientific community and other public interests would develop and adopt a formal set of earthquake prediction cases for the region to be instrumented. This might be the Los Angeles — southern California region, the San Francisco — northern California area, Charleston, South Carolina, or any other area that would be instrumented and considered geologically and seismologically to be a "prediction unit." To enhance the participation and co-

* From Weisbecker, L. W., Stoneman, W. C., Ackerman, S. E., Arnold, R. K., Halton, P. M., Ivy, S. C., Kautz, W. H., Kroll, C. A., Levy, S., Mickley, R. B., Miller, P. D., Rainey, C. T., and van Zandt, J. E., *Earthquake Prediction, Uncertainty, and Policies for the Future,* Stanford Research Institute, Menlo Park, Calif., 1977, 28. With permission.

operation of the public in this process, the set of possible earthquake-prediction cases should be formally published — perhaps in the *Federal Register.*

All levels of government and private interests will have to participate in the process of evaluating alternative tactics for responding to the promulgated range of possible earthquake-prediction cases for the region. This will require detailed assessments of costs and benefits for reducing deaths, injuries, and property damage. An assessment will have to be made of the number of "false alarms" that can be tolerated for each risk. General criteria will have to be established for responding to residual risk which could change over time. Such studies will require the accumulation of regional data bases that are not presently available, and will in fact prove useful for many kinds of socioeconomic studies.

After the available damage-reduction tactics have been evaluated and ranked according to explicit, but multiple, and not necessarily compatible criteria (e.g., deaths and property damage), the public through its appropriate legislative bodies and through its political process will adopt those alternatives that are deemed most appropriate to the existing goals and objectives of society.

In concert with the process described above a predictive capability will have been established in the area. Accordingly, there is the possibility that earthquake premonitors may be detected. It would then seem logical that the same Earthquake Prediction Case Development and Evaluation Committee that developed the formal set of earthquake-prediction cases should also validate and translate the scientific data into a formal announcement that these scientifically identified precursors best fit one or the other earthquake-prediction cases. The decision maker is required only to respond with the appropriate, previously adopted alternative. Those alternatives may, of course, provide broad bands of discretion in certain areas, according to the will of the people in the legislature.

233

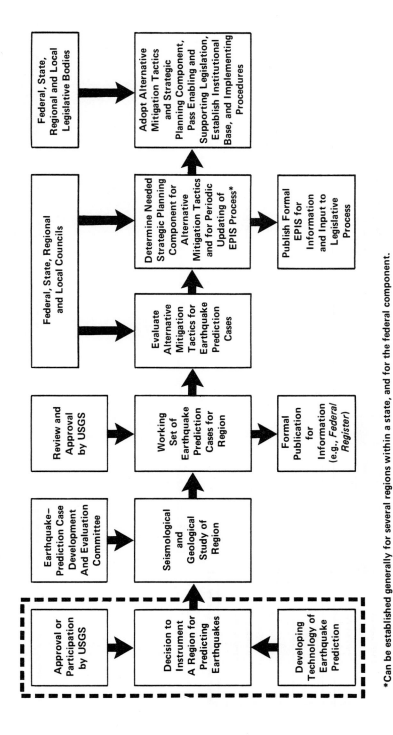

FIGURE 1. The Earthquake-Prediction Impact-Statement Process.

*Can be established generally for several regions within a state, and for the federal component.

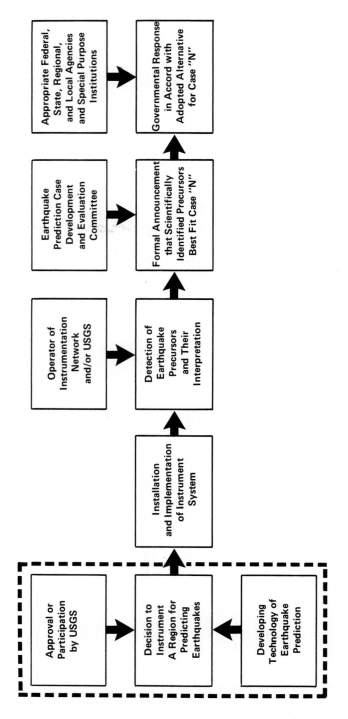

FIGURE 2.   Application of the Earthquake-Prediction Impact-Statement Process.

Appendix L

TAILORING MITIGATION MEASURES TO EARTHQUAKE WARNINGS*

A planning and operations guide could be developed to identify measures to be taken for various types of warning (short term versus long term) in places inside and outside the predicted damage area. The guide could be prepared and periodically updated as earthquake prediction is improved and as changes occur in enabling legislation and other factors that influence the preparedness program. If and when a damaging earthquake is predicted, appropriate guidance could be given to the concerned agencies as part of the warning process.

Case 1: Short-Term Warning

The first situation for which guidance could be prepared is that resulting from the prediction that a damaging earthquake will occur within a period of days. During such a period, it would be too late for preparedness measures that require a long lead time. The recommended actions that might be included in a warning to communities within the predicted damage area are the following:

Short-Term Warning: Damaging Earthquake Highly Probable (Risk Areas Specified, Time Insufficient for Extensive Preparedness Measures)
Broadcast public information and advice for this situation.

- Order evacuation of known hazardous structures and restrict access to known hazardous locations.
- Advise public and private organizations to tie down equipment for security against shock or displacement and protect shelf items from falling.
- Urge public through all mass media to make final preparations without delay (e.g., cleaning up trash or filling water containers); advise them to stay out of specified areas and specific types of structures.
- Disseminate through mass media information on fire prevention, self-help fire fighting, and medical self-help.
- Order shutdown of hazardous industrial operations.
- Direct operating departments to suspend all nonemergency functions, alert personnel, check equipment and supplies, and prepare for deployment of forces if ordered.
- Mobilize all available organized forces and deploy to preassigned emergency duty stations.
- Fully man all control centers and establish 24-hour operations.
- Establish and maintain communications with other jurisdictions and service facilities.
- Activate staging areas and make final preparations there.
- Take actions to ensure the safety of institutionalized persons.
- Discontinue all elective surgery, release all hospital patients except those who are critically ill, and take other actions to expand bed capacity and to protect remaining patients.

---

* From Weisbecker, L. W., Stoneman, W. C., Ackerman, S. E., Arnold, R. K., Halton, P. M., Ivy, S. C., Kautz, W. H., Kroll, C. A., Levy, S., Mickley, R. B., Miller, P. D., Rainey, C. T., and van Zandt, J. E., *Earthquake Prediction, Uncertainty, and Policies for the Future,* Stanford Research Institute, Menlo Park, Calif., 1977, 18. With permission.

- Deploy assigned personnel, equipment, and supplies to designated staging areas.
- Advise utilities and industry to shut down nonessential services throughout the emergency area.
- Deploy field units and maintain them on standby so that they can rapidly survey area for damaging and other earthquake-induced problems.
- Move fire-fighting and other emergency equipment and supplies outside the stations.
- Deploy engineering and other equipment.

Case 2: Long-Term Warning

The second situation for which guidance could be prepared is a longer prediction that provides sufficient time to implement measures to reduce seismic risk and substantially improve capability for disaster operations. The general character of the emergency measures that might be recommended in an initial warning to threatened communities is indicated below. The specific measures would depend on the nature of the prediction (weeks, months, years) and the characteristics of the threatened community.

Long-Term Warning: Damaging Earthquake Highly Likely (Risk Areas Specified, Time Sufficient for Preparedness Measures)

Establish public policy for long-term situation. Brief key government and nongovernment officials on situation and basic emergency plan and earthquake response plan. Review, update, or, if necessary, develop listed items:

- Legislation and local ordinances dealing with this type of situation.
- Organization and assignment of responsibility to emergency service units.
- Mutual aid agreements with other local jurisdictions and state agencies.
- Plans for informing the public during emergency.
- Preparedness plans for hospitals, other institutions, and organizations that operate essential utilities (power, water, natural gas, sanitation, communications, and transportation, including food and fuel distribution).
- Staffing and operation of emergency operating center and other headquarters; communications with emergency service units and with other localities.
- Maps indicating risk areas — fires, potential dam flood areas, landslides, structures that are susceptible to damage, etc.
- Procedures for determining (1) distribution of earthquake damage and ensuing hazards and (2) postearthquake capability of hospitals, water systems, and other vital facilities and services.

Conduct planning workshops for each service. Review checklist of postearthquake actions.

- Prepare instructions for service units and personnel, assign responsibility for specified actions, and indicate when, where, how, and with what resources the actions are to be accomplished, and by whom.
- Evaluate existing capability for performing the listed actions and, where appropriate, identify measures and resources that would improve capability.
- Identify measures that will reduce earthquake losses.
- Determine what normal activities and services could be deferred or curtailed to free funds for emergency preparations.
- Develop detailed plans for actions to be taken if a short-term warning is issued.
- Determine requirements and prepare standby procurement orders for needed equipment and supplies.

- Determine requirements and prepare standby procurement orders for needed equipment and supplies.

Identify and mark hazardous structures and locations in the risk area. Consider actions to reduce risk (e.g., removal, strengthening, prohibition of occupancy).

Expand fire prevention programs and abate fire hazards.

- Augment fire-fighting resources; prepare mobilization instructions.
- Survey community for current fire risk, modifying or confirming fire contingency plans as appropriate.

Begin actions to expand cadre and improve capability of emergency operations.

- Recruit, train, and assign personnel as needed to increase service capabilities for rescue, first aid, fire fighting, fire prevention, sanitation, etc.
- Prepare mobilization instruction.
- Bring emergency operating center and other headquarters to full readiness; provide for auxiliary power and augment communications.
- Arrange for use of facilities selected for staging areas, mass care, and other purposes, and prepare them for use.
- Procure previously identified and needed equipment and supplies.

Improve readiness in potential dam flood areas.

- Complete evacuation plans, warning system.
- Transfer key facilities.
- Develop engineering procedures to determine damage.
- Consider lowering water level.

Improve readiness and capability of lifeline organizations, resource agencies, essential industries.

- Identify measures to reduce earthquake losses and disruption of services.
- Activate standby agreements for transportation and other lifeline services.
- Activate standby agreements for utilization of commercial and educational facilities.
- Consider moving up resources from locations outside risk area.

Improve readiness and capability of hospitals, medical and allied professionals, and public health agencies.

- Prepare instructions for mobilizing personnel and resources.
- Expand stocks of drugs, medicines, and sanitation supplies.
- Check readiness of hospitals to discharge or move patients and expand bed capacity, consider deferring elective surgery.
- If appropriate, begin moving in resources from locations outside risk area.

Appendix M

RESEARCH ELEMENTS OF THE U.S. EARTHQUAKE HAZARDS REDUCTION PROGRAM*

FUNDAMENTAL EARTHQUAKE STUDIES

Objectives

- To obtain a comprehensive understanding of the natural phenomena involved in the earthquake process.
- Improve global networks of seismograph stations to provide a sound data base for studies in observational seismology and provide associated data services.

Activities

a.   The Earthquake Process — Develop a fundamental understanding of the earthquake process
   1) Develop theoretical models based on laboratory data and field observations. Study physical properties of rocks under conditions similar to those in the earth's crust and upper mantle. Determine seismic source parameters from field observations.

b.   The Implications of Plate Tectonics for Earthquake Hazards Reduction — Determine how stress is accumulated, distributed, and released along boundaries of moving plates and in plate interiors.
   1) Determine relative motions of plates. Refine definitions of plate boundaries. Determine deep-crustal and upper-mantle structure. Identify seismic gaps, measure stress and deformation of plate boundaries and in intraplate regions. Study the relationship of seismicity to geologic structures in intraplate regions.

c.   Global Seismology — Collect and disseminate seismological data from around the world.
   1) Operate the Worldwide Standarized Seismograph Network (WWSSN) and reestablish a maintenance program for the stations that lapsed several years ago.
   2) Operate the data acquisition and processing capability of the National Earthquake Information Service, including use of satellite telecommunications, issuance of new seismicity maps, and routine computation of the parameters of the earthquake mechanism.
   3) Upgrade about half of the WWSSN and establish the capability to produce integrated tapes of digital seismic data.
   4) Acquire and operate a ten-station array of transportable broad-band seismographs for global seismic studies.
   5) Operate an integrated digital network consisting of High-Gain Long-Period stations, Seismological Research Observatories, and the upgraded WWSSN stations called for in activity 3), and produce integrated tapes of digital seismic data.
   6) Acquire, install, and operate 10 ocean-bottom seismographs.

* From National Science Foundation and U.S. Geological Survey, Earthquake Prediction and Hazard Mitigation — Options for USGS and NSF Programs, U.S. Government Printing Office, Washington, D.C., 1976, 25.

These activities are related to programs of the National Aeronautics and Space Administration, the National Oceanic and Atmospheric Administration, and the Defense Advanced Research Projects Agency.

PREDICTION

Objectives

- Predict moderate to large earthquakes in time and space and estimate limits of probabilities of their occurrence on the basis of detailed observation.
- Issue earthquake predictions in a timely manner through development of automated data analysis systems.
- Reduce unreliable predictions of earthquakes through development of a sound understanding of the physical basis for the precursory phenomena observed.

Activities

a. Deformation Monitoring Instrumentation (Purchase and Installation) — Measure ground deformation in active seismic regions to monitor the long-term accumulation of strain, determine the physics of the seismic source, and observe precursors.
   1) Deploy continuously recording strain meters, tiltmeters, tide gauges, gravimeters, water-well level monitors, etc., in selected areas of high or unique seismicity.
b. Seismic-Monitoring Instrumentation (Purchase and Installation) — Determine the patterns of seismicity in time and space, the physics of the seismic source, and the variation over time of seismic source and seismic wave parameters.
   1) Deploy narrow and broad-band seismic instruments in selected areas of high or unique seismicity.
c. Geochemical, Magnetic, Electrical and Other Instrumentation (Purchase and Installation) — Study other types of phenomena that have been reported as earthquake precursors.
   1) Deploy geochemical sensors, magnetometers, resistivity arrays, telluric sensors, self-potential sensors, etc., and carry out studies with animals in selected areas of high or unique seismicity.
d. Operations — Operate installed networks of instruments and provide bulletins and computer files of uniformly processed data to provide bases for development of a theoretical and empirical framework for earthquake prediction.
   1) Operate networks of instruments including maintenance and routine data processing in selected areas of high or unique seismicity.
e. On-Line Computer Processing Capability — Develop and implement automated analysis techniques of all network data to allow rapid identification of earthquake precursors for timely formulation of predictions.
   1) Develop software and hardware for monitoring all instrument networks online. Deploy these computers in central and southern California first and then eventually at the hub of any major network.
   2) Develop computer models incorporating known and hypothesized physical behavior of fault zones to allow predictions of future behavior from rapid inclusion of newly gathered data.
f. Land Deformation Surveys — Determine the large-scale distribution of vertical and horizontal land deformation in seismically active regions to evaluate the stress and strain regime, and to observe precursors.

1) Repeat trilateration and leveling surveys.
2) Repeat gravimetric, magnetic, and other surveys.

g. Intensive Field Studies in Strategic Areas — Conduct field experiments of a topical nature to collect observations and test hypotheses regarding the physics of the failure process and the origin and nature of precursors.
   1) Conduct a wide variety of topical field studies such as after-shock studies, refraction or reflection surveys, seismic velocity monitoring, detailed surface deformation studies, and short-term studies using a portable network of any one or a combination of sensors used to monitor precursors.

h. Detailed Analyses of Data and Theoretical Modeling — Analyze field and laboratory data, and develop and test hypotheses concerning the physics of the failure process and precursory phenomena.
   1) Detailed analysis, theoretical modeling and synthesis of results from:
      • Strain Data
      • Seismic Data
      • Other Data and Syntheses

i. Instrument Development — Develop and improve instruments for field use that show significant promise of detecting earthquake precursors.
   1) Develop or complete development of instruments, such as a portable multi-wavelength laser-ranging devices, portable long-base tiltmeters, broad-band seismometers, data telemetry systems, absolute gravemeters, stress detectors, etc., and improve the reliability and sensitivity of instruments already utilized.

j. Laboratory Studies — Determine the physical behavior of rocks near rupture and model earthquake processes in the laboratory with specific application to earthquake prediction.
   1) Examine the properties of rocks, the physics of fracture, and the occurrence of precursors prior to fracture on rock samples in the laboratory. Model earthquake processes in the laboratory.
   2) Conduct laboratory experiments using large-size samples and small-scale field experiments.

k. Controlled Earthquake Experiments — Examine the feasibility of stimulating small earthquakes and creep along natural fault zones by artificial means to study premonitory effects.
   1) Begin search for site with appropriate geologic, tectonic, and environmental conditions to carry out experiments.
   2) Conduct supporting theoretical and laboratory studies on fluid flow in natural fault zones and its influence on earthquakes.

These activities are related to programs of the National Aeronautics and Space Administration, National Oceanic and Atmospheric Administration, and Defense Advanced Research Projects Agency.

INDUCED SEISMICITY

Objectives

• Develop techniques for diagnosing in advance whether reservoir impoundment or fluid injection in wells at a particular site holds the potential for triggered earthquakes.
• Develop techniques to permit development and operation of large reservoirs and deep injection wells including a basis for remedial action should earthquakes be triggered.

- Determine to what extent theories and models are successful in explaining how induced or triggered earthquakes apply to natural earthquake processes.
- Determine the feasibility of artificially modifying natural seismicity by measuring the physical properties of fault zone materials in drill holes. Select a site for a field experiment in an uninhabited area.

Activities

a. Reservoir Induced Earthquakes — Determine the effects of reservoirs on seismic activity and the reasons for these effects.
  1) Monitor in detail the seismicity, the strain field, and the fluid pressure at depth around major reservoirs before, during, and after they are constructed.
b. Drilling into Fault Zones — Determine the physical properties of fault zones and fault gouge.
  1) Drill several holes into major faults and measure such properties as permeability, porosity, elastic parameters, temperature, fluid pressure and stress.

These activities are focused on the process of reservoir-induced earthquakes and are considered to be supplementary to monitoring and baseline studies by agencies responsible for reservoirs, e.g., the Bureau of Reclamation and the Corps of Engineers.

HAZARDS ASSESSMENT

Objectives

- Determine the expected location, size, frequency, and characteristics of earthquakes and of associated surface faulting for various regions of the United States.
- Develop a physical basis for predicting the character of damaging ground motion as a function of distance from a postulated earthquake and varying geologic site conditions.
- Develop a physical basis for predicting the incidence, nature and extent of earthquake-induced ground failure and flooding.
- Delineate geographical variations in the nature and likelihood of occurrence of earthquake hazards.
- Evaluate earthquake risk (e.g., hazard assessment and vulnerability analysis) on a nationwide and regional basis.

Activities

a. Earthquake Potential — Determine the expected location, size, frequency, and characteristics of surface faulting and associated earthquakes for various regions of the United States.
  1) Improve the location, accuracy and completeness of historic earthquake data, including a remapping of poorly located historic (both instrumental and felt) events.
  2) Delineate seismically active faults by monitoring regional earthquake activity in selected zones of the U.S. by determining accurate epicenters, focal depths, and focal mechanisms.
  3) Investigate earthquake recurrence from analysis of Quaternary history of individual faults.

4) Delineate seismic source areas on the basis of seismic, geologic, and geophysical characteristics. Estimate rates of activity and evaluate upper bound earthquakes.

    5) Delineate active faults and seismic source areas and monitor earthquake activity in selected areas of the outer continental shelf.

b. Geologic Factors Influencing Ground Motion — Develop a physical basis for predicting the character of damaging ground motion as a function of distance from a postulated earthquake and varying geologic site conditions.

c. Geologic and Hydrologic Effects — Develop a physical basis for predicting the incidence, nature and extent of earthquake-induced ground failure and flooding.

    1) Investigate mechanisms of earthquake-induced liquefaction and mass movements; refine criteria for predicting the occurrence of ground failure.

    2) Develop improved methods to predict inundation and the consequences of flooding caused by subsidence, tectonic downwarping, massive landslides into water, tsunamis, and other secondary earthquake effects.

    3) Conduct geologic hazards evaluations of the effects of damaging earthquakes.

d. Earthquake Hazards — Delineate geographical variations in the nature and likelihood of occurrence of earthquake hazards.

    1) Prepare refined probabilistic ground motion maps for entire United States.

    2) Expand state-of-the-art evaluation and mapping of earthquake hazards (ground shaking and failure, surface faulting, elevation changes, and inundation) in areas of high seismic risk; develop new methods for probabilistic hazard evaluation, including faulting, ground failure, and tsunami effects.

e. Earthquake Risk — Evaluate earthquake risk on a nationwide and regional basis.

    1) Develop and improve methods for estimating damage and loss based on probabilistic maps of earthquake hazards. Apply methods to estimate risk on a regional and nationwide basis.

These activities are related to programs of the Department of Housing and Urban Development, the Nuclear Regulatory Commission, the Department of Transportation, and the Veterans Administration.

ENGINEERING

Objectives and Activities

This section outlines the objectives of earthquake engineering research and the activities required to accomplish those objectives. An activity common to all subelements of this research is the dissemination of the results to practicing engineers and architects, and to code and regulatory organizations.

Subelement a:
Characterization of ground motion for structural analysis and design

Objective:
Develop methods to characterize the nature of the input motions and corresponding response of simple systems for use in engineering analysis, planning, and design.

Activities

1. Develop analytic models to estimate the special characteristics of ground motion and the acceleration, velocity, and displacement time-histories of this motion for use as input motion in structural analysis and design. Such models will include

the effects of the earthquake source, the transmission path, the amplification caused by local site conditions, and the influence of the presence of a structure on this motion (soil-structure interaction).

2. Develop techniques for measuring the severity of earthquake effects based on parameters significant in engineering analysis and design. Apply these techniques in post-earthquake investigations and pilot studies of zoning.

Subelement b:
Acquisition of Strong-Motion Data

Objective:
Obtain comprehensive data on the nature of the earthquake motions at typical sites and in representative structures.

Activities:
1. Improve the national strong-motion instrumentation network by:

    (a) Replacing obsolete instruments,
    (b) Installing adequate instrumentation arrays in all seismic regions,
    (c) Developing arrays to measure the two and three dimensional distributions of ground motion.
    (d) Instrumenting representative types of structures, particularly in the more active parts of the country. A high priority must be placed on obtaining an adequate number of records from the next major earthquake. A predicted earthquake provides an opportunity to gather valuable, comprehensive ground motion and structural response data.

2. Develop new instruments with a view toward minimizing total costs of instrumentation, data acquisition, and data processing.
3. Expand data gathering activities including studies in those active areas in other parts of the world where data of importance to the U.S. can be obtained, possibly at a greater frequency.

Subelement c:
Investigation of Dynamic Soil Properties and Failures

Objective:
Develop in-situ and laboratory methods to determine the dynamic properties of soils and analytical procedures including the potential for failure of slopes, embankments and foundations.

Activities

1. Develop techniques for determining the dynamic properties of soils, both in-situ and in laboratories.
2. Develop analytical methods to evaluate soil failures (liquefaction and landslides) and to assess the possibility of controlling such failures.
3. Investigate the dynamics and design of various types of foundations (include spread footings, mats or rafts, piles or caissons, and deep foundations), and develop criteria for selecting a type of foundation appropriate to various settings and soil conditions.

Subelement d:
Investigations of Structural Response

Objective:
Develop analytical procedures for characterizing the earthquake response of structures and structural elements based on both analytical and experimental studies.

Activities:

1.   Investigate the dynamic behavior of structures and components experimentally to determine performance characteristics up to ultimate capacity and to provide a basis for the formulation, development, and validation of analytical methods of analysis and design. This may require the development of substantial new experimental facilities.
2.   Develop analytical methods to characterize the earthquake response of structures and structural components, with an emphasis on three-dimensional, nonlinear, and inelastic behavior to ultimate capacity. Simplify these analytical procedures for computer-aided structural design.
3.   Develop methods to assess the hazard vulnerability of existing structures and to upgrade their performance when subjected to earthquake motions.

Subelement e:
Studies of Special Structures and Systems

Objectives:
Develop analytical methods to evaluate the earthquake response of special types of structures (dams, critical facilities, bridges and other extended structures) and of interconnected structures and systems (pipelines, transmission lines, etc.).

Activities:

1.   Develop design, construction and analysis methods to minimize the impact of earthquakes on critical facilities such as hospitals, power plants, other emergency facilities, and structures housing or storing hazardous materials.
2.   Improve the procedures for analyzing the effects of contained liquids on dams, reservoirs, tanks, etc. This should include the effects of waves generated by seiches or landslides and the design of systems to perform adequately during tsunamis.
3.   Develop methods of planning and construction of utility and public service structures, including facilities for storing emergency supplies to minimize disruptions caused by earthquakes. Develop criteria for systems design that are compatible with other comprehensive urban and regional planning considerations.

Subelement f:
Post-earthquake Investigations

Objective:
Obtain information for engineering analysis and design from observations of damage (or lack of damage) following earthquakes that support the development of improved U.S. engineering practices and construction techniques.

Activities

1. Investigate all potentially damaging earthquakes in the U.S. with an emphasis on correlation of damage or lack of damage with design and construction practices.
2. Investigate earthquakes in foreign countries that are likely to provide information that can improve engineering design and construction practices in the U.S.

These activities are related to programs of the National Bureau of Standards, the Department of Transportation, the Department of Housing and Urban Development and other agencies concerned with earthquake-resistant design of structures.

RESEARCH FOR UTILIZATION

Objectives

*   Define options for the mix of measures to mitigate earthquake hazards by considering research, social, economic, legal, and political barriers and incentives to policy implementation.
*   Assess public and private regulation impacts and develop alternatives where necessary.
*   Facilitate the beneficial utilization of earthquake hazard mitigation measures by developing effective techniques for communicating information to the public and decision makers.
*   Increase the capability of public officials to implement earthquake hazard mitigation measures through land-use planning, preparedness planning, building regulation, and disaster response.
*   Define alternatives the private sector could adopt for mitigating earthquake hazards.

Activities

a. Allocation of Earthquake Mitigation Resources: Develop comprehensive cost-benefit methods of analysis to provide a basis for choosing among possible earthquake mitigation actions.
    1) Evaluate how people and organizations establish acceptable levels of risk for low-probability and potentially catastrophic events.
    2) Develop a prototypical economic model for earthquake-prone regions for estimating the interactions among the public and private sectors for various earthquake mitigation measures (e.g. financial sector, building codes, land use, prediction).
    3) Study the economic incentives and disincentives to correct or eliminate existing hazardous conditions, including buildings. This includes the availability of public and private financing.
    4) Examine comprehensively the national implications of regional and local earthquake mitigation practices, and the local implication of regional and national practices and policies.
    5) Develop cost-benefit analyses applicable to decisions at the individual, group, and community levels through case studies.
    6) Study alternate strategies of mitigation based on comprehensive planning, statutory regulation, etc.
b. Preparedness: Develop a basis for preparedness planning in anticipation of an impending earthquake.

1) Investigate the division of functions and responsibilities between public and private sectors and develop plans for the coordination of preparedness and response activities.

2) Establish socioeconomic monitoring to develop baseline data to evaluate the impact of earthquake predictions and other mitigation procedures.

3) Examine the social, economic, legal, and political aspects of earthquake predictions and develop recommendations for maximizing the benefits of prediction.

4) Initiate comprehensive investigations of the legal issues likely to be encountered in the application of earthquake mitigation procedures.

5) Investigate the likely political consequences of alternative preparedness programs through case studies.

6) Develop model policies for implementing preparedness activities on local, state, and regional bases.

c. Relief and Rehabilitation: Assess and develop means to provide for the relief and rehabilitation of disaster-struck communities.

1) Develop and implement a comprehensive program to evaluate relief and rehabilitation programs; develop program guidelines to hasten community recovery and decrease future vulnerability to earthquake and other hazard agents.

2) Examine the trade-offs between the provision of post-disaster relief and rehabilitation and financial assistance for predisaster hazard reduction.

3) Conduct long-term, longitudinal studies on the return of the disaster-struck community, family, public agencies and utilities to normalcy. Such studies should include all aspects of the pre- and post-disaster periods as well as very long response.

4) Systematically conduct post-audits to collect information on the consequences of major disasters (including non-earthquake occurences).

5) Prepare model legislation to implement relief and rehabilitation.

d. Information Flow: Develop effective methods for communicating earthquake hazard mitigation information to decision makers and the public.

1) Investigate the flow of information within institutions and develop alternative ways to facilitate this flow.

2) Conduct training programs (e.g., seminars, continuing education to institutionalize earthquake mitigation measures in State and local government).

3) Establish workshops with representatives of the private sector (e.g., engineers, architects, bankers, model code agencies) on methods of reducing earthquake losses.

4) Initiate an information program to acquaint the public with earthquake hazard mitigation measures.

5) Examine alternative information strategies for informing the public of services and facilities available to reduce the disaster's impact.

e. Regulation and Assessment: Assess public and private regulation impacts on the achievement of disaster mitigation.

1) Assess the impact of earthquake mitigation measures on public and private attitudes and practices.

2) Evaluate the effectiveness of physical regulation (e.g. building code, land-use controls, occupancies) to achieve given levels of earthquake protection.

3) Evaluate the effectiveness of financial regulations and practices (e.g. insurance, mortgage and financial regulations, taxation policies) to achieve given levels of earthquake protection.

4) Evaluate the effectiveness of regulatory, operation and investment policies

of public utilities (e.g., water, communications, transportation) in hazard prone areas to provide short and long term essential public services.

5) Prepare model legislation for different mitigation strategies based on matrix of seismic hazard and mitigation/benefit.

6) Analyze the feasibility and impact of extensive local microzonation.

7) Evaluate regulation and zoning changes to modify hazard of existing buildings.

These activities are related to programs of all governmental agencies concerned with reducing earthquake losses.

Appendix N

EARTHQUAKE HAZARDS REDUCTION ACT OF 1977*

PUBLIC LAW 95-124 — OCT. 7, 1977

91 STAT. 1098

Public Law 95-124
95th Congress

An Act

Oct. 7, 1977
[S. 126]

To reduce the hazards of earthquakes, and for other purposes.

Be it enacted by the Senate and House of Representatives of the United States of America in Congress assembled,

Earthquake
Hazards
Reduction Act of
1977.
42 USC 7701
note.
42 USC 7701.

SECTION 1. SHORT TITLE.

That this Act may be cited as the "Earthquake Hazards Reduction Act of 1977".

SEC. 2. FINDINGS.

The Congress finds and declares the following:

(1) All 50 States are vulnerable to the hazards of earthquakes, and at least 39 of them are subject to major or moderate seismic risk, including Alaska, California, Hawaii, Illinois, Massachusetts, Missouri, Montana, Nevada, New Jersey, New York, South Carolina, Utah, and Washington. A large portion of the population of the United States lives in areas vulnerable to earthquake hazards.

(2) Earthquakes have caused, and can cause in the future, enormous loss of life, injury, destruction of property, and economic and social disruption. With respect to future earthquakes, such loss, destruction, and disruption can be substantially reduced through the development and implementation of earthquake hazards reduction measures, including (A) improved design and construction methods and practices, (B) land-use controls and redevelopment, (C) prediction techniques and early-warning systems, (D) coordinated emergency preparedness plans, and (E) public education and involvement programs.

(3) An expertly staffed and adequately financed earthquake hazards reduction program, based on Federal, State, local, and private research, planning, decisionmaking, and contributions would reduce the risk of such loss, destruction, and

---

* From U.S. 95th Congress, Earthquake Hazards Reduction Act of 1977, in Earthquake Hazards Reduction Program — Fiscal Year 1978 Studies Supported by the U.S. Geological Survey, U.S. Geological Survey Circular 780, 1978, 31.

disruption in seismic areas by an amount far greater than the cost of such program.

(4) A well-funded seismological research program in earthquake prediction could provide data adequate for the design, of an operational system that could predict accurately the time, place, magnitude, and physical effects of earthquakes in selected areas of the United States.

(5) An operational earthquake prediction system can produce significant social, economic, legal, and political consequences.

(6) There is a scientific basis for hypothesizing that major earthquakes may be moderated, in at least some seismic areas, by application of the findings of earthquake control and seismological research.

(7) The implementation of earthquake hazards reduction measures would, as an added benefit, also reduce the risk of loss, destruction, and disruption from other natural hazards and manmade hazards, including hurricanes, tornadoes, accidents, explosions, landslides, building and structural cave-ins, and fires.

(8) Reduction of loss, destruction, and disruption from earthquakes will depend on the actions of individuals, and organizations in the private sector and governmental units at Federal, State, and local levels. The current capability to transfer knowledge and information to these sectors is insufficient. Improved mechanisms are needed to translate existing information and research findings into reasonable and usable specifications, criteria, and practices, so that individuals, organizations, and governmental units may make informed decisions and take appropriate actions.

(9) Severe earthquakes are a worldwide problem. Since damaging earthquakes occur infrequently in any one nation, international cooperation is desirable for mutual learning from limited experiences.

(10) An effective Federal program in earthquake hazards reduction will require input from and review by persons outside the Federal Government expert in the sciences of earthquake hazards reduction and in the practical application of earthquake hazards reduction measures.

42 USC 7702.

SEC. 3. PURPOSE.

It is the purpose of the Congress in this Act to reduce the risks of life and property from future earthquakes in the United States through the establishment and maintenance of an effective earthquake hazards reduction program.

42 USC 7703.

SEC. 4. DEFINITIONS.

As used in this Act, unless the context otherwise requires:

(1) The term "includes" and variants thereof should be read as if the phrase "but is not limited to" were also set forth.

(2) The term "program" means the earthquake hazards reduction program established under section 5.

(3) The term "seismic" and variants thereof mean having to do with, or caused by earthquakes.

(4) The term "State" means each of the States of the United States, the District of Columbia, the Commonwealth of Puerto Rico, the Virgin Islands, Guam, American Samoa, the Commonwealth of the Mariana Islands, and any other territory or possession of the United States.

(5) The term "United States" means, when used in a geographical sense, all of the States as defined in section 4(4).

42 USC 7704.

SEC. 5. NATIONAL EARTHQUAKE HAZARDS REDUCTION PROGRAM.

(a) ESTABLISHMENT. — The President shall establish and maintain, in accordance with the provisions and policy of this Act, a coordinated earthquake hazards reduction program, which shall —

(1) be designed and administered to achieve the objectives set forth in subsection (c);

(2) involve, where appropriate, each of the agencies listed in subsection (d); and

(3) include each of the elements described in subsection (e), the implementation plan described in subsection (f), and the assistance to the States specified in subsection (g).

(b) DUTIES. — The President shall —

(1) within 30 days after the date of enactment of this Act, designate the Federal department, agency, or entity responsible for the development of the implementation plan described in subsection (f);

Plan, submittal to congressional committees.

(2) within 210 days after such date of enactment, submit to the appropriate authorizing committees of the Congress the implementation plan described in subsection (f); and

(3) by rule, within 300 days after such date of enactment—

(A) designate the Federal department, agency, or interagency group which shall have primary responsibility for the development and implementation of the earthquake hazards reduction program;

(B) assign and specify the role and responsibility of each appropriate Federal department, agency, and entity with respect to each object and element of the program;

(C) establish goals, priorities, and target dates for implementation of the program;

(D) provide a method for cooperation and coordination with, and assistance (to the extent of available resources) to, interested governmental entities in all States, particularly those containing areas of high or moderate seismic risk; and

(E) provide for qualified staffing for the program and its components.

(c) OBJECTIVES. — The objectives of the earthquake hazards reduction program shall include —

Earthquake resistant construction.

(1) the development of technologically and economically feasible design and construction methods and procedures to make new and existing structures, in areas of seismic risk, earthquake resistant, giving priority to the

development of such methods and procedures for nuclear power generating plants, dams, hospitals, schools, public utilities, public safety structures, high occupancy buildings, and other structures which are especially needed in time of disaster;

Earthquake prediction.

(2) the implementation in all areas of high or moderate seismic risk, of a system (including personnel, technology, and procedures) for predicting damaging earthquakes and for identifying, evaluating, and accurately characterizing seismic hazards;

Model codes.

(3) the development, publication, and promotion, in conjunction with State and local officials and professional organizations, of model codes and other means to coordinate information about seismic risk with land-use policy decisions and building activity;

Earthquake-related issues, understanding.

(4) the development, in areas of seismic risk, of improved understanding of, and capability with respect to, earthquake-related issues, including methods of controlling the risks from earthquakes, planning to prevent such risks, disseminating warnings of earthquakes, organizing emergency services, and planning for reconstruction and redevelopment after an earthquake;

(5) the education of the public, including State and local officials, as to earthquake phenomena, the identification of locations and structures which are especially susceptible to earthquake damage, ways to reduce the adverse consequences of an earthquake, and related matters;

Research.

(6) the development of research on —

(A) ways to increase the use of existing scientific and engineering knowledge to mitigate earthquake hazards;

(B) the social, economic, legal, and political consequences of earthquake prediction; and

(C) ways to assure the availability of earthquake insurance or some functional substitute; and

(7) the development of basic and applied research leading to a better understanding of the control or alteration of seismic phenomena.

(d) PARTICIPATION. — In assigning the role and responsibility of Federal departments, agencies, and entities under subsection (b) (3) (B), the President shall, where appropriate, include the United States Geological Survey, the National Science Foundation, the Department of Defense, the Department of Housing and Urban Development, the National Aeronautics and Space Administration, the National Oceanic and Atmospheric Administration, the National Bureau of Standards, the Energy Research and Development Administration, the Nuclear Regulatory Commission, and the National Fire Prevention and Control Administration.

(e) RESEARCH ELEMENTS. — The research elements of the program shall include —

(1) research into the basic causes and mechanisms of earthquakes;

(2) development of methods to predict the time, place, and magnitude of future earthquakes;

(3) development of an understanding of the circumstances in which earthquakes might be artificially induced by the injection of fluids in deep wells, by the impoundment of reservoirs, or by other means;

(4) evaluation of methods that may lead to the development of a capability to modify or control earthquakes in certain regions;

(5) development of information and guidelines for zoning land in light of seismic risk in all parts of the United States and preparation of seismic risk analyses useful for emergency planning and community preparedness;

(6) development of techniques for the delineation and evaluation of the political effects of earthquakes, and their application on a regional basis;

(7) development of methods for planning, design, construction, rehabilitation, and utilization of manmade works so as to effectively resist the hazards imposed by earthquakes;

(8) exploration of possible social and economic adjustments that could be made to reduce earthquake vulnerability and to exploit effectively existing and developing earthquake mitigation techniques; and

(9) studies of foreign experience with all aspects of earthquakes.

(f) IMPLEMENTATION PLAN. — The President shall develop, through the Federal agency, department, or entity designated under subsection (b) (1), an implementation plan which shall set year-by-year targets through at least 1980, and shall specify the roles for Federal agencies, and recommended appropriate roles for State and local units of government, individuals, and private organizations in carrying out the implementation plan. The plan shall provide for —

(1) the development of measures to be taken with respect to preparing for earthquakes, evaluation of prediction techniques and actual predictions of earthquakes, warning the residents of an area that an earthquake may occur, and ensuring that a comprehensive response is made to the occurrence of an earthquake;

(2) the development of ways for State, county, local, and regional governmental units to use existing and developing knowledge about the regional and local variations of seismic risk in making their land use decisions;

(3) the development and promulgation of specifications, building standards, design criteria, and construction practices to achieve appropriate earthquake resistance for new and existing structures;

(4) an examination of alternative provisions and requirements for reducing earthquake hazards through Federal and federally financed construction, loans, loan guarantees, and licenses;

(5) the determination of the appropriate role for insurance, loan programs, and public and private relief efforts in moderating the impact of earthquakes; and

(6) dissemination, on a timely basis, of —

(A) instrument-derived data of interest to other researchers;

(B) design and analysis data and procedures of interest to the design professions and to the construction industry; and

(C) other information and knowledge of interest to the public to reduce vulnerability to earthquake hazards.

Report, filing with congressional committees.

When the implementation plan developed by the President under this section contemplates or proposes specific action to be taken by any Federal agency, department, or entity, and, at the end of the 30-day period beginning on the date the President submits such plan to the appropriate authorizing committees of the Congress any such action has not been initiated, the President shall file with such committees a report explaining, in detail, the reasons why such action has not been initiated.

(g) STATE ASSISTANCE. — In making assistance available to the States under the Disaster Relief Act of 1974 (42 U.S.C. 5121 et seq.), the President may make such assistance available to further the purposes of this Act, including making available to the States the results of research and other activities conducted under this Act.

(h) PARTICIPATION. — In carrying out the provisions of this section, the President shall provide an opportunity for participation by the appropriate representatives of State and local governments, and by the public, including representatives of business and industry, the design professions, and the research community, in the formulation and implementation of the program.

Program plan review. Report to Congress.

Such non-Federal participation shall include periodic review of the program plan, considered in its entirety, by an assembled and adequately staffed group of such representatives. Any comments on the program upon which such group agrees shall be reported to the Congress.

Measures developed pursuant to paragraph 5(f) (1) for the evaluation of prediction techniques and actual predictions of earthquakes shall provide for adequate non-Federal participation. To the extent that such measures include evaluation by Federal employees of non-Federal prediction activities, such measures shall also include evaluation by persons not in full-time Federal employment of Federal prediction activities.

42 USC 7705.
Submittal to
congressional
committees.

SEC. 6. ANNUAL REPORT.

The President shall, within ninety days after the end of each fiscal year, submit an annual report to the appropriate authorizing committees in the Congress describing the status of the program, and describing and evaluating progress achieved during the preceding fiscal year in reducing the risks of earthquake hazards. Each such report shall include any recommendations for legislative and other action the President deems necessary and appropriate.

42 USC 7706.

SEC. 7. AUTHORIZATION OF APPROPRIATIONS.

(a) GENERAL. — There are authorized to be appropriated to the President to carry out the provisions of sections 5 and 6 of this Act (in addition to any authorizations for similar purposes included in other Acts and the authorizations set forth in subsections (b) and (c) of this section), not to exceed $1,000,000 for the fiscal year ending September 30, 1978, not to exceed $2,000,000 for the fiscal year ending September 30, 1979, and not to exceed $2,000,000 for the fiscal year ending September 30, 1980.

(b) GEOLOGICAL SURVEY. — There are authorized to be appropriated to the Secretary of the Interior for purposes of carrying out, through the Director of the United States Geological Survey, the responsibilities that may be assigned to the Director under this Act not to exceed $27,500,000 for the fiscal year ending September 30, 1978; not to exceed $35,000,000 for the fiscal year ending September 30, 1979; and not to exceed $40,000,000 for the fiscal year ending September 30, 1980.

(c) NATIONAL SCIENCE FOUNDATION. — To enable the Foundation to carry out responsibilities that may be assigned to it under this Act, there are authorized to be appropriated to the Foundation not to exceed $27,500,000 for the fiscal year ending September 30, 1978; not to exceed $35,000,000 for the fiscal year ending September 30, 1979; and not to exceed $40,000,000 for the fiscal year ending September 30, 1980.

Approved October 7, 1977.

Index

INDEX

unreinforced, earthquake damage
    susceptibility, II: 105, 106, 110, 123, 124,
    126
mechanical systems, damage to, II: 133
mobile homes, vulnerability to ground motion,
    II: 130
mortar mixes, II: 125
nonstructural elements, damage to, II: 108, 131
parapets, damage to, II: 127, 128, 129
pilasters, effect on stability, II: 121
repair of damage, II: 133
responses to ground motions, II: 105—108
roofs, detachment from end walls, II: 120
seismometers, placement of, II: 166
split-level, damage to, II: 128, 131
steel-frame construction, earthquake damage
    susceptibility, II: 106, 110, 114
steel structures, damage to in Kern County
    earthquake, II: 109
strain-gauge platforms, use in testing responses
    to force, II: 166
stress loads from ground motion, I: 156
strong-motion seismograph placement in, II:
    162
structural continuity requirement, II: 117
tilt-up walls, earthquake resistance, II: 119, 122
torsion as cause of failure, II: 113, 117, 129
transient excitations, testing of response to, II:
    157—166
Turkey earthquake, damage by
    earthquake-resistant houses, II: 142
    reinforced concrete buildings, II: 139
    stone buildings, damage to, II: 139, 140
    unreinforced masonry buildings, II: 139
unit masonry walls, earthquake resistance, II:
    119, 122
water tanks, damage to, II: 133
wood-frame
    earthquake resistance, II: 105, 106, 107, 109,
    127
    exception from seismic hazard legislation,
    III: 2
    masonry veneered, damage to, II: 129, 132
Bulgaria, southern region earthquake (1928),
    quick condition failure, I: 183
Busch fault, earthquake on, II: 16
Butler Valley dam project, image study of region,
    III: 21—23

C

Cabot fault as extension of Great Glen fault, I:
    100
Calaveras fault, prediction model based on creep
    along, II: 9
California
    active classification of faults, I: 26
    Antioch fault, creep observation, I: 28
    Banning fault as branch of San Andreas, I: 21
    Bear Valley earthquake (1972), seismicity

    preceding, II: 12
Big Pine fault, displacements on, I: 21
Borrego Mountain earthquake (1968)
    fault creep following, I: 29
    photographic study of fault area, III: 13
    seismicity preceding, II: 12
    surface breaks corresponding to thermal IR
    image, III: 25
building code development, II: 80
Busch fault, earthquake on, II: 16
Butler Valley dam project, image study of
    region, III: 21—23
Calaveras fault as branch of San Andreas fault,
    I: 21
Concord fault, creep observation, I: 28
Corralitos earthquakes, seismicity preceding,
    II: 12
Coyote Creek fault, I: 29; III: 13
Crescent City, tsunami damage to, I: 190, 191
    Danville earthquake swarm (1970), delineation
    of fault area, III: 26
    Division of Mines and Geology
    telluric measuring network on San Andreas
    fault, II: 19
    tiltmeters used by, II: 13
    earthquake disaster planning, III: 39—50, 104,
    105
    earthquake hazard response studies, III: 67—74
    earthquake prediction evaluation council
    evaluation guidelines, Appendix H, III:
    213—216
Eureka-Arcata earthquake (1954)
    death count, I: 6
    property damage, I: 7
fault map, I: 19
Field Act, provisions of, II: 80, 94
freeway system, damage to by San Fernando
    earthquake, II: 154
Ft. Tejan earthquake (1857), inactive zone at
    previous rupture zone, II: 3
Garlock fault
    creep observation, I: 28
    displacements on, I: 21
    photographic study of, III: 12, 13
Garrison Act, provisions of, II: 95
Hayward earthquake (1868), death count, I: 6
Hayward fault
    as branch of San Andreas, I: 21
    creep observation, I: 28
    hospital location on, III: 4, 53
    study of potential damage and casualties, III:
    51—54
    urban development along, I: 141
historical development of seismic regulations,
    II: 78, 79
Hollister earthquake (1974), see Hollister
    earthquake
hospital building construction legislation, II: 96
Humboldt County earthquake (1932), death
    count, I: 6
Humboldt earthquake (1975), property